Energy Internet

Ahmed F Zobaa · Junwei Cao
Editors

Energy Internet

Systems and Applications

 Springer

Editors
Ahmed F Zobaa
College of Engineering, Design,
and Physical Sciences
Brunel University London
Uxbridge, Middlesex, UK

Junwei Cao
Tsinghua University
Beijing, China

ISBN 978-3-030-45455-5   ISBN 978-3-030-45453-1   (eBook)
https://doi.org/10.1007/978-3-030-45453-1

This Springer imprint is published by the registered company Springer Nature Switzerland AG
The registered company address is: Gewerbestrasse 11, 6330 Cham, Switzerland

Contents

Part III   Information and Communication for Energy Internet

Part IV   Energy Management Systems for Energy Internet

Part I
Architecture and Design of Energy Internet

Chapter 1
Foundation and Background for Energy Internet Simulation

Shuqing Zhang, Shaopu Tang, Peter Breuhaus, Zhen Peng, and Weijie Zhang

Abstract The energy internet has been proposed and utilized to alleviate existing environment, sustainability, efficiency and security problems and satisfy increasing energy demand in energy applications, by coupling the energy flows of various energy types with information flow. The energy internet mainly involves the energy flow systems of various energy types and is an energy-coupling system composed of a physical energy system and corresponding controllers. Then, using an in-depth study of dynamic processes and the interaction among energy production, transmission and consumption, this chapter concludes that the design, planning, operation and control of the energy internet rely heavily on analytical tools. Next, the major components, the architecture, and the technical characteristics of the energy internet are described from the perspective of simulation. Finally, application scenarios for dynamic simulation of the energy internet and the existing simulation technical foundation for various energy types are introduced.

S. Zhang (✉) · S. Tang · Z. Peng · W. Zhang
Tsinghua University, Beijing, China
e-mail: zsq@mail.tsinghua.edu.cn; zsq@tsinghua.edu.cn

S. Tang
e-mail: tsp1988@mail.tsinghua.edu.cn

Z. Peng
e-mail: pengzhenctgu@sina.cn

W. Zhang
e-mail: 459592598@qq.com

P. Breuhaus
IRIS, Stavanger, Norway
e-mail: pebr@norceresearch.no

© Springer Nature Switzerland AG 2020
A. F. Zobaa and J. Cao (eds.), *Energy Internet*,
https://doi.org/10.1007/978-3-030-45453-1_1

1.1 Concept, Components, Structure and Characteristics of the Energy Internet

This sub-chapter briefly describes the definition of the energy internet, provides a brief analysis of the composition and classification of the energy internet from the view of simulation and considers the coupling relationships between various energy sources and related equipment in the system from a microcosmic point of view.

1.1.1 Concept and Main Body of the Energy Internet

The energy internet is a multi-energy system based on a power system or a smart grid and is designed to fully utilize renewable energy and promote comprehensive energy efficiency. In addition, advanced information and communication technology and power electronics technology are employed to improve the system performance and energy quality, as shown in Fig. 1.1. Through a distributed intelligent energy management system (IEMS), wide-area coordinated control of distributed energy equipment is implemented to realize the complementarity of cold, heat, gas, water, electricity and other energy sources and to establish safe, intelligent and efficient energy systems.

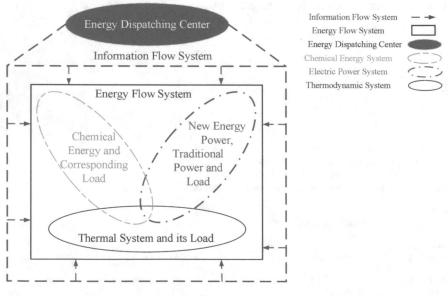

Fig. 1.1 Energy internet

The National Development and Reform Commission of China and the National Energy Administration of China summarized two modes of energy internet systems: regional and wide-area. This book focuses on the regional energy internet.

To meet the demands of customers according to local geography and natural features, an energy internet system cooperatively utilizes traditional and renewable energy; optimizes the facilities of an integrated energy internet system; and takes advantage of distributed resources, combined cooling, heating and power (CCHP) and smart micro-grids (MGs) to realize hierarchical utilization of various kinds of energy.

Additionally, the energy internet takes advantage of large power plants, where different kinds of energy are combined and energy of multiple forms are generated. In the future, it seems that distributed gas stations, which can reduce the influence of the intermittence of renewable energy, will contribute as reserves after the interconnection of various forms of energy. Moreover, electric vehicles also support energy internet systems.

In the future, more coupling of electrical power systems and thermal power systems will emerge with the high penetration of low-carbon emission techniques, which leaves more degrees of freedom for system planning and dispatch optimization. Conversely, the complexity of system topology will create new challenges in control and protection. Additionally, dynamic characteristics also need to be redefined. Therefore, modelling and simulation of energy internet systems are required [1].

The energy internet is the main link in an energy flow system that contains a coupling transformation among at least three different forms of energy: electrical energy, thermal energy and chemical energy, as shown in Fig. 1.1. Based on the traditional power grid and an MG, the energy internet expands and brings in new energy sources, power electronic devices, thermal systems and gas systems and includes the conversion, transmission, storage, distribution and consumption of various energy types. Additionally, it involves energy device design and manufacturing, system construction, control and system operation. Due to the differences in energy types of different regions, the forms of energy internet also vary. Figure 1.2 shows a typical energy internet.

As proposed above, the energy internet is a new concept. At present, most research on the energy internet has concentrated on system design and planning, which are constrained by static characteristics, and research on the dynamic processes of the system is still in the preliminary stages. From the perspective of the system dynamic analysis and simulation in the energy field, research on the energy internet has mainly focused on energy flows, which involve coupled electricity, cold/heat and fuel networks, the dynamic energy transformation among the electrical power system, renewable energy sources, the thermodynamic system and chemical fuel pipelines, and system-level and equipment-level control. The dynamic processes of components, devices and networks in response to system control, faults and disturbances, and performance tests are concerns in dynamic simulations.

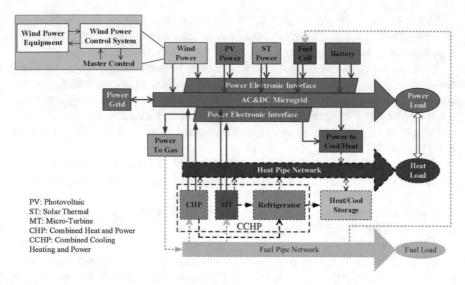

Fig. 1.2 A typical energy internet system structure

1.1.2 Backbone and Main Components of a Typical Energy Internet

1.1.2.1 Categories of Key Techniques

The energy internet involves and employs various techniques. Generally, the key techniques can be summarized in the following categories.

(1) According to the Energy Source

The energy flows in the system take different forms, so devices dealing with different forms of energy can be separated. Thus, the techniques can be sorted into renewable, electrical, chemical and thermal techniques. In this case, researchers may look for the diverse features of different forms of energy:

- Renewable sources (including solar, wind and tidal power);
- Electrical power (devices such as electrical storage, electric vehicles and heat pumps);
- Chemical sources (including natural gas, diesel, coal, biomass, nuclear, etc.);
- Thermal sources (including solar heat, thermal/cooling storage, hot water, steam, etc.)

(2) According to Energy Production

A special feature of an energy internet system is that grids of various forms of energy are interconnected to form a system that can produce energies of specific forms to meet various demands:

- Electricity,
- Heat,
- Cool,
- Chemical fuels.

Furthermore, CHP and CCHP can provide more than one form of energy, and the combined systems are always more efficient.

(3) According to the Prime Mover

The energy internet system is based on an electrical power system that is connected to a thermal system through prime movers, so the combined system can also be divided by the prime mover:

- Turbines: micro-turbines, gas turbines (GTs), steam turbines, internal combustion engines and combined-cycle gas turbines.
- Reciprocating internal combustion engines.
- New emerging techniques: fuel cells, stirling engines, and the organic Rankine cycle.

(4) According to the Energy Sequence

Due to the location of the system, the energy sequence can be determined by the local energy structure and categorized into the following:

- Topping cycle system,
- Bottoming cycle system.

In a topping cycle system, electrical power is the priority production and is widely used as a common operation condition. Thermal energy is the major production in a bottoming cycle system. In this case, the system supplies an industry that requires a very large thermal demand.

(5) According to Scale

The scale of energy internet systems is varying. Generally, these systems can be separated into micro-systems, regional systems and large systems. Moreover, they can also be classified into distributed systems and centralized systems [2].

1.1.2.2 Backbone and Main Components

Interconnected energy internet grids and the corresponding control system constitute the backbone of the energy internet.

(1) Electrical energy

- **Generator**

Electrical energy, as a secondary energy source, originates from different forms of energy and different energy conversion equipment. Most of this energy source comes from generators, which are driven by mechanical energy and realize direct energy conversion. With the development of technology, the types of generators have become diversified, such as micro-gas turbine power generation, wind power generation, photothermal power generation, biomass power generation and tidal power generation, and some energy is directly generated by energy conversion equipment, such as solar panels that directly convert solar energy into electricity and fuel cells in which the chemical energy stored in fuel is directly converted into electricity. Due to the different features of primary energy sources, electrical energy production varies greatly from daytime to evening and from summer to winter. Power generation devices have different capacities, voltages and frequencies, and most of them are distributed. All of these factors make the characteristics of the energy internet complex. In addition, the combined use of multiple power sources can improve the system performance, such as wind and solar complementarity, which can make the power supply more reliable, and fuel cell and micro-gas turbine combined power generation, which can improve efficiency.

- **Flexible alternating current transmission system (FACTS) device and converter**

The electrical power from generators is transmitted to alternating current (AC) and direct current (DC) MGs and then distributed to users through the power grid, distribution transformers and converters. Converters are mainly power electronic devices. Because electrical voltage and frequency vary among different power sources, they must be transformed into consistent AC or DC power through power electronic devices to be sent to a unified MG. The characteristics and control of power electronic devices have a considerable influence on the performance of power grids.

- **Electrical power load**

In addition to conventional power loads, there are also new types of loads and loads coupled with different forms of energy in the energy internet, such as charging piles and stations for electric vehicles, heat pumps, inverter air conditioners and electrical conversion devices. Charging piles and stations for vehicles have a considerable impact on the power grid and other energy sources.

- **Electrical energy storage**

Electrical energy storage plays an important role in the energy internet. When there is surplus energy generation, the storage devices can store it, and when the energy generation is insufficient, the storage devices can output electrical energy as sources. Energy storage supports the operation of the energy internet with high penetration of renewable energy and improves the flexibility and reliability of system operation. There are several main types of electrical energy storage devices, such as electro-chemical energy storage, mechanical energy storage and electromagnetic energy storage. At present, mature devices are mainly based on batteries.

(2) **Thermal energy**

- **Prime mover**

As a linkage between an electrical power system and thermal network, a prime mover plays a significant role in the energy internet system. For this reason, it is necessary to analyse the characteristics of prime movers. To date, several types of prime movers have been applied in the energy internet, including micro-turbines, gas turbines, steam turbines, combined-cycle gas turbines, reciprocating internal combustion engines, Stirling engines, fuel cells and the organic Rankine cycle. They have been modelled, and some typical examples have been given [3–11].

Steam turbines are widely used in the energy internet and are also one of the oldest typical prime movers. The history of steam turbines traces back to 100 years ago. In addition, the Rankine cycle is the principle of thermodynamic models of steam turbines. There are two kinds of steam turbines, backpressure and extraction-condensing, and researchers need to choose according to heat quality, the quantity of power and heat, and economy.

- **Thermally activated technologies**

The key to distinguishing CCHP from other traditional combined heat and power (CHP) technologies is that CCHP not only produces heating and power but also cooling capacity because of cooling or dehumidification components. During the cooling procedure, thermally activated technologies, which are involved in the cooling or dehumidification components, are triggered. Evidence from recent studies has inferred that thermally activated technologies can improve total efficiency, which makes them popular. In addition, thermally activated technologies also satisfy low emissions and cost reductions [12–16].

- **Thermal energy storage (TES)**

The different energy production of thermal system devices and electrical power generators gives a time difference between the peak demands. The unsynchronized peak demand allows the energy internet system to reduce the net peak demand by suitably using diversity. TES, as a reserve and heat recovery device, is competent for decoupling generation and demand [17]. Thus, TES not only improves the discontinuity of the energy internet system but also lengthens the operating time. Moreover, it may also defer investment.

The duration of the storage cycle is determined both by the load demand and by the storage mechanisms. Practical heat storage normally involves liquid tanks. Future heat storage will employ phase change materials (PCM) that can achieve high energy densities; thermochemical materials; and nanometre materials.

(3) **Gas energy**

Natural gas is one of the most significant chemical energy types in the energy internet. Gas energy systems usually consist of starting stations, gas pipelines, intermediate pressure stations and gas distribution stations.

The main tasks of a starting station are to keep the gas transmission pressure stable, automatically adjust the gas pressure, measure the gas flow and remove the droplets and solid impurities in fuel gas.

Intermediate pressure stations are an important part of a gas network and should be installed every certain distance in gas energy systems. The number of intermediate pressure stations and the outlet pressure should be determined through technical and economic calculations.

A gas distribution station is located at the end of a long pipeline and is the gas source for a small city or industrial gas distribution network. The distribution station receives gas from the long-distance pipeline. After dust extraction and pressure regulation, the gas purity and pressure meet the requirements of an urban distribution ring network or industrial pipeline network.

The main task of gas storage and distribution stations is to make gas pipelines and networks preserve the required pressure and maintain a balance between gas supply and consumption according to the instructions of the gas dispatching centre.

In dynamic simulations, a gas energy system mainly includes pipelines, compressors and networks.

1.1.3   Basic Structure of the Energy Internet

In the energy internet, power energy is transformed from renewable energy by wind/photovoltaic power generation, and a power network is coupled with a thermal energy network by steam turbines, gas turbines and electrical heating equipment, and is coupled with a chemical energy network by biomass power generation, fossil energy power generation and electrical hydrogen generation.

Due to variations in climate and energy types among different regions, there are some differences in composition among energy internet systems. However, the main structure of the energy internet remains similar. As shown in Fig. 1.2, the energy internet mainly contains networks with at least three different forms of energy: a power grid (containing AC and DC networks), a thermal system and a chemical fuel pipeline grid, through which energy interactions among various energy devices can be established.

In the energy internet, an electrical power grid containing renewable power generation is coupled with a thermal system by synchronous generators, gas turbines and electrical heating equipment, and coupled with a chemical fuel pipeline grid by biomass power generation, fossil energy power generation and electrical hydrogen generation. The thermal system is coupled with the chemical fuel pipeline grid by boilers.

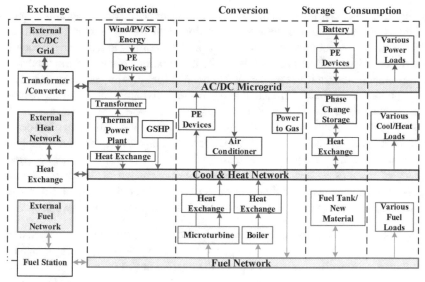

PV: Photovoltaic ST: Solar Thermal PE: Power Electronic GSHP: Ground Source Heat Pump

Fig. 1.3 Typical structure of a multi-energy flow system

The structure of the energy internet mainly reflects the energy links and interactions among energy systems, as shown in Fig. 1.3.

The electrical power grid, thermal system and chemical fuel pipeline grid individually form their own networks. These networks are coupled through several key devices (energy conversion devices), including micro-turbines, boilers and air conditioners. Each kind of key device has its own dynamic response characteristics and corresponding time constants. When the networks are coupled through these devices and a dynamic excitation occurs in any network, the dynamic processes are filtered by the key devices before they penetrate other networks. Therefore, studying the dynamic response characteristics of key coupling devices is essential for studying the cross-energy-coupling problem in energy internet systems.

1.1.4 Technical Characteristics

1.1.4.1 Wide Time Scale Characteristics

Existing research has summarized the range of the response time constant for dynamic devices in a conventional power system, an MG with interference and control intervention. By omitting the rapid and detailed physical processes of some device levels, e.g. the switching transient process of power electronic devices, we can summarize the physical processes involved in electrical energy under disturbances and their control intervention, and we can approximately divide the response time into a transient scale from microseconds to several seconds and a medium- to long-term scale from tens of seconds to tens of minutes. With further supplementation of the time scales of the dynamic and transient processes of heat and fuel pipeline network equipment, we can obtain the dynamic, transient and long-term dynamic ranges of the response time constant for a typical energy internet system, as shown in Fig. 1.4.

The results of an induction [18–21] show that the response time span of all devices in the energy internet is very wide, from the microsecond level in the electromagnetic transient of the high-frequency switching circuit in a power system converter to dozens of minutes to several hours for the heat exchange in the thermodynamic system. Overall, the response time constants of dynamic devices and the components of electrical energy, thermal energy and chemical energy overlap in two scopes:

- For the time scale from several seconds to a couple of minutes, the physical processes involve the generator motive, speed-regulating device, generator mechanical inertia, protection and restriction on the overheating of the equipment, action of

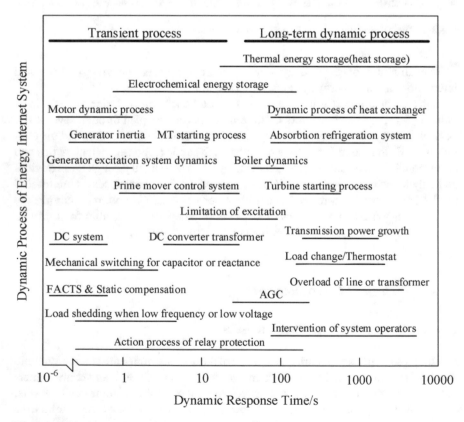

Fig. 1.4 Wide time scale dynamics of an energy internet system

the mechanical switch, automatic generation control, and transient load shedding. The middle and long processes of the system involve the start-up of the micro-gas turbine and the boiler dynamics in the thermal system. The heat exchange process between the heat storage and the heat exchanger involves the dynamic process of the fuel supply for the fuel network.

- The time scale ranging from tens of minutes to several hours mainly involves load adjustment and operational intervention processes in power and thermal systems.

Coupling of equipment for the same form of energy is not difficult to understand, and coupling among different types of energy systems and equipment in the above two time scales is further explored in the following ways:

- The electrical power grid, thermal system and gas pipeline network are tightly coupled on the source side. A part of the thermal energy of the boiler and micro-gas turbine rotates the turbine to drive the generator to produce electricity, and the other part of the thermal energy enters the heating, cooling or storage of the thermal network.
- Thermoelectric, fuel cell and other power generation devices are coupled on the fuel supply side, and these devices are coupled with the power grid and the heating network.
- The above two forms of coupling are in a time scale from several seconds to a few minutes and belong to coupling of fast processes.
- The coupling between thermal power and the power network belongs to the coupling of slow processes in the scale from tens of minutes to several hours caused by the change in thermal and electrical load power.

The coupling of an energy internet system in a wide time scale creates difficulties in solving simulation models. Generally, a small step is fundamental to solving devices and subsystems containing fast dynamics to ensure convergence and accuracy of the algorithm. For a power system with small time scale dynamics, calculation of the device and component models requires dozens of microseconds or even smaller step sizes. If the mathematical model of the whole system is solved by the same step length, the computational burden increases significantly, the whole simulation decelerates and the scale of simulation is limited. If a multistep approach is adopted, stable, accurate and fast, numerical algorithms are necessary to guarantee the accuracy and validity of the model solution.

1.1.4.2 Fundamental Differences in Modelling and Solving for Various Forms of Energy

1. Modelling and calculation of coupling among equipment and between equipment and networks

There are essential differences in the coupling mechanisms within energy grids, and different modelling and calculation methods of coupling among equipment and between equipment and networks pose difficulties to dynamic simulation of the energy internet. Specifically, there are several cases as follows.

The main body of the electrical power system is basically composed of generators with exciters and speed regulators, transmission and distribution networks, power electronic devices and loads. The grid is the coupling link among equipment components of the electrical power system. In electromagnetic and electromechanical transient simulations, the device elements are also coupled with the grid and incorporated into a whole grid through the "component-network interface", realizing simultaneous solution of the equipment and power grid models. Taking a generator as an example, the terminal voltage and current are transformed and connected to the network equation in synchronous rotating coordinate and solved together. The rotor angle is used for the interface between the electric quantity in the dq coordinates of the generator and the electric quantity in the synchronous coordinates of the network in the xy coordinates.

In comparison, the scale of the thermodynamic system is relatively small, the number of equipment components is limited, and the coupling among equipment components is more direct. In the simulation, dynamic models of components and devices are joined by associated equations or associated parameters.

Different coupling methods among components and between systems and components lead to differences in model solution algorithms. In each time step of the simulation, each component and the dynamic element model in the electrical power grid are separately solved, and then all the models of components and dynamic elements are solved jointly through the component–network interface. For the thermodynamic system, the models of each component and dynamic component are directly connected to form a higher dimensional model group, and the whole system is solved simultaneously.

2. Strong non-linear algebraic equations

Dynamic models of an energy network and elements include two sections: algebraic equations and differential equations. Non-linear factors mainly exist in algebraic equations, which make it difficult to solve energy system models. The convergence and the convergence speed depend on many factors, such as the initial values of unknown variables in the equations, the iterative flow and the method for accelerating convergence. There are obvious differences in non-linearity sources and processing methods of non-linearities in different energy systems.

For electrical power systems, the grid is often modelled by sparse linear equations or linear network equations, which are deduced by differentiating the models of dynamic network components of lumped parameters. The non-linearity is mainly embodied in the change in the power network topology, particularly in the power electronic converter. When the topology changes frequently, the network equations are reconstructed in the simulation. A switching action depends on the state of the switch circuit, but the switching action time is not exactly at the end of each solution step. Therefore, it is necessary to introduce interpolation or a compensation mechanism. The other equipment models described by non-linear algebraic equations, such as non-linear static load, are generally converted to current injections on the basis of the bus voltages. Then, they are combined with the network equation and solved iteratively.

The non-linear generation equations of dynamic models in the thermodynamic system mainly include the performance of the compressor and the turbine components, the heat transfer equation of the heat exchanger (the area multiplied by the heat transfer temperature difference), the total volume equation of the system and equipment, and the physical equation. In general, these equations are connected to a non-linear equation group of high order and are solved by the Newton algorithm and the improved algorithm.

3. **Long dynamics from radical state change of working medium and multi-physics integration**

In the thermodynamic system, a change in the state of the working medium (energy transfer medium), such as a transformation among solid, liquid and gas phases, exhibits a non-smooth and continuous change or a significant change. For example, when the absorption refrigeration system starts, the generating fluid is heated until the saturation temperature is reached and natural convection heat transfer occurs. Then, steam begins to appear. In this case, it is very challenging for only one set of model equations to correctly describe multiple stages of the physical processes.

The overlapping of multi-time-scale physical processes of some dynamic components in electrical power systems creates substantial difficulties in modelling and model parameter acquisition. For example, large synchronous machines produce long dynamics containing super transients, transient processes after a disturbance. The electrical equation of the synchronous generator is built based on the Park transformation in $dq0$ coordinates and is used in the system simulation, and the stator and rotor resistance and inductance in $dq0$ coordinates are parameters of the equation. Although they basically determine the electrical characteristics of the motor, these parameters are difficult to determine by terminal tests.

Long dynamics can also be observed in energy components when radical change occurs in the working fluid state. The overlapping fusion of multi-scale physical processes increases the difficulty of system modelling and simulation, and it is difficult to establish a set of models that can be adapted to the whole dynamic processes in the equipment-system-level simulation.

1.2   Importance and Application of System Simulation

Based on the understanding of an energy internet system and the characteristics of simulation techniques, this chapter highlights the importance of modelling, simulation and analysis of complex integrated energy grids in the energy internet by first introducing an example and then giving application scenarios for dynamic simulation. The development background of simulation techniques in the energy industry is summarized, which lays a foundation for the development of simulation models and solution algorithms.

1.2.1   Necessity of Dynamic Simulation

The energy internet is a complex system involving a variety of forms of energy coupling and containing a variety of different types of systems and devices, including fuel gas transportation systems, heat supply systems, refrigeration systems, photovoltaic systems, wind energy systems, biomass power generation systems and energy storage systems. To improve energy production and consumption efficiency and achieve real-time adjustment of the energy supply and consumption in the situation where a large number of renewable energy stations are connected to the power grid, basic research on the dynamic characteristics of the coupling and interaction among these complex energy systems is particularly important. An energy internet simulation based on the system combination and mathematical models can be used in the study of the dynamic relationship among the internal equipment and energy grids of an energy internet system, providing a theoretical basis and practical guidance for equipment selection, system planning and design, operation assistance and control strategy optimization, etc.

To date, thermoelectric simulations have been widely used in thermal power plants. Thermoelectric simulations are vital to skill training and anti-accident manoeuvres. The commonly used modelling software packages include the Real-time Object-oriented Simulation Environment (ROSE), Engineering Analysis System V5 (EASY5), 3KEYmaster and Professional TRansient Analysis eXpert (Pro-TRAX). Electromagnetic and electromechanical transient simulations of the power grid have good basic research foundations and practical applications. The commonly used simulation software packages include Bonneville Power Administration (BPA), Power Systems Computer-Aided Design (PSCAD), Network Torsion Machine Control (NETOMAC) and Power System Analysis Software Package (PSASP). The simulation of natural gas transmission systems also has a solid research foundation and wide practical applications. These simulations are mainly used in pipeline leakage and detection, optimization of operation, gas storage and peak-shaving. The commonly used modelling software packages include Stoner Pipeline Simulator (SPS), Transient Gas Network (TGNET), PIPEPHASE and Real Pipe-Gas. However, research on dynamic simulations of the energy internet started much later, and no publicly available simulation software package has yet been reported.

To emphasize the significance of modelling, simulation and analysis of the energy internet, a case in which the lack of doing so resulted in severe problems is presented here.

An existing energy system of a large building was extended by integrating a micro-gas turbine (MGT) in which the thermal energy of the exhaust gas via a heat exchanger was used to contribute to heating the building. The nominal power output of the MGT was 100 kW, and the turbine was controlled by prescribing a setpoint for power output. The heat extraction in the heat exchanger was performed by water and controlling the inlet temperature to the heat exchanger to ensure that the limit values for the water temperature were not exceeded.

Fig. 1.5 Overheating in the combustor resulted in melting metal, deformation and further damage

The system was in operation for some weeks before it could not be restarted after an unexpected shutdown. Taking a closer look at the MGT showed that a complete failure had occurred (Figs. 1.5 and 1.6). The damage pictures indicate that flame flashback might have occurred, which damaged the combustor (Fig. 1.5). Due to overheating and embrittlement of the material, combustor parts entered the high-speed turbine, which was damaged. Parts of the destroyed rotor penetrated the inlet spiral to the turbine (Fig. 1.6).

A root cause analysis was performed, which included evaluating the history of the installation, analysis of operational data and modelling and simulation.

When the unit was installed, it was necessary to place the unit in a room distant from the room hosting the heating system of the building. Because the power and heat output matched the needs of the building, no further in-depth analysis was performed. The only connections between the control systems of the building were that the required power setpoint was sent to the MGT controller and the operating data were visualized on the control screen for the building system.

Processing the collected historical operational data resulted in graphs similar to those shown in Figs. 1.7 and 1.8. The data were collected by the control system and had a resolution of approximately 1/30 s. The power output that follows the setpoint, the relative rotational speed $\left(rpm/rpm_{100\%}\right)$ and the water temperature in the waste heat recovery heat exchanger are visualized.

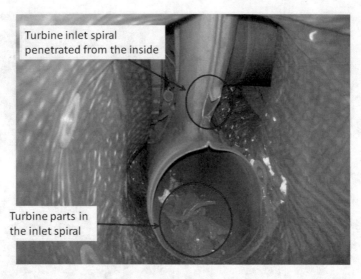

Fig. 1.6 Damage at the inlet to the turbine (inlet spiral)

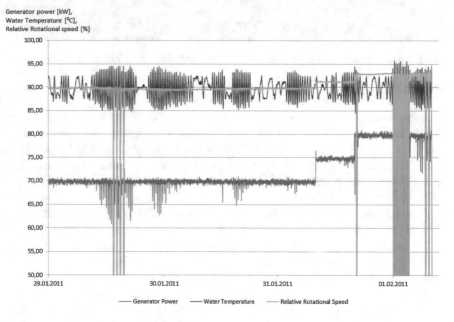

Fig. 1.7 Measurements demonstrating the impact of sensor location on the control and operational stability of a micro-gas turbine CHP system

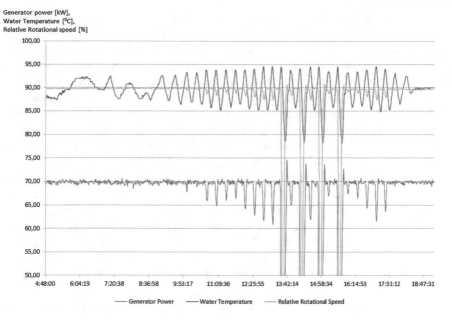

Fig. 1.8 Zoomed-in view of Fig. 1.7 for data collected on 29.01.2011

It is interesting to observe the fluctuations in the water temperature, which are, after exceeding a certain amplitude, followed by changes in power output and rotational speed. In some cases, the rotational speed decreased well below 50%, and the engine shut down, followed by an automatic restart. The failure of the engine was evaluated to be caused by flashback of the flame during the large fluctuations in the rotational speed, which was possible because the combustor was of a lean premixed type, and reduced airflow might have resulted in flame speeds higher than the flow velocity. Such conditions allowed the flame to propagate back into the premixing zone. The flashback led to overheating and melting material, material embrittlement and material breaking and entering the turbine, which caused complete failure. However, the question remained about the cause of the fluctuations because the MGT was considered robust and stable in several other applications.

The answer was found when analysing the integrated installation, considering the locations of sensors and valves as well as the control setup. It was found that the temperature sensor for the water was located more than 20 metres from a control valve for regulating the fluid temperature towards the heat exchanger by mixing cold and hot streams to maintain a target temperature. This distance was the result of restrictions in the area, as the integration of the MGT was not foreseen when the building was planned. When the control system detected a need to adjust the heat extraction, a corrective signal was sent to the control valve to adjust the inlet temperature of the heat exchanger. The adjusted temperature was checked via the aforementioned sensor. Due to the given tube dimensions (distance and diameter), water with the adjusted temperature took a while to reach the sensor. This time delay resulted in further adjustments of the valve because the control system did not register a changed water temperature shortly after sending corrective signals to the valve. When the sensor was finally exposed to the adjusted temperature, the control system registered an over-corrected value exceeding the threshold within the control system, which led to immediate counter corrections, which again suffered from the same problem of time delay. It can easily be seen that the system ended up operating with oscillating temperatures. When the amplitude exceeded a certain level, the power output was adjusted, ignoring its required setpoint. Power adjustments were predominantly performed by adjusting the rotational speed while the hot gas temperature was targeted to be kept constant for MGT efficiency.

The variations combined with exceeding the inbuilt threshold values to protect the MGT forced the control system in some instances to shut down and restart the engine, which resulted in relative speed values decreasing to well below 50%.

In addition, some system internal control and safety algorithms were in conflict because of the setup of the system. The main setpoint of the engine was the electrical power output, which was maintained via the rotational speed (i.e. mass flow in the GT) and the temperature in the turbine, while the water temperature/heat extraction was controlled separately. Additionally, the restrictions in place to protect the system from failures, which were related to, for example, the exit temperature of the turbine, created issues. These sometimes-conflicting requirements then triggered the shutdown of the engine followed by a fast restart.

The main reason that the failure occurred was the integration of the MGT without considering dynamic processes and the inbuilt control system. If these considerations had been accounted for in an early stage, corrective actions could have been taken. For example, corrective actions could have included either changing the position of the sensor relative to the valve or considering the distance within the control system.

1.2.2 Application Scenarios

Dynamic simulation techniques can also be applied in the field of the energy internet, just as they can be applied in electrical power systems to achieve scheme verification, accident analysis, countermeasure making, skill and operation training and so on. Several application scenarios are as follows:

- Basic theory research and validation

Dynamic simulation can be used to support basic theory research and validation. Under complicated external conditions, dynamic simulations are performed to study unstable forms and key factors, control features, energy quality, voltage fluctuations in system nodes after a certain disturbance/fault, analysis of harmonic content in the system, analysis of the mechanism of commutation failure, pressure in the pipeline network after a certain disturbance/fault, variation in flow velocity, temperature in the heat transfer process and so on. Basic theory research can lay a foundation for the development of energy internet techniques and the planning and construction of practical projects.

- Verification of the planning and design scheme

According to planning and design drawings and the imported parameters of the equipment and system, the system model can be established. Based on this information, a dynamic simulation platform can provide comprehensive data analysis for verification of the feasibility of the planning scheme, rationality of equipment capacity and location allocation.

- Research and optimization of control strategy

Based on the dynamic simulation platform, we can effectively use the portability, reusability, security and efficiency of the simulation model to verify the effectiveness of the control strategy designed in an actual case of the energy internet.

- Adjustment to the configuration and operation mode

For an established industrial park energy internet, we can study the energy transmission of energy attributes in the system based on the simulation platform. By combining existing control strategies or developing new control strategies in the platform, we can obtain the best matching of various attributes of energy strive for the optimization of the energy efficiency, environment and system economy of the whole park.

- Other application scenarios

Hardware-in-the-loop, online system security assessment, training of operation and maintenance staff, fault reappearance and countermeasure formulation.

1.2.3   Foundation of an Energy Internet Dynamic Simulation

This section briefly reviews the basis of the development of energy internet simulation techniques, namely, the development status of system simulation techniques in power grid, MG and CCHP systems.

1.2.3.1   Power System and MG Simulation

- Power System Simulation

Currently, power system simulation techniques have become relatively mature. Many simulation approaches have been formed, such as electromagnetic transient simulation, electromechanical transient and quasi-steady-state simulation, middle and long-term dynamic simulation, steady-state simulation and electromagnetic-electromechanical simulation interface techniques, to meet various demands and conditions.

Electromagnetic transient simulation (EMTP) can accurately simulate the dynamic response of each element and simulate HVDC (high voltage direct current) and FACTS power electronic devices. Compared with electromechanical transient analysis, it is necessary to apply a more precise method to analyse the electromagnetic transient phenomenon of the power system. For example, the system must consist of inductance, resistance and capacitance, which are described by differential equations with transient variables. The range of the frequency analysed is wide, from low-frequency behaviour of dynamic overvoltage and dynamic control to high-frequency behaviour of the first wave of thundershock, which covers almost all physical phenomena that affect the system. The elements of the system can be classified as elements with lumped parameters and distributed parameters. A wave propagation process must be considered with distributed parameters. During the electromagnetic transient calculation, a transmission line with distributed parameters is often simulated using a cascaded Pi-type equivalent circuit with lumped parameters. If the process of wave propagation is considered, the transmission line can be described by the network method, the characteristic line method (also called the Bergeron model and the Dommel model) and the improved Fourier transform method. The network topology of a DC system changes frequently when the detailed controller model and the action process of power electronic switches are considered.

Electromechanical transient simulation mainly analyses the transient stability during large disturbances and steady-state stability during small disturbances in the power system. The differential equations and algebraic equations of the whole system according to the topological structure are obtained. Then, using the initial value, which is the power flow without disturbances, the time variations of state variables after a disturbance are calculated. Only the positive sequence phasor with fundamental frequency is considered to describe the whole AC system and system response, while the circuits are described by impedances.

Hybrid simulation of electromechanical and electromagnetic transient methods separates the network into two parts to be solved: one is an electromagnetic transient computing network, and the other is an electromechanical transient computing network. Then, the system realizes the integrated simulation by exchanging data through the interface. The hybrid simulation expands the scale of the electromagnetic transient simulation and simultaneously provides the necessary background for the simulation analysis of the electromagnetic transient network.

Based on power grid simulations, simulations of MGs have been developed through modelling and introduced to some applications [21, 22]. An MG simulation provides a basic model and algorithm for simulating the energy internet system.

• Distributed Resource (DR) and MG Simulation

Some achievements have been made in the study of the influence of DR and MG penetration on a distribution network and its models. A simulation platform and simulation software package are effective methods for simulating the system operation, which provides important support for the implementation of DR and MG projects.

The simulation of an MG includes steady-state analysis, transient stability analysis and fine simulation of dynamic behaviours of equipment [22].

Steady-state analysis includes two parts: solving the steady-state operating point of the system according to the given operating mode of the DR system; obtaining the fault current by short-circuit fault analysis, which provides the basis for selecting equipment and switching capacity.

Transient stability simulation focuses on the relatively slow dynamic processes of the system, with simplified network components, power electronic devices, distributed power supply and various controller models, to model and simulate the system (using a quasi-steady-state model to describe the system).

Fine simulation of device behaviours pays more attention to the faster and more precise features of the MG or equipment. It uses electromagnetic transient simulation as the core and considers the MG power quality, system controller design, protection and emergency control system design, short-term load tracking characteristics, transient short-circuit current and fault ride-through capability of DRs.

In [23], electromechanical and electromagnetic transient simulations of the same MG were carried out by using DigSilent and Matrix Laboratory (MATLAB) Sim-PowerSystems, and the similarities and differences of the multi-dynamic-process simulations were compared. Yan [24] used the fundamental positive sequence component method and load identification techniques to build a MG load modelling system by MATLAB. Lin [25] proposed the difficult point and key techniques for solving a real-time simulation of an MG.

With the continuous penetration of DRs and changes in energy structure, smart grids have gradually become a new generation of power grids. Power grid simulation, combined with computer, information and communication techniques, has developed and greatly improved in accuracy, rapidity and flexibility, laying a foundation for the simulation of energy internet systems [25–27].

Simulation methods expanded from those of conventional grids are not enough to calculate the new generation of power grids with MGs. Due to the complexity of new patterns in the power grid, new standards are required to determine the stable domain. Furthermore, it is difficult to apply electromagnetic transient simulation to an actual MG system, which contains a large number of converters and electrical nodes. Electromechanical transient simulation ignores the electromagnetic transient processes of the equipment and power grid, especially converters with fast response characteristics and distributed generation equipment.

1.2.3.2   Non-electrical Power System Simulation

A thermodynamic system is a typical non-electrical power system and is a significant part of the planning and construction of an energy internet system. Thermodynamic system equipment is diverse, and its dynamic response crosses over a wide time scale. The model is described through a set of algebraic equations, ordinary differential equations and partial differential equations with distributed parameters. These equations with their solving algorithms can cover the key technologies of energy system modelling and simulation, which are representative.

Modelling and simulation of thermodynamic systems can be broadly divided into 2 levels: simulation analysis for devices and for the whole system.

- Simulation Analysis for Devices

Simulation analysis for devices includes optimization of the internal structure of equipment, design and verification of all parts of equipment, performance analysis, operation simulation with varying conditions or disturbances, control design and analysis of control characteristics and control design. It does not need to consider the correlation and coupling between devices. According to the main components involved in a thermodynamic process, models with different levels of detail are established based on various requirements. According to the results of model computation, the dynamic behaviours of equipment components at different levels are analysed. Then, the modelling and simulation can be divided into static lumped parameters, static distributed parameters, dynamic lumped parameters and dynamic distributed parameters. A lumped parameter model given by K.E. Herold in [28] is representative and can be used for calculation of the thermodynamic system and simplified calculations. A static distributed parameter model can calculate the distributions of refrigerant and solution in each component. Considering the differences in the heat capacity and working fluid state of the system, [29–31] set up a more detailed model of the generator, absorption chiller and absorption exchanger and described the operating characteristics of the system, which greatly improved the accuracy.

At present, considerable work has been performed in research on non-electrical power devices. In gas turbine simulation research, some notable achievements have been made. Weng [31] with Shanghai Jiao Tong University performed a detailed analysis on the components and overall performance of a gas turbine, especially under variable working conditions. The paper expounded on a simplified method of gas turbine modelling and considered that the model should be simplified according to the physical process rather than the convenience of solving the mathematical equation. The basic dynamic and static characteristics of gas turbines should be preserved as much as possible. Badyda [32] reviewed the gas turbine techniques and development trends of other advanced power technologies.

Modelling of absorption chillers is relatively mature. Zinet [33], based on the Nusselt falling film theory, established and resolved an absorption process model of an absorber and generator. A three-dimensional model of a condenser was established, and other heat transfer models were established by the simplified effectiveness-number of transfer units (NTU) method. Then, the dynamic characteristics during the start-up/shut down and the variable working conditions were studied. Wang [34] modelled a direct-fired lithium bromide absorption chiller unit. A neural network structure with high precision and generalization performance was obtained by a feed forward neural network, a large quantity of field data was used as the input and output parameters and the internal weights were constantly adjusted. Then, the neural network was used to simulate the operation of the unit.

- Simulation Analysis of the Whole System

Simulation analysis of the system mainly focuses on the performance of part of the system or the whole system, including system operation characteristic analysis, equipment matching and coordinated operation, influence of equipment parameters on the system, system performance analysis and system and equipment control characteristic analysis and design. Simulations at this level do not consider the differences in the component parameters and the working condition, and response characteristic modelling is mostly adopted for the components/equipment. An equipment model is connected by the correlation state variables or the input–output [31]. Overall, simulation analysis for the system is divided into static simulation and dynamic simulation.

(1) Static simulation considers the stability of the system in a certain condition. A static model can be used to determine the system configuration plan, operation mode and operation plan for a period according to the conditions of the climate and the properties of buildings. Moreover, according to the electricity price and power and heat (cooling) load changes in different periods of the day, the model can optimize the operation plan or coordinate control of the system.

A lumped parameter steady-state model of a boiler and 3-cylinder steam turbine was established by Zhang [35], which was used for system calculation during fixed conditions and variable working conditions. Dong [36] established a steady-state model of a thermodynamic system described in an algebraic equation, and the model was used for energy efficiency diagnosis and economic evaluation. A complete thermodynamic system was established by Zhang [37] using more than 60 non-linear algebraic equations, including a boiler, superheater, reheater, steam turbine, and condenser, and was solved by the Powell method. A steady-state model of a ship thermodynamic system was established in [38], which combined mechanism modelling and characteristic curve fitting modelling. Considering the characteristics of climate and energy prices in southern China, Liu [39], from the Advanced Energy Power Key Laboratory of the Chinese Academy of Sciences, investigated the applicability of an integrated energy system composed of micro-gas turbines and small gas turbines in typical urban buildings, including office buildings, shopping malls, hospitals and hotels. Zheng [40] from Shanghai Jiaotong University analysed four typical waste heat recovery methods and noted that the major factors affecting the primary energy efficiency of the system were the total amount of waste heat recovery of the system and the conversion process of high-temperature heat to high-quality power.

(2) Dynamic simulation is used to analyse transient and dynamic processes, the control strategy of a research and development system, and the security and stability performance analysis of a thermal system. A dynamic simulation model of

a steam combined-cycle unit and a heavy burning unit with a lumped parameter model was established [41, 42], based on which system dynamic characteristics were obtained. The model can be used for unit operation analysis, fault diagnosis, control development and operation training. Luo [43] built a dynamic model of solar thermal power systems, including receivers, oil tanks, shell-and-tube heat exchangers, steam accumulators, and steam turbines, and developed dynamic characteristic research. In addition, some special research has been conducted on equipment-system simulation modelling and algorithms, such as [44, 45], which mentioned several methods for improving the accuracy of lumped parameter dynamic modelling and simulation of thermodynamic systems.

Compared with an electrical power system, a thermodynamic system is smaller in scale, and the number of pieces of equipment and components is limited; therefore, the coupling of the equipment is more direct in system simulation. By combining the input/output and associated state variables, a set of algebraic equations or differential equations of high dimension is formed. For a model equation, direct and conventional solutions are generally used. For example, for static model algebraic equations, the Newton method or Newton's improved method [33–35] is used; for dynamic differential equations, a low-order algorithm with better stability, such as the implicit trapezoidal method [46, 47], is often used to solve the problem.

Modelling and simulation of thermodynamic systems have been carried out for many years. A relatively mature model for the corresponding equipment components and working medium changes has been developed, which lays a good foundation for system-level simulation. However, the combined simulation has technical difficulty in the application for the thermal system and the electrical power system due to the wide differences in the response time of the dynamic processes.

1.2.3.3 Energy Internet System Simulation

Considering steady-state modelling simulations and their applications, the corresponding research is relatively abundant. An operation control strategy that reduces the operating cost was proposed for an electric-driven heat pump regional CCHP station using a component input–output characteristic model [48]. Operation optimization and configuration optimization of the distributed CCHP system were studied [49], and economic operation was set as the goal. Ono [50] simulated the optimized operation of a power supply to improve the energy utilization efficiency of a regional system. The optimized operation strategy reduced the energy consumption and cost of the system by 11% and 8.2%, respectively. Guo [51] discussed the application of a water source heat pump system with seawater as a heat source and cold source for a regional CCHP system in Dalian, China. The research analysed the economic efficiency, energy efficiency and environmental impact of the system and verified the feasibility of the system both technically and economically.

Thus far, the dynamic analysis of energy internet systems is still developing. Most dynamic simulations of energy internet systems have focused on CHP or CCHP. Wang [52] analysed the CCHP of energy islands. A dynamic mathematical model was built and verified by pre-measured data. Moreover, this research analysed the system response during a step change in working conditions. Shi [53] from the EPRI (Electric Power Research Institute) of China modelled and simulated a cooperative power generation system with micro-GTs and fuel cells. The research analysed the influence of the system on the distribution system with which it is connected, including voltage distribution and system stability. A dynamic model of a Stirling engine (SE) micro-CCHP system was published by Rosato et al. [54]. The system was used for residential utilization. The research proved the dynamic comparison of traditional separated generation approaches and energy internet systems. Yang [55] built a model according to demonstration projects that were a CCHP model with a micro-gas turbine, absorption chiller and low-temperature adsorption chiller. The system was a bottoming cycle system, which means that heat was a priority. The model was built on the simulation platform of EASY5. Steady-state simulation and dynamic simulation were illustrated in that research. Then, the author analysed the system response with the various working conditions. That study covered research on both characteristic analysis and control system analysis. Song [56] built a distributed energy system model in MATLAB/Simulink according to an air conditioning system and CCHP system of a hotel. The author researched the optimal supply option and system performance. The simulation of this CCHP system provided the influence of the gas turbine heat recovery, loading rate and efficiency of the heat recovery boiler on the system performance.

1.3 Conclusion

The energy internet takes full advantage of cold, heat, gas, electricity and other types of energy sources and can be used to establish an intelligent energy system that can achieve high energy efficiency. The production, transfer, distribution and consumption of various types of energy with smart control and protection constitute the main body of the energy internet.

The techniques of the energy internet include the conversion, transmission, storage, distribution and consumption of various energy types and involve energy device design and manufacture, system planning and construction, control and system operation. These key techniques can be categorized by energy source, energy production, prime mover, energy sequence and system scale. The structure of the energy internet mainly contains three types of energy networks: electrical, thermal and fuel networks, through which interactions and coupling among various energy devices can be established.

Two aspects make the new forms of energy systems different from traditional systems. First, complex long dynamic processes exist and are intertwined in the energy system: wide time scale and multi-phase long dynamic processes. Second, various energy devices and grids of different energy types are coupled and combined through multiple paths. Large amounts of energy conversion and transmission devices and energy grids compose the energy internet. These characteristics make the systems much more complicated than traditional energy systems.

Thus, dynamic process-based research, study, diagnostic analysis and energy production elicit the requirements of powerful analysis methods and tools. Many activities make dynamic simulations essential: system operation behaviours, system control characteristics, unforeseen behaviours and characteristics of new types of devices and system-level control strategies, and training of operation and maintenance staff. Some examples also illustrate the importance of dynamic simulation of the energy internet, where the lack of prior simulation analysis results in severe problems. Potential application scenarios include verification of the planning and design scheme, research and optimization of the control strategy, hardware-in-the-loop testing, adjustment of the configuration and operation mode of the park energy grid, online system security assessment, training of operation and maintenance staff, fault reappearance and countermeasure formulation.

As summarized in this chapter, dynamic simulation of traditional energy systems has laid a solid foundation for dynamic simulation of the energy internet. First, traditional dynamic simulation provides abundant practical dynamic models for energy devices and networks, which can be directly employed in dynamic simulation of the energy internet. Second, we can refer to the solution algorithms for traditional dynamic devices and energy systems when simulating the energy internet. However, this approach is still far from the target: to carry out dynamic simulation of the energy internet. The simulation techniques for a single energy type have become mature and are applied in energy industries. At present, no software or platform is available for dynamic simulation of the energy internet. Most research and practice only deal with steady-state problems, such as the allocation of energy sources according to the energy users, balancing energy flow and calculating energy loss. Simulation of CCHP systems has only initialized dynamic simulation of the energy internet.

References

1. I.V. Beuzekom, M. Gibescu, J.G. Slootweg, A review of multi-energy system planning and optimization tools for sustainable urban development. Powertech, IEEE Eindhoven. IEEE, 2015
2. S. Tian, W. Luan, D. Zhang et al., Technical forms and key technologies on energy internet. Proc. CSEE **35**(14), 3482–3494 (2015)
3. P.J. Mago, A.K. Hueffed, Evaluation of a turbine driven CCHP system for large office buildings under different operating strategies. Energy Build. **42**(10), 1628–1636 (2010)
4. S.M.H. Mohammadi, M. Ameri, Energy and exergy analysis of a tri-generation water-cooled air conditioning system. Energy Build. **67**, 453–462 (2013)
5. M.M. Korouyeh, M.H. Saidi, M. Najafi et al., Evaluation of desiccant wheel and prime mover as combined cooling, heating, and power system. Int. J. Green Energy **16**(3), 1–13 (2019)
6. P. Ahmadi, M.A. Rosen, I. Dincer, Greenhouse gas emission and exergo-environmental analyses of a trigeneration energy system [J]. Int. J.Greenh. Gas Con. **5**(6), 1540–1549 (2011)

7. Y. Wang, Y. Huang, E. Chiremba et al., An investigation of a household size trigeneration running with hydrogen. Apply Energy **88**(6), 2176–2182 (2011)
8. F.A. Al-Sulaiman, I. Dincer, F. Hamdullahpur, Energy analysis of a trigeneration plant based on solid oxide fuel cell and organic Rankine cycle. Int. J. Hydrogen Energy **35**(10), 5104–5113 (2010)
9. G. Mulder, P. Coenen, A. Martens et al., The development of a 6 kW fuel cell generator based on alkaline fuel cell technology. Int. J. Hydrogen Energy **33**(12), 3220–3224 (2008)
10. H. Al Moussawi, F. Fardoun, H. Louahlia-Gualous, Review of tri-generation technologies: design evaluation, optimization, decision-making, and selection approach. Energy Convers. Manag. **120**, 157–196 (2016)
11. D. Xie, A. Peng, Modeling of KW-scale fuel cell combined heat and power generation systems. Paper presented at Asia-pacific power and energy engineering conference (Chengdu, China, 2010), pp. 28–31
12. Z. Aminov, N. Nakagoshi, T.D. Xuan et al., Evaluation of the energy efficiency of combined cycle gas turbine. Case study of Tashkent thermal power plant, Uzbekistan. Appl. Therm. Eng. **103**, 501–509 (2016)
13. M.N. Hidayat, F. Li, Impact of distributed generation technologies on generation curtailment. Power & Energy Society General Meeting, IEEE, 2013
14. X. Liu, J. Li, Y. Qu et al., Overview of modeling of combined cooling and heating and power system. Power Syst. Clean Energy **28**(7), 63–68 (2012)
15. X. Zhang, J. Zhang, C. Xiao et al., Dynamic simulation research progress of absorption chiller unit. Gas Heat **34**(3), 16–20 (2014)
16. X. Wang, D. Zhang, J. Xu, et al., Research review of structural type and operating characteristic of waste heat boiler. China Nonferr. Metal. **A**, 40–45 (2012)
17. E.S. Barbieri, F. Melino, M. Morini, Influence of the thermal energy storage on the profitability of micro-CHP systems for residential building applications. Appl. Energy **97**, 714–722 (2012)
18. Y. Tang, New progress in research on multi-time scale unified simulation and modeling for AC/DC power systems. Power Syst. Technol. **33**(16), 1–8 (2009)
19. Y. Du, G. Li, J. Meng, Multi time-scale dynamic models for doubly-fed induction generator in DIgSILENT. 2013 IEEE PES Asia-Pacific Power and Energy Engineering Conference (APPEEC), IEEE, 2013
20. S. Zhang, S. Tang, Y. Zhu et al., Key techniques for phased and multi-mode hybrid simulation of multi-energy micro grid. J. Comput. Res. Dev. **54**(4), 683–694 (2017)
21. J. Mírez, A modeling and simulation of optimized interconnection between DC microgrids with novel strategies of voltage, power and control, in *IEEE Second International Conference on Dc Micro grids*, IEEE, 2017
22. C. Wang, *Analysis and Simulation Theory of Micro Grid* (Science Press, China micro grid, 2013)
23. X. Yang, J. Su, Z. Lv et al., Overview on micro-grid technology. Proc. CSEE **34**(1), 57–70 (2014)
24. Y Yan, The research on digital simulation method and development of load modelling system of micro grid. University of Hunan, China micro grid, 2012
25. Z. Lin, *Research on key technologies of distributed real-time simulation for Micro-grid* (Beijing Institute of Technology, Beijing, 2015)
26. J. Wang, Modeling of micro grid and simulation platform development. Beijing University of Civil Engineering and Architecture micro grid, Beijing, 2013
27. G. Han, Micro grid modeling and simulation of its operation and control. Shandong University, 2014
28. K.E. Herold, R. Radermacher, S.A. Klein, *Absorption Chillers and Heat Pumps* (Crc Press, Florida, 1996)

29. M. Ogasawara et al., Dynamic behavior of absorption water chiller. **46**(520), 1–10 (1998)
30. C. Zhuo, Absorption heat transformer with TFE-Pyr as the working pair. The Netherland, 1998
31. S. Weng, Performance analysis of gas turbine. Shanghai Jiaotong University, 1998
32. K. Badyda, State and prospects of gas turbine and combined cycle technology development (2014)
33. M. Zinet, R. Rullierer, P. Haberschill, A numerical model for the dynamic simulation of a recirculation single-effect absorption chiller. Energy Convers. Manag. **62**(5), 51–63 (2012)
34. C. Wang, J. Jia, The application of neural network to the performance simulation of gas-fired lithium bromide apsorption chiller. Refrig. Air Cond. **8**(1), 49–52 (2008)
35. W. Zhang, H. Zhang, M. Su, Modeling and emulation analysis of units mixedly burning blast furnace gas and caol in thermal power plant. Therm. Power Gener. **41**(4), 59–64 (2012)
36. K. Dong, Thermodynamic system simulation and energy efficiency diagnosis of off-design condition. Dalian Institute of Technology, 2012
37. C. Zhang, Simulation on the thermodynamic system of a 300 MW coal fired power plant. J. Chin. Soc. Power Eng. **32**(9), 705–711 (2012)
38. S. Yao, X. Yan, Y. Ye, et al., Simulation study on thermal response of a certain type of LNG storage tank for ship. J. Jiangsu Univ. Sci Technol (Natural Science Ed.) **30**(2), 136–141 (2016)
39. A. Liu, Research on gas turbine based distributed CCHP system in southern China. Instate of Engineering Thermophysics China Academy of Sciences, 2009
40. M. Zheng, W. Wang, Optimization for waste heat in CCHP system. Shanghai Energy Conserv. **5**, 12–17 (2009)
41. B. Wang, N. Cui, C. Zhang et al., Dynamic simulation model for gas-steam combined cycle thermodynamic system. J. Syst. Simul. **20**(12), 3107–3113 (2008)
42. N. Cui, B. Wang, Y. Deng et al., Dynamic simulation model for the heavy duty gas turbine thermodynamic system. Proc. CSEE **28**(2), 110–117 (2008)
43. B. Luo, Dynamic simulation of thermodynamic system of the solar thermal power plant. Chongqing University, 2014
44. J. Guo, F. Liu, D. Liu, The improving of modeling methods at thermal system simulation based on the parameters. CD Technol. **5**, 50–52 (2008)
45. F. Fang, J. Zhang, Modeling and simulation of thermal system in power station based on system dynamics. Proc. CSEE **31**(2), 96–103 (2011)
46. W. Ma, Simulation research on GAS turbine and its combined cycle in part load. Shanghai Jiao Tong University, 2009
47. M. Su, Novel technique for simulation of large flows into and out of small volume in power system. J. Shanghai Jiao Tong Univ. **4**, 11–13 (1998)
48. M. Murai, Y. Sakamoto, T. Shinozaki, An optimizing control for district heating and cooling plant, in *Proceedings of the 1999 IEEE International Conference on Control Applications*, 1999
49. D. Buoro, M. Casisi, P. Pinamonti et al., Optimal lay-out and operation of district heating and cooling distributed trigeneration systems. Industrial and cogeneration; Microturbines and small turbomachinery; oil and gas applications. Wind Turbine Technol. **5**, 157–166 (2010)
50. E. Ono, H. Yoshida, F.L. Wang, Retro-commissioning of a heat source system in a district heating and cooling system, in *Proceedings of International Building Performance Simulation Association 2009. United States: International Building Performance Simulation Association* (2009), pp. 1546–1553
51. Z. Guo, X. Zhong, M. Gu et al., Research on the energy saving of ocean water heat pump system. J. Shanghai Univ. Electr. Power **4**, 331–334 (2010)
52. Z. Chong, Scheme design and Multil-criteria comparative evaluation of CCHP system. University of Guangdong for Technology, 2016

53. H. Shi, Study on modeling and simulation for distributed generation system based on Gas turbine/fuel cell combined generation. China Electric Power Research Institute, 2011
54. A. Rosato, S. Sibilio, G. Ciampi, Energy, environmental and economic dynamic performance assessment of different micro-cogeneration systems in a residential application. Appl. Therm. Eng. **59**(1–2), 599–617 (2013)
55. J. Yang, Integration and performance simulation study of combined cooling heating and power system based on micro turbine. Shanghai Jiao Tong University, 2009
56. S. Song, The scheme research and performance simulation on the distributed heating-cooling-power energy supply system. Dalian Univsersity of technology, 2013

Chapter 2
Modelling, Simulation and Analysis

Shuqing Zhang, Peter Breuhaus, Shaopu Tang, Zhen Peng, Xianfa Hu,
Ning Liu, Yaping Zhu, and Jinxin Liu

Abstract Although there are many simulation tools in various energy fields, due to
the differences in technical characteristics in various energy fields, simulation techniques cannot be directly integrated to meet the development needs of the energy
internet. Facing different application requirements, different modelling ideas and
methods as well as solution algorithms are proposed in this chapter. For modelling,
simulation and analysis of quasi-steady-state and long-term issues of the energy internet, a standard procedure is suggested for developing concepts or modifying existing
systems, with underlying issues discussed. For the study of dynamics and transients,
modelling methods of components and networks under different energy conditions
are first discussed. The node equation in matrix form is used in the power grid, and
the node pressure equation and branch flow equation are adopted in a thermal network and gas pipeline network. Then, considering the interweaving and interaction
of long dynamic processes of wide time scale and phased evolution in the energy
internet, this chapter proposes a three-layer multi-mode phased hybrid simulation
framework to solve the dynamic and transient coupling and interactions between
devices and adjacent networks of different energy types in the energy internet. Then,
some existing simulation software suites and tools in energy engineering fields and
academic circles are introduced in detail. Possible problems are discussed, including
determining how to select software suites and tools with embedded models, combining different tools or developing and integrating new modules or tools. Finally,
this chapter enumerates some typical component models in the energy internet and
presents simulation results based on several cases. A brief analysis of a park energy
internet case shows the correctness of the modelling and solution considerations
mentioned above.

S. Zhang (✉) · S. Tang · Z. Peng · X. Hu · N. Liu · Y. Zhu · J. Liu
Tsinghua University, Beijing, China
e-mail: zsq@mail.tsinghua.edu.cn

P. Breuhaus (✉)
IRIS, Stavanger, Norway
e-mail: pebr@norceresearch.no

© Springer Nature Switzerland AG 2020
A. F. Zobaa and J. Cao (eds.), *Energy Internet*,
https://doi.org/10.1007/978-3-030-45453-1_2

2.1 Modelling, Simulation and Analysis of Quasi-steady-State and Long-Term Issues of the Energy Internet

2.1.1 Introduction

Modelling, simulation and analysis of the energy internet should be a standard procedure when developing concepts or modifying existing systems, especially when elements with different dynamic characteristics are integrated. This approach is usually closely connected with algorithms and the underlying numerical schemes. However, before using any kind of modelling and simulation tools, it is essential to evaluate the purpose of evaluation as well as the system characteristics and their main components. This approach is similar to the case analysis presented in **part 1.2.1**, in which a not properly executed procedure that resulted in severe damage.

2.1.2 Elements of Integrated Energy Systems

The main elements of any energy system can be classified as follows:

- **Energy sources** usually represent primary energy sources, such as wind, solar, biomass, natural gas and coal. They are also divided into different categories according to their characteristics (fluctuating/non-fluctuating), local availability and other criteria. Most commonly, they are divided into two categories: new energy and traditional energy.
- **Energy conversion units** convert incoming energy into other forms of energy for further distribution via energy grids either on the transmission or distribution level. Conversion systems may include PV cells or solar collectors, wind turbines, power plants, heaters, electrolysers and other power-to-gas conversion technologies.
- **Energy grids** are diverse types of transmission and/or distribution grids that transport energy over some distances. Examples are electricity transmission and distribution grids, the gas grid, grids for heating and cooling systems and the water grid.
- **Energy storage systems** usually target the storage of a single form of energy, and the main purpose is to timewise decouple energy production and use. Energy storage systems also contribute to peak demand shaving, to support grid stability and power quality, and to balance disturbances resulting from integrated fluctuating power generation sources (e.g. PV and wind) or sudden changes in energy demand. It should be noted that energy grids may also decouple energy production and use. For example, the energy content in the grid may increase or decrease (e.g. via the pressure in the gas grid or temperature-level variations in thermal grids).

- *Energy sinks* usually cover all so-called end users and are also referred to as the demand side. This could be any kind of consumer, such as motors, computers, homes and offices.

The size of an integrated energy system heavily depends on the definition of the boundaries and could vary largely between smaller units (e.g. a PV panel plus inverter, cables, battery and end user) and large ones that cover entire districts or countries and include several energy vectors. When modelling modern systems, it needs to be mentioned that energy fluxes are not necessarily continuously fixed and defined in terms of magnitude or direction. Examples in this context are batteries (charging or releasing energy); single homes having, for example, smaller local energy production units included so that they are partly releasing energy to the grid or dragging energy out of the grid; and even industries.

Given the often-varying use of elements in an integrated energy system, it is necessary to generate a clear picture of the characteristics and use cases of the elements before beginning any modelling.

2.1.3 Component Characteristics

The characteristics of the different elements in a system need to be known, especially when the dynamic behaviour of integrated systems will be analysed. The characteristics of a component include the efficiency and performance for the anticipated operating range as well as its dynamics.

The efficiency and performance need to cover aspects of partial load behaviour and restrictions during operation. For example, it may be the case that power cannot be varied between 0 and 100% but is limited to variation between 50 and 100%. Furthermore, changes in ambient conditions may impact performance. These are parameters such as temperature, humidity and pressure, with the magnitude of the impact being dependent on the component itself. This information is usually provided with the technical description of the components. The background of the changes is that most components are designed for specific operating conditions. For example, gas turbines and motors are mostly designed to be placed at an altitude of sea level, 1,013 bar ambient pressure, 15 °C temperature and 60% relative humidity. Any deviation in these conditions is expected to result in reduced or increased performance and efficiency from the values that are usually given as design values.

The dynamics of the components are connected to how fast components start-up (may be different for so-called cold or warm start conditions) or shut down as well as on how fast they may react to changes in the operational set point during operation. These dynamics are usually influenced by design details and restrictions resulting from the required minimum component lifetime, considering the material properties, dimensions, paring of material, etc. Examples in the area of combined heat and power generation are summarized in Table 2.1 [1]. Differences in start-up times for a simple cycle gas turbine (GT) and combined gas turbine/steam turbine

Table 2.1 Examples of transients in different CHP units [1]

Plant type	Warm start (h)	Cold start (h)	Load variation (%/min)
Simple cycle GT	0.25	0.50	20
Combined GT/ST	1.00	2.50	15
Coal	–	–	4
Solid oxide fuel cell	0.025	25	–

(GT/ST) plants originate from different levels of complexity and the extended warm-up times of the components of the bottoming steam cycle. These are due to thick walls in the steam turbine and the heat recovery steam generator. The lower load variation in a coal plant (ST-based) is due to similar reasons, and these plants usually have large boilers that have a high thermal inertia. The very long cold start-up time in a Solid Oxide Fuel Cell (SOFC) results from material combinations (ceramics and metals). The materials used have quite different thermal expansion rates, so a careful and slow heat-up proceeding from ambient temperature to the operating temperature of 650–1000 °C is required.

While the examples shown in Table 2.1 highlight component characteristics, the characteristics of connecting elements, such as the dimensions of connecting tubes and response times of valves, also need to be considered, as indicated in the example in **part 1.2.1**. Their impact can be reduced to pressure losses only when steady-state analysis is the target of the evaluation.

2.1.4 Time Resolution

A key element in modelling the dynamic behaviour and performance of energy systems is connected to the required time resolution of the modelling. To balance the necessary effort, an analysis should be performed in terms of the following:

- Integral systems or steady-state analysis, such as analysing the yearly energy balance of systems that operate nearly stably and continuously.
- Timewise balancing of energy production, distribution and demand.
- Safety-relevant issues such as emergency shutdown of the system or elements of it.

The first type of modelling and tools to be applied fits systems having low variation in production and demand. These tools are often describing, for example, continuously operating industrial processes and their energy demand. These systems have relatively constant operating parameters for extended periods, especially when operated on a 24/7 basis. If the energy demand of these systems is supplied by dispatchable energy production units, it indicates that these systems tend to be operating in steady state. However, specific attention must be paid to operating conditions such as start-up, shut down and other transient processes. Large industrial plants and processes are often operated continuously with usually only small changes in production. An example of such an energy system may be offshore oil and gas platforms. These are systems usually without connection to onshore systems and that need to operate continuously (i.e. 24/7 basis) for economic reasons. Figure 2.1 shows a possible system on a platform that was used as a base for an analysis of energy fluxes, exergy destruction and the evaluation of possibilities for process optimization. However, it needs to be noted that energy systems may differ from platform to platform. The graph below does not consider the energy used for a residential area due to the purpose of the project. Even though production is considered to be stable (neglecting planned and unplanned changes in operation due to service and maintenance requirements), these components are exposed to long-term variations due to changes within the reservoir.

These changes result from the depletion of the oil and gas field and the connected reduced pressure in the reservoir and are long-term changes. Systems usually have a period of some years until the maximum production rate is reached, followed by decreasing production a few years later, which again lasts for 10 or more years. The production time may be extended by introducing enhanced oil recovery measures, which in turn increase the local energy need and may change the composition of

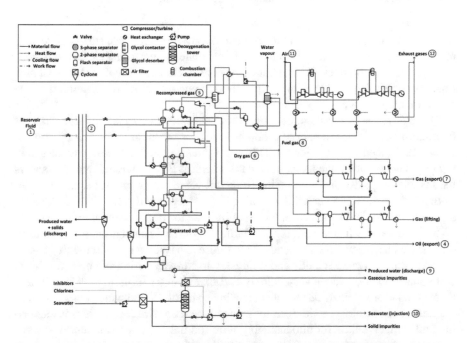

Fig. 2.1 Example of a simplified flow diagram of the energy system and flux model of an offshore platform [39]

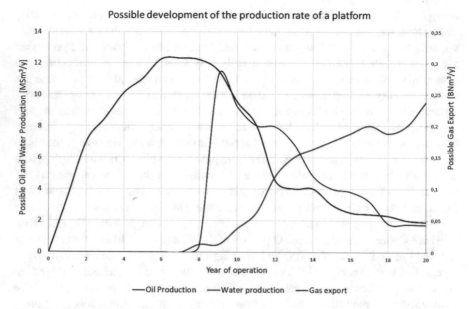

Fig. 2.2 Theoretical example of possible long-term changes in the yearly production rate on a platform from the start of production until the expected end of life

the produced oil and gas. Given the connected long-term changes, steady-state simulations are sufficiently accurate enough to evaluate the performance in the case of different production rates. To illustrate longer-term changes, Fig. 2.2 shows an artificial example of a possible changing production rate on a platform, assuming a time frame of approximately 20 years.

Quite different is the situation when it is necessary to consider a large share of fluctuating renewables in energy production and/or when large fluctuations on the demand side need to be considered. This is of increasing importance due to the growing implementation of fluctuating renewables such as PV and wind. Figure 2.3 shows the installed capacity of PV and wind in recent years. The graph is based on recent statistics from the International Renewable Energy Agency IRENA [2] and indicates a growth in installed wind power capacity by a factor of approximately 4.4 and by 26 for PV capacity. In addition, Fig. 2.4 indicates the rapid changes in solar intensity on a day with several passing clouds. This figure is based on measurements of solar intensity during a mixed cloudy and sunny day in Norway during summer and shows that solar irradiation varies very quickly and that changes of up to 80% within less than a second are typical. Due to the increasing share of PV and wind power and their fluctuating characteristics, the need for unsteady and transient evaluations is increasing and forms the base for dimensioning required balancing power, energy storage and the management and control of energy grids and energy fluxes. Depending on the system to be analysed and the targeted result, it is necessary to choose the timewise resolution and the level of detail to which the characteristics of the components must be known. For example, in the case of balancing power generation and demand, it

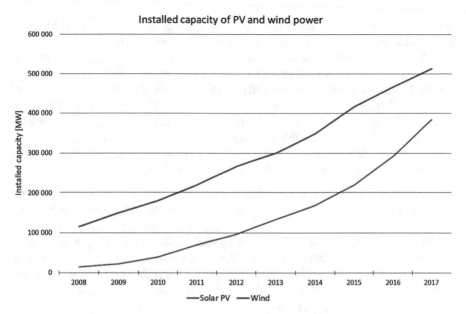

Fig. 2.3 Worldwide installed capacity of solar PV and wind-based power [3]

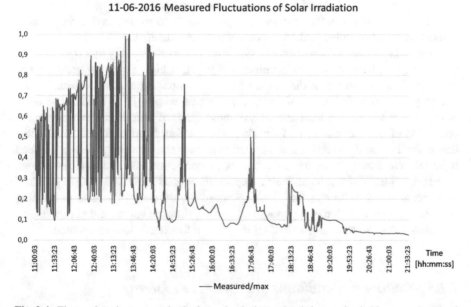

Fig. 2.4 Fluctuations in measured solar intensity during a cloudy/sunny day in Norway; visualized are the measured values divided by the maximum value in the period of measurement

is necessary to define the resolution according to the component having the highest gradient and level of fluctuations during transient operation. Transients also impact aspects of the stability of the electrical grid and/or the power quality to be considered [3–6]. The main impact is related to harmonics, performance characteristics and voltage fluctuations of PV inverters and potential imbalance between phases. In these cases, therefore, a high timewise resolution is required. The same criteria should be applied to the resolution of collected experimental data so that they can be used for verification and validation of the models and the results of simulation.

The required timewise resolution when modeling and analysing an energy system with solar energy as a source may depend on the following:

- The scale of the system and the distribution of the solar panels. In the case of a large system with geographically relatively widely distributed solar panels, the gradients of transients are expected to be smaller, as not all panels are exposed to the sun at the same point in time (also depends on how fast clouds may pass).
- When solar energy is converted into electricity, a higher resolution is required, as the entire conversion system (panel, cabling and power electronics) has a very low inertia. In contrast, a solar thermal system reacts more slowly due to a relatively high thermal inertia (heat carrier material, etc.).

A similar approach may be applied to the integration of other technologies. Consequently, the timewise resolution needs to be chosen based on the system to be modelled (e.g. electrical, thermal, or gas), phenomena to be detected and evaluated and components included in the system and their characteristics.

To evaluate the responses of energy systems towards emergency and safety events with the aim of evaluating their vulnerability [7], a high timewise resolution is required. The background is the evaluation of the propagation of large disturbances due to problems and failures of some components within the systems. The main motivation in this case is the balancing of these kinds of unforeseen events and the protection of other components from damage resulting from exposure to a disturbance. Furthermore, modeling and simulation of such events might indicate the need to install additional protective components or corrections for control. Additional information on the systems and their components beyond those available in standard documents needs to be acquired. Examples are the moment of inertia of rotating equipment, sizes of volumes, opening and closing times of valves and several more. The details depend to a large extent on the system that needs to be evaluated.

2.1.5   Boundary Conditions and System Properties

Boundary conditions and system properties are important for evaluation and consideration in advance of starting modelling.

- *Geometrical boundary conditions* are defined to identify interfaces with other systems, and are therefore key for defining the characteristics required for modelling. An example might be modelling an electrical system with electrical input from a PV unit. In this case, it is possible to either define a boundary/interface at the location where the PV unit is connected to the electrical grid (i.e. behind the inverter and power electronics) or outside the PV panel (i.e. before solar radiation hits the PV cell). In the first case, electricity (i.e. amperes as a function of time) is delivered by the inverter. The second case has solar radiation (i.e. W/m^2 as a function of time) as input but also requires the characteristics of the PV cell, the inverter and possibly other elements (e.g. in the case of local battery energy storage on the DC side). These additional components increase the complexity of the model, the stability of the iteration process during simulation and the required computing time and power. However, where the boundary condition is placed may also depend on the quality of the available characteristics of the components or energy characteristics at the boundary. A similar approach also applies to the demand side and the level of detail in its modelling as well as available information. Information on energy needs in industrial processes is often better available than that of businesses or public and private consumers.
- *Time frame* of the simulation: Depending on the time frame that needs to be simulated, the level of detail can vary. This can be as follows:

  Long-term energy system developments and simulations may target basic future trends in energy system development, and therefore cover a time frame of several years or perhaps even a few decades. In this kind of simulation, general information is required, such as the growth in yearly energy consumption and the exchange or replacement of components with the goal of developing and comparing a set of scenarios or concepts.

  Medium-term evaluation covers between one and a few years as well as aspects of seasonal balancing of energy systems. Here, a resolution of approximately 1/hour is required and sufficient, which may be covered by some generic consumption and production patterns, such as those used and available in the open-source code Energy Plan [8]. The main goal in these cases is to plan well-balanced energy systems with boundary conditions covering villages, cities, areas, countries and even interconnecting across countries. However, these cases are not well suited for evaluating the stability of a grid, even though they give some indications.

  Short-term evaluation covers a time of up to a few hours but has a fine resolution in time. This simulation requires relatively detailed information on the characteristics of the components at least at the 1/s level. When considering grid stability, the power quality of electrical or emergency events in energy systems sometimes requires a resolution of even below 1/s.

  In all three cases, validation and verification data with a similar resolution are necessary.

Single energy carrier or hybrid systems may require different modelling and simulation. This is due to differences in the responses to sudden changes on the production or demand side of the system. In comparison with other systems such as thermal systems, electrical systems have a relatively low internal inertia, meaning that without any implemented correction (e.g. via real or virtual energy storage or advanced control), a disturbance directly influences the grid and power quality [9].

2.1.6 Sensors, Data and Accuracy

As already indicated in previous paragraphs, measured data and their availability are essential for modelling and validation. The quality of the measured data is essential for the accuracy of the tool and for any activity connected to monitoring and diagnosing such systems as well as their control. It may, therefore, make sense to express some general thoughts about sensors, measured data and accuracy.

The *location* of a sensor needs to be considered for interpreting the collected values. Depending on where it is positioned within the energy system, it may represent a condition in the system. An example is a temperature sensor in a room (e.g. office or home) and the extent to which it represents the temperature inside the room, as it measures only at a single point. The positioning of the sensor may result in the measured value being influenced, for example, by direct solar radiation (it will, therefore, not represent the temperature of the air in the room) or an airflow via a partly or fully opened window or door. The first indicates a temperature higher than that of air, and the second indicates a temperature that is lower. A connected control system for air conditioning would, in the first case, reduce the temperature or increase cooling, and in the second case, increase the temperature; in both cases, these changes would most likely lead to values outside of the comfort zone of the person(s) in the room. The same also applies to installations that, after having been in operation for some time, are equipped with additional components. Research on optimized placement of sensors in, e.g. buildings, is ongoing and might eliminate uncertainties when applied in the future [10].

The *timestamp* of measured and stored values is important when modelling and analysing the dynamic behaviour of energy systems. A key is the synchronization often built-in clocks in the different components of the energy system. The necessary accuracy depends on the required sampling rate and the dynamics of the system. The higher the sampling rate and the faster the changes in transient processes are, the more accuracy that is required in synchronization. In this context, it needs to be noted that in many systems, measurements are not collected simultaneously but collected one after the other via a scanner (e.g. a scanner for measuring pressures [11]). This will cause slight differences in time between the values and will, therefore, result in problems when processing and evaluating the data, especially in transient conditions. An example is the analysis of pressure in a flow field in a relatively large channel. In these cases, a standard approach is to distribute sensors within the channel and, after collecting the data, generate a representative average value [12]. In the case of fast transients and when pressures are measured via a scanner (e.g. if the local temperature exceeds the local applicability of fast pressure transducers), generating an average value is challenging, if not impossible.

The *sensor application* may also impact the possibility of accurate modelling, simulation and analysis, especially when connected to transients with steep gradients. While a delay in transmitting electrical signals is in most cases negligible, it does not apply to sensors for pressure, temperature and concentration of substances (e.g. organic matter, pH value and turbidity of the water), which might control the relevant parameters to be considered in some systems, such as water treatment plants (see, e.g. [13]). In addition, the specific response time of a sensor can impact additional aspects of the implementation and influence the time delay between the occurrences of a change at the location where it should be measured until it is registered by the sensor. The abovementioned example of a pressure scanner might again serve to highlight this aspect. This technology is often used when, for example, temperatures exceed the limit values of, e.g. piezo-electrical measurements, at locations where the pressure should be measured (e.g. in high-temperature parts of gas turbines or industrial processes). In these cases, a hosepipe or a small tube is installed to transmit the local pressure to the location of the scanner. The length of this connection defines the time that a change in pressure at the targeted location needs to reach the scanner with the pressure sensor. Other aspects of sensor design and application contribute and might need to be considered on a case-by-case basis. However, these may be negligible if the whole chain is calibrated (e.g. [14]) and considered within the documentation, data acquisition and control system.

We should also pay attention to the impact of the *replacement* of sensors, e.g. during service and maintenance. The recorded data might then differ from those before the replacement. The main reasons for deviation may be the following:

- A different *type of sensor* or *production series* may result in the need to adjust sensor characteristics. The parameters that most likely need to be adjusted are the response rate of the sensor (inertia), its sensitivity (e.g. calibration curve) and perhaps its accuracy.
- The *changed position* of a sensor after its replacement can also result in different measured values. The impact depends mainly on the gradient of the parameter close to the location of the sensor. An example is the measurement of the material temperature in a cooled gas turbine blade, as a kind of extreme case. In these cases, gradients of approximately 100 K/mm are not unusual, indicating that only slightly different locations can, even at identical other boundary conditions, result in a different measured temperature. This, in turn, can result in a misleading interpretation of a situation of not having accurate information on the sensor position.

Another issue may result from information flow. In many cases, sensors are replaced in standard service and maintenance processes but sometimes also as a corrective action for a disturbance. While it is obvious in the second case that the existing sensor was replaced, as a corrective action was performed for exactly this purpose, in the first case, it might happen that the replacement was mentioned in a service report but not necessarily marked in the database. In such a case, it might be possible that when the systems are restarted and connected, the recorded data differs from data collected before service. This might then, in turn, lead to a false interpretation of the data. A possible sensor replacement should, therefore, also be considered a reason for deviating measurements when restarting a system after a shut down for service.

2.2 Modelling, Simulation and Analysis of Dynamics and Transients in the Energy Internet

In this section, modelling methods for the dynamics and transients of components and networks of different energy forms are introduced. Accordingly, a multi-network hierarchical coupled hybrid simulation solution algorithm is proposed with a "trunk-branch-leaf" three-tier bidirectional interface collaborative interaction structure.

2.2.1 Modelling of Conventional Power Equipment and Units

2.2.1.1 Electrical Power Grid

Conventional equipment and units in the electrical power grid comprise generators, transformers, electrical power loads, reactive power compensation devices, DC converters, generator speed regulation systems, generator excitation systems and so on. To accurately simulate an electrical power system and the responses of the equipment in the electrical power grid, detailed models of the equipment and units must be established.

- **Modelling Method**

Electrical power equipment models are generally obtained through theoretical analysis or system identification methods. Since an electrical power system is a continuous system, existing mathematical models of conventional electrical power equipment and units are usually established by differential equations, transfer functions and state-space descriptions.

Differential equations are widely used to describe the relationships between quantities in electrical power system modelling. For devices with inductors and capacitors, there is a differential relation between electrical quantities due to electromagnetic induction. There is also a differential relation between the mechanical quantities of a rotating machine, such as a motor. The general form of ordinary differential equations is as follows:

$$a_0 \frac{\mathrm{d}^n y}{\mathrm{d}t^n} + a_1 \frac{\mathrm{d}^{n-1} y}{\mathrm{d}t^{n-1}} + \cdots + a_{n-1} \frac{\mathrm{d}y}{\mathrm{d}t} + a_n y = a_1 \frac{\mathrm{d}^{n-1} y}{\mathrm{d}t^{n-1}} + c_2 \frac{\mathrm{d}^n - 2u}{\mathrm{d}t^{n-2}} + \cdots + c_n u$$

$$(2.1)$$

Fig. 2.5 Model of a
generator

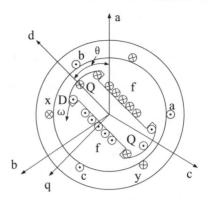

where n is the system order, a_i ($i = 1, 2, \ldots, n$) is the system structural parameter
and c_i ($i = 1, 2, \ldots, n$) is the structural parameter of the input function, which are
all real constants; y is the system output; and u is the system input.

Through theoretical analysis, equations of physical quantities are established, and
a mathematical model is obtained. Here, an example of a synchronous generator is
presented (Fig. 2.5).

According to Faraday's law of electromagnetic induction and the relationship
between the electrical quantities, the voltage equation and flux linkage equation of
the a-phase winding of the generator stator are obtained.

$$u_a = p\psi_a - r_a i_a \tag{2.2}$$

$$\psi_a = L_{aa}i_a - L_{ab}i_b - L_{ac}i_c + L_{af}i_f + L_{aD}i_D + L_{aQ}i_Q \tag{2.3}$$

In these formulas, $p = d/dt$, Ψ is the flux linkage through the winding, f represents
the excitation winding, D and Q represent the equivalent damping winding on the d
and q coordinate axes, respectively, L_{aa} is the self-inductance of a one-phase winding,
and L_{ab}, L_{ac}, $\ldots$, L_{aQ} are the mutual inductances between windings of two phases.
By using the same method, the voltage equations of rotor windings and other stator
windings can be obtained.

Based on Newton's law, the motion equation of the rotor can be obtained as shown
in Formula (2.4).

$$\begin{cases} J\frac{d^2\theta_m}{dt^2} = T_m - T_e \\ \frac{d\theta_m}{dt} = \omega_m \end{cases} \tag{2.4}$$

where

T_m is the mechanical torque obtained by the generator,
T_e is the electromagnetic torque of the generator,
ω_m is the mechanical angular displacement,

θ_m is the mechanical angular velocity and J is the rotational inertia of the generator rotor.

By combining all these equations, we obtain a mathematical model of the generator. The electrical quantities are coupled with each other. To achieve decoupling of the electrical quantities, **coordinate transformation** is needed. For different electrical power equipment models, different coordinate transformations are carried out according to the characteristics of the models. A generator model is usually expressed by **Park's equations** in the *dq0* coordinate system. A model of an asynchronous motor that rotates at the rated frequency usually adopts the *xy0* coordinate system. A three-phase symmetrical static component is modelled in a static rectangular coordinate system, that is, the *αβ0* coordinate system. In addition, partial differential equations are also used in modelling, such as for the establishment of distributed parameter models of transmission lines.

Transfer functions are also used in component modelling. A transfer function is the ratio of the output to the input of a linear system in the zero state after a Laplace transformation. Transfer functions are usually expressed as follows:

$$G(s) = \frac{Y(s)}{U(s)} = \frac{\sum_{j=1}^{n} c_{n-j} s_j}{\sum_{j=0}^{n} a_{n-j} s_j} \tag{2.5}$$

This formula can also be obtained from a Laplace transformation of formula (2.1), and the variables refer to the same physical quantities. When a model has the characteristics of module structure, the transfer functions are very practical and easy to modify when the model changes. For example, in generator excitation systems and speed regulation systems, the control units have different functions. Adding a PSS control unit to an excitation system model fully demonstrates the advantage of the transfer function model.

State-space analysis methods are also useful for modelling electrical power systems. The target system is described by the state-output equation as follows:

$$\begin{cases} \dot{x} = Ax + Bu \\ y = Cx \end{cases} \tag{2.6}$$

In this formula, u and y are the input and output of a system, respectively, and x is the state quantity. Compared with the node equation, a state-space method provides more system information, such as system eigenvalues, and the solution is very flexible. However, the biggest drawback of this method is that the modelling process is complex and is not easily programmed. In recent years, some scholars have proposed an automated state-space model generation algorithm for electromagnetic transient simulation of active distribution networks that overcomes the shortcomings and is compatible with nonlinear links, such as a distributed power supply and controller.

For modelling with discontinuous characteristics, such as discrete control systems, the main problem is determining how to search for breakpoints quickly and prevent computation errors from exceeding the requirements.

- **Model Selection**

On the premise that the requirements for the accuracy of the model be satisfied, we properly simplified the model to achieve rapidity of simulation. For different simulation conditions and requirements, different mathematical models are selected for the same device according to different simulation purposes.

A lumped parameter model is adopted in electromechanical transient simulation. The electrical components are represented by algebraic equations based on complex impedance. The electrical power load is usually modelled by a static model, but in regard to determining how different load characteristics affect the behaviour of power systems, an induction motor model is always applied. In electromagnetic transient simulation, we calculate the changes in instantaneous types of electrical variables at the microsecond level for the study of the dynamic responses of various components. The influence of nonlinearity, electromagnetic coupling and distributed parameters should be considered. In a study of the influence of a transformer in the power system, we should take the saturation of the transformer core into consideration and use the saturated transformer model.

Modelling of conventional electrical power equipment and components has been mature for many years. In recent years, micro-grid and distributed power supply systems have developed rapidly. Power electronic equipment, all kinds of new energy generation equipment and energy storage equipment have been widely used in micro-grids and interact with each other. The transient characteristics are more complex and special, and higher requirements for the accuracy of the model are put forward. Research has made some progress on determining how distributed power sources and micro-grids impact distribution networks and the corresponding models, but the requirements of practical applications cannot be satisfied.

2.2.1.2 Thermodynamic System

- **Modelling Principles**

A thermal energy system is a typical non-electrical energy system and is an important part of an energy internet system in domestic planning and construction. A thermal system contains a variety of complex physicochemical processes ranging from simple mechanical movements to complex thermal phenomena, mass transfer and chemical reactions. For example, there are a variety of complex physical and chemical processes in a furnace (combustion chamber) at the same time, such as fuel combustion, gas turbulent diffusion and heat and mass transfer and the various processes restrict and influence each other. The universal laws that reflect the dynamic processes of a thermodynamic system are listed below.

- *Law of conservation of mass*: The mass of a given fluid can neither be created nor destroyed. An increase or decrease in the fluid mass inside the control volume is equal to the sum of the fluid mass flowing in or out through the surface, and its mathematical expression is a continuity equation.

- *Newton's laws of motion, including the momentum law and momentum moment law*: Momentum law: for a given fluid, the rate of the momentum change in the control volume is equal to the total external forces applied on it. The mathematical expression is an *N-S* equation. Momentum moment law: For a rotating part, the derivative of the momentum moment versus time for a certain point O is equal to the vector sum of the moments at which the external force of the particle system is subjected to the point.
- *Law of energy conservation*: An energy increase in the control volume is equal to the sum of the net energy brought in when the working fluid flows, the net heat introduced from the outside through the interface, and the net heat released by the internal heat source minus the net work done by the working fluid.
- *State equation*: A constitutive equation that characterizes the intrinsic relationship among the dynamic variables of the working fluid. The thermodynamic variables of the working fluid include pressure, temperature, density, enthalpy, etc. Any two thermodynamic parameters can determine a thermodynamic equilibrium state, and the rest of the thermodynamic parameters can be determined by functions. The fluid is generally not in a strictly balanced state. According to the thermo-equilibrium hypothesis, the thermodynamic state at each point in the flow field is infinitely close to the equilibrium state in each moment.

  By combining the above four types of equations, the various processes of the thermal system can be described in the form of mathematical equations.

- **Modelling Methods**

Generally, thermal energy system modelling can be divided into two levels: the component-device level and component-device-system level.

- *Component-device level*. It mainly focuses on the optimization of equipment internal structure parameters, performance analysis, simulation of equipment operating conditions under variable conditions or disturbances, analysis of control characteristics, control design, etc. Component-device level modelling always omits the correlation and coupling among devices. Different models are created according to different requirements.
- *Component-device-system level*. Its focus is on equipment-system level processes and issues, including equipment selection, matching and coordinated operation in the system environment; simulation and analysis of system and equipment operating characteristics; system performance analysis; analysis and design of system and equipment control; etc. At this level of modelling, device models are associated with each other by associated state quantities or inputs and outputs.

Thermal system modelling can be divided into dynamic and static models according to the time dimension; it can also be divided into distributed parameter and lumped parameter models according to spatial dimensions. A static model assumes that the parameters do not change with time and mainly studies the relationship among each component when the system is stable under different operating conditions. Therefore, a static model is suitable for optimizing the system configuration scheme, operation mode and scheme. In a practical running system, the states and parameters are time-variant. Especially for simulation studies of transient conditions such as start-up, shut down and variable load, dynamic models are indispensable for analysing the dynamic and transient characteristics of the thermal system after a disturbance, researching the control strategy and analysing the safety and stability of the system. The calculation results of a static model can generally be used as the initial conditions of a dynamic model. A lumped parameter model treats the entire system or component as a "black box", considering only the changes in the system's inlet and outlet parameters. In system-level simulation studies, to improve the accuracy, some components that are quite important and whose parameters vary widely in space should be treated as distributed parameters.

General modelling steps
Any modelling is just a finite abstraction of an actual system and generally goes through four steps:

- **Mathematical model establishment.** This process cannot and need not consider the full details of the system but focuses on the major, essential content that reflects the behavioural characteristics of the system. According to different research purposes, different main research contents are selected, and reasonable assumptions and simplifications are made for the system. Then, based on the three conservation laws, heat transfer equations and state equations, the system's mathematical description equations are established. For dynamic simulation of thermodynamic systems, partial differential equations are often used.
- **Discretization of mathematical models.** Differential equations or partial differential equations describing actual continuous processes are transformed into discrete equations that can be computed by computers.
- **Algorithm development and programming.** According to the system operation process and the characteristics of the group of equations, an appropriate solution sequence and algorithm are determined. The stability, precision and speed of the solution should satisfy the needs of the simulation. The algorithm is converted into a computer-executable program.
- **Model verification and application.** This process is indispensable. On the one hand, it finds algorithm defects through program debugging and modifies and perfects the algorithm. On the other hand, it verifies whether the accuracy and reliability of the model meet the requirements.

Example of modelling complex thermodynamic equipment: Gas Turbine
The power system and thermal system are coupled through energy conversion devices in an energy internet system, e.g. the gas turbine, which is also the core of the CCHP system and is mainly composed of a compressor, combustion chamber, turbine and generator. The compressor, turbine and generator are installed on one shaft and rotate together. The compressor compresses the air to increase its pressure and temperature. Then, the compressed air enters the burner, mixes with the fuel and burns to form high-temperature gas. This gas enters the turbine and expands in the turbine to do work. In addition to overcoming the compression work of the

compressor, the other expansion work is used to promote the rotor of the generator
to rotate and generate high-frequency alternating current for external output. The
temperature and pressure of the flue gas decrease after expansion, and the exhaust
gas enters the waste heat recovery and utilization system after leaving the turbine.
According to the component characteristics of the gas turbine, a dynamic model of
the main components is established by a lumped parameter-based modelling method.
The compressor is a strong nonlinear component, and its operating characteristics are
generally represented by general characteristic curves, including the pressure ratio
π, converted rotating speed $n/\sqrt{T}$, converted flow $G\sqrt{T}/p$ and efficiency η. When
two of these parameters are determined, the compressor has a defined working state,
so the relationship between these parameters can be fitted as follows:

$$\begin{cases} G_c = f_{c1}(n, \pi_c) \\ \eta_c = f_{c2}(n, G_c) \end{cases} \tag{2.7}$$

Similarly, the relationship among the expansion ratio π, converted rotating speed
$n/\sqrt{T}$, converted flow $G\sqrt{T}/p$ and efficiency η of the turbine can be fitted based
on the general characteristic curves as follows:

$$\begin{aligned} G_t &= f_{t1}(n, \pi_t) \\ \eta_t &= f_{t2}(n, G_t) \end{aligned} \tag{2.8}$$

When modelling the combustion chamber, the following factors are considered:
capacity inertia, thermal inertia, combustion efficiency and pressure loss, ignoring
the heat dissipation to the outside world. Therefore, the differential equations for
outlet pressure and enthalpy are

$$\frac{dH_4}{dt} = \frac{k\left(G_f LHV\eta_b + G_a H_3 - G_4 H_4 - \left(H_4 - \frac{\sigma_b p_4}{\rho}\right)(G_f + G_a - G_4)\right)}{\rho V_b}$$

$$\frac{dp_4}{dt} = \rho \frac{k-1}{k} \cdot \frac{dH_4}{dt} + \frac{R_g T_4}{\sigma_b V_b}(G_f + G_3 - G_4) \tag{2.9}$$

When the output power of the turbine is not equal to the power consumption of
the compressor and generator, the rotor torque becomes unbalanced. This imbalance
drives the whole system into a dynamic transition process, manifested as changes in
the speed as follows:

$$\frac{dn}{dt} = \frac{900}{I \cdot n \cdot \pi^2}(W_t \cdot \eta_{mt} - W_c - W_e) \tag{2.10}$$

In addition, the physical properties of gas and air are supplemented to complete
a gas turbine mathematical model, which is mainly related to the specific heat at
constant pressure, specific enthalpy and temperature.

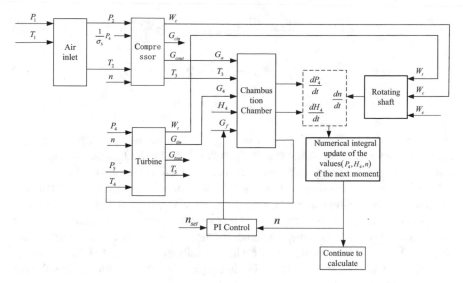

Fig. 2.6 Single-shaft gas turbine dynamic simulation model

According to the operating characteristics of the system, a single-shaft gas turbine dynamic simulation model was established using each component model, as shown in Fig. 2.6. After setting the boundary parameters and iterative initial values, dynamic simulation of the model can be achieved.

2.2.1.3 Fuel Pipe

A natural gas pipeline network consists of several main elements, including a gas source, negative pipeline, pressure station, valve and pressure regulating valve. Valves and pressure regulating valves are used to control the flow or cut off natural gas in pipelines.

Steady-state one-dimensional gas flow through a pipe is described as follows:

$$f_r = K_r s_{ij} \sqrt{s_{ij}(p_i^2 - p_j^2)} \tag{2.11}$$

where K_r is the pipeline constant, p_i and p_j are the pressures at nodes i and j, respectively, and s_{ij} is the direction of gas flow. If $p_i > p_j$, then $s_{ij} = 1$; else, $s_{ij} = -1$.

Dynamic one-dimensional gas flow through a pipe is described by Newton's motion equation, the continuity equation of mass, the state equation and the energy equation. Because gas pipelines are buried underground, it is typically assumed that the gas temperature is constant. The first three equations are shown as follows:

$$\frac{\partial(\rho v)}{\partial t} + \frac{\partial(\rho v^2 + P)}{\partial x} + \rho g \sin \alpha + \lambda \rho \frac{|v| v}{2D} = 0 \qquad (2.12)$$

$$\frac{\partial \rho}{\partial t} + \frac{\partial(\rho v)}{\partial x} = 0 \qquad (2.13)$$

$$P = \rho z R_g T \qquad (2.14)$$

By linearizing (2.11)–(2.13) under steady-state conditions and simplifying these equations under the assumption that the gas flow speed is far less than the speed of sound in the gas, ordinary differential equations (ODEs) are obtained and can be solved for various boundary conditions. The method of the characteristic curve is one way to numerically solve such equations. However, the stability of the solution process is guaranteed only if the time step is strictly limited, which always leads to an unacceptably small time step. Another method is the centred implicit difference method, the computational stability of which is fully guaranteed. A disadvantage of this method is that the computational burden rapidly increases as the scale of the gas network increases.

2.2.2 Modelling of Long Dynamics

New dynamic characteristics have been shown in multi-energy systems, including two kinds of long dynamic processes: wide time scale long dynamic processes and phased evolution long dynamic processes.

2.2.2.1 Wide Time Scale Long Dynamics

One type of long dynamics originates from the complex energy conversions in power equipment. The response time constants of the dynamic process cover a wide time range, as shown in Fig. 1.4.

For example, in a synchronous machine, two wide time scale processes are observed after the machine is disturbed by a short-circuit fault.

- A sudden change in armature current causes induced current in the damper winding (or rotor core). The time constants of these induced currents are usually less than 0.1 s, which is generally called a sub-transient effect.
- Because of the sudden change in the armature current, the induced current is induced in the excitation winding. This transient time constant is several seconds, which is called a transient effect.
- If a change in the generator terminal voltage causes a change in the exciter voltage, it leads to transients in the damping windings on the rotor. Both sub-transient and transient effects can be observed.

To model this type of long dynamics, four fundamental methods are introduced:

- **Procedure of decomposition and modelling**

For the situation where the entire equipment contains wide time scale processes, we can decompose the processes into several fast and slow sub-processes and model them separately.

- **Component decomposition and modelling**

When wide time scale processes are scattered across different components, the equipment can be broken up into components. Then, the components are modelled and linked together to obtain the equipment model.

- **Time scale compression and model simplification**

For system-level simulation, it is sometimes not necessary to build a model covering the entire time scale. Thus, the wide time scale can be compressed, and the model can be simplified.

- **Linearization and superimposition**

Due to distributed parameter characteristics, many devices and components, such as fuel pipes and thermal pipelines, are typical wide time scale dynamic elements. A practical method for modelling them is linearization and superimposition. A series of linear models of several orders are employed to approximate the wide time scale characteristics.

2.2.2.2 Phased Evolution Long Dynamics

The state of the working medium often changes during operation and start-stop of the thermal system, such as the melting process of the phase change material when absorbing heat, the solidification when storing cold, the liquid vaporization and boiling of the liquid in the water wall of the power station, the vapour condensation in condensers and the vaporization in evaporators and generators of the refrigerating equipment.

When the state of the working fluid changes, significant changes take place in its physical parameters, such as density, specific heat capacity, thermal conductivity and viscosity. On the other hand, changes in the state of the working fluid change the heat and mass transfer processes to some extent. Hence, the corresponding physical property parameters should be selected according to the different states of the working fluid during the modelling. Taking a state change of liquid as an example, the changes in its physical parameters are introduced.

- **Melting and solidification**

For a phase change energy storage material, in the early stage of the heat storage process, the material is in the solid phase before the temperature reaches its corresponding melting point. The heat flow is transmitted by heat conduction. The control

Fig. 2.7 Melting process of
a phase change material in a
vertical cavity

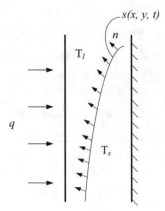

equation is a heat conduction differential equation, and its two-dimensional model
expression is

$$\rho c \frac{\partial T}{\partial \tau} = \lambda \left(\frac{\partial^2 T}{\partial x^2} + \frac{\partial^2 T}{\partial y^2} \right) \tag{2.15}$$

where ρ is the density of the solid material; c is the heat capacity; λ is the thermal
conductivity of the solid material; and T is the material temperature.

After absorbing a certain amount of heat, the phase change material is in a solid–
liquid two-phase state, as shown in Fig. 2.7. In the solid phase region, the heat flow is
transferred by conduction. In the liquid phase region, the heat flow transfer method
is not only heat conduction but also may include heat convection. At present, the
description of liquid phase heat transfer can be solved by the heat source moving
method, which establishes a conduction equation with a certain plane heat source
or radiator. According to the continuity equation and energy equation, the control
equations at the solid–liquid interface is

$$\lambda_s \frac{\partial T_s}{\partial n} - \lambda_l \frac{\partial T_l}{\partial n} = \rho r \frac{\mathrm{d}s(x, y, \tau)}{\mathrm{d}\tau} \tag{2.16}$$

$$t_s(x, y, \tau) = t_l(x, y, \tau) = T_m \tag{2.17}$$

In these equations, n represents the normal direction of the interface; $s(x, y, \tau)$ is
the solid–liquid interface function; r is the phase transition enthalpy; T_m is the phase
transition temperature; and subscripts s and l represent the solid phase and liquid
phase, respectively.

If the internal temperature distribution of the material is excluded, it is not necessary to track the position and shape of the solid–liquid interface. In this situation,
the effective heat capacity method and the enthalpy method are generally used for
modelling. The phase change process is reflected by the changes in the material's

physical property parameters, such as the expression of specific heat capacity as follows:

$$c = \begin{cases} c_s & (T < T_{M1}) \\ \frac{c_s + c_l}{2} + \frac{r}{T_{M2} - T_{M1}} & (T_{M1} \leq T \leq T_{M2}) \\ c_l & (T > T_{M2}) \end{cases} \qquad (2.18)$$

Heat is continuously absorbed, and the phase change material completely turns into liquid. Then, the heat flow is transferred by heat conduction and natural convection. The heat transfer method must be judged by the Grash of number G_r. When $G_r \leq 2860$ for the vertical interlayer and $G_r \leq 2430$ for the horizontal interlayer (the underside is hot), the heat transfer in the interlayer depends on heat conduction. When G_r exceeds the above values, natural convection begins to form within the interlayer, and convection develops increasingly violently as G_r increases. When G_r reaches a certain value, natural convection transits from laminar to turbulent flow. The key to the convective heat transfer problem is the solution to the heat transfer coefficient h.

The temperatures of the hot and cold surfaces of the cavity are T_h and T_l, respectively, the characteristic temperature is taken as $(T_h + T_l)/2$, and the characteristic length is taken as the distance δ between the hot and cold surfaces; then

$$Gr = \frac{g\alpha_V(T_h - T_l)\delta^3}{\nu^2} \qquad (2.19)$$

$$Pr = \frac{\nu\rho c}{\lambda} \qquad (2.20)$$

$$Nu = C(Gr\,Pr)^n \qquad (2.21)$$

$$Nu = \frac{h\delta}{\lambda} \qquad (2.22)$$

where $\alpha_v = -\frac{1}{\rho}\left(\frac{\partial \rho}{\partial t}\right)_P \approx -\frac{1}{\rho}\frac{\rho_\infty - \rho}{T_\infty - T}$ is the expansion coefficient; ν is the kinematic viscosity; and C and n are determined by experiments and related to the heat transfer surface shape and location, thermal boundary conditions and different flow patterns.

- **Vaporization and condensation**

Vaporization has two forms: evaporation and boiling. Evaporation occurs on the surface of a liquid and can be described using only a set of mathematical equations. Boiling takes the form of bubbles in a liquid. This process is quite complicated and cannot be completely described by a set of mathematical equations. During the boiling of a large vessel, as the superheat Δt (temperature deviation between the surface and the saturated liquid) increases, heat exchange between the heating surface and the liquid will take place along with the processes of natural convection, nucleate boiling, transition boiling and film boiling, and each stage can be represented by the mathematical equations shown in Table 2.2.

There are two kinds of condensation: film condensation and bead condensation. For film condensation, the cold surface is always covered by a liquid film, and the liquid film layer becomes the main thermal resistance. For bead condensation, the thermal resistance on the cold surface is negligibly small. Therefore, heat transfer studies of the condensation process are primarily directed towards film condensation. The flow pattern in the liquid film is divided into laminar and turbulent flow, so it is necessary to use the film R_e to determine the flow pattern, as shown in Table 2.3.

2.2.3 Modelling of Networks

2.2.3.1 Electrical Power Grid

All the components of the electrical power system are connected to the grid through the transmission lines and electrical power transformers, and all kinds of compensation devices are in the electrical transmission network. The complex interactions among massive components in the network make system analysis very difficult. Consequently, the transient process of each component needs to be considered simultaneously. In regard to the establishment of transient models of all network components, the equivalent circuit of transient solutions is derived based on the differential equations of each component.

• **Network modelling**

The transient model of a transmission line is based on the π equivalent circuit of concentrated parameters (if the wave process is considered, a distributed parameter model is used) and is composed of resistance, inductance, capacitance and other factors. Among them, the ground capacitance model for transmission lines is shown in Fig. 2.8.

The electromagnetic transient model in the abc coordinate system can be described as follows:

$$i_{abc} - i'_{abc} = C_{abc} p u_{abc} \tag{2.23}$$

$$C_{abc} = \begin{bmatrix} 2C_m + C_g & -C_m & -C_m \\ -C_m & 2C_m + C_g & -C_m \\ -C_m & -C_m & 2C_m + C_g \end{bmatrix} \tag{2.24}$$

where i is the current injected into the node; i' is the current out of the node; u is the voltage of the node; p is the differential operator; C_g is the wire-to-ground capacitance; C_m is the wire-to-wire capacitance; and i_{abc}, i'_{abc} and u_{abc} are all instantaneous values. After the $\alpha\beta0$ coordinate transformation, the model is transformed into the following equation:

Table 2.2 Different stages of the boiling process and the corresponding mathematical equations

Different stages of the boiling process	Mathematical equations		
Natural convection zone	The degree of superheat is small ($\Delta t < 4$ °C for water at one atmosphere), and no bubbles are generated on the wall surface	e	t_w is the wall temperature, t_∞ is the ambient temperature and l is the characteristic length
Nucleate boiling zone	When the degree of superheat $\Delta t \geq 4$° C, bubbles begin to form at the vaporization centre of the heating surface, Δt increases, vaporization cores increase, and steam blocks and steam columns are synthesized. The end of this process corresponds with the peak heat flux density	$h = Cq^{0.67}M_r^{-0.5}p_r^m(-\lg p_r)^{-0.55}$ $C = 90(W^{0.33}/(m^{0.66} \cdot K))$ $m = 0.12 - 0.21\lg\{R_p\}_{\mu m}$	Mr is the relative molecular mass; p_r is the ratio of liquid pressure to its critical pressure; R_p is the surface average roughness; and q is the heat flux density
Transition boiling zone	Δt continues to increase, and the heat transfer law is very unstable. The heat flux density decreases with increasing Δt. This process continues until the heat flux density reaches the minimum value q_{min}		
Film boiling zone	A stable vapour film layer forms on the heating surface, the generated steam is regularly discharged from the film layer, and q increases as Δt increases	$h_c = C\left[\frac{gr\rho_v(\rho_l-\rho_v)\lambda_v^3}{\eta_v d(t_w-t_s)}\right]^{1/4}$ $h_r = \frac{\varepsilon\sigma\left(T_w^4-T_s^4\right)}{T_w-T_s}$ $h^{4/3} = h_c^{4/3} + h_r^{4/3}$	d is the characteristic length; η is the dynamic viscosity; $C = 0.62$ for a transverse tube, and $C = 0.67$ for a spherical surface; σ is the Stefan–Boltzmann constant; and ε is the emissivity

Table 2.3 Determining the flow pattern and mathematical equations of film condensation

Determining the flow pattern		Mathematical equations
Vertical wall	$Re = \frac{4hl(t_s - t_w)}{\eta r}$	$Re \cdot 1600,$ laminar flow	$h = 1.13\left[\frac{g\rho^2\lambda^3 r}{\eta(t_s - t_w)l}\right]^{1/4}$
		$Re > 1600,$ turbulence	$Nu = Ga^{1/3}\dfrac{Re}{58\,\mathrm{Pr}_s^{-1/2}\left(\mathrm{Pr}_w/\mathrm{Pr}_s\right)^{1/4}\left(Re^{3/4} - 253\right) + 9200}$ $Nu = hl/\lambda,$ $Ga = gl^3\big/v^2,$ $Pr = v\rho c/\lambda$
Horizontal pipe	$Re = \frac{4\pi hd(t_s - t_w)}{\eta r}$	Smaller diameter, laminar flow	$h = 0.729\left[\frac{g\rho^2\lambda^3 r}{\eta(t_s - t_w)d}\right]^{1/4}$

Fig. 2.8 Ground capacitance model of a transmission line

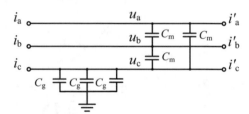

$$\begin{bmatrix} i_\alpha \\ i_\beta \\ i_0 \end{bmatrix} - \begin{bmatrix} i'_\alpha \\ i'_\beta \\ i'_0 \end{bmatrix} = \begin{bmatrix} C_\alpha & 0 & 0 \\ 0 & C_\beta & 0 \\ 0 & 0 & C_0 \end{bmatrix}\begin{bmatrix} \rho u_\alpha \\ \rho u_\beta \\ \rho u_0 \end{bmatrix} \qquad (2.25)$$

where $C_\alpha = C_\beta = C_g + 3C_m,\ C_0 = C_g$. After differentiation of the equation by an implicit trapezoidal method, we can obtain the following:

$$\begin{bmatrix} i_\alpha(t) \\ i_\beta(t) \\ i_0(t) \end{bmatrix} - \begin{bmatrix} i'_\alpha(t) \\ i'_\beta(t) \\ i'_0(t) \end{bmatrix} = \begin{bmatrix} G_{C_\alpha} & 0 & 0 \\ 0 & G_{C_\beta} & 0 \\ 0 & 0 & G_{C_0} \end{bmatrix}\begin{bmatrix} u_\alpha(t) \\ u_\beta(t) \\ u_0(t) \end{bmatrix} + \begin{bmatrix} I_{C_\alpha}(t - \Delta t) \\ I_{C_\beta}(t - \Delta t) \\ I_{C_0}(t - \Delta t) \end{bmatrix} \qquad (2.26)$$

According to this equation, we can draw the equivalent circuit diagram of the equivalent terminal model of the transmission line, as shown in Fig. 2.9. The differential equations and transient equivalent circuits of other network components can be obtained by using the same method.

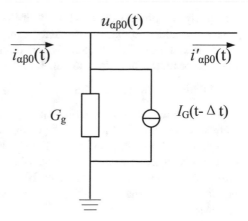

Fig. 2.9 Equivalent terminal model of a transmission line

After transient models of all components are obtained, according to the differential equations of each branch in the network and Kirchhoff's current law, each node equation of the single-phase circuit in the n-node network can be derived as follows:

$$G_{i1}u_1(t) + G_{i2}u_2(t) + \cdots + G_{ij}u_j(t) + \cdots + G_{in}u_n(t) = i_i(t) - I_i \quad (2.27)$$

where G_{ij} is the equivalent mutual conductance between nodes i and j, G_{ii} is the equivalent self-conductance of node i and I_i is the sum of the currents of the equivalent current sources of all branches connected to node i. Combining all node equations, we can obtain the following:

$$\begin{bmatrix} G_{11} & 0 & \vdots & 0 \\ 0 & G_{22} & \vdots & 0 \\ \vdots & \vdots & \vdots & \vdots \\ 0 & 0 & \vdots & G_{nn} \end{bmatrix} \begin{bmatrix} u_1(t) \\ u_2(t) \\ \vdots \\ u_n(t) \end{bmatrix} = \begin{bmatrix} i_1(t) \\ i_2(t) \\ \vdots \\ i_n(t) \end{bmatrix} - \begin{bmatrix} I_1 \\ I_2 \\ \vdots \\ I_n \end{bmatrix} \quad (2.28)$$

This equation can be simplified into the following form: $G \cdot u(t) = i(t) - I$. Extending this equation to the three-phase system, we obtain an expression in $\alpha\beta0$ coordinates as follows:

$$G_{\alpha\beta0}u_{\alpha\beta0}(t) = i_{\alpha\beta0}(t) - I_{\alpha\beta0} \quad (2.29)$$

The basis for solving the nodal equation is the acquisition of a companion admittance matrix. In an electrical power network, because the neutral points of some transformers are grounded, there are some differences between the positive sequence equivalent circuit and zero sequence equivalent circuit: some nodes do not exist in the zero sequence equivalent circuit. Since the $\alpha\beta$ component equations are decoupled from the zero sequence component equations, their admittance matrixes are formed separately. To prevent the singularity of the admittance matrix, a virtual equation $U_{k0} = 0$ is added to the zero sequence component equation of node K, which does not exist in the zero sequence equivalent circuit, so that the diagonal elements in the admittance matrix of node K are all 1, and it has no effect on the equation.

- **Machine–network interface**

After obtaining the node equations of the network, we connect them with the equations of other electrical power equipment, such as generators and electrical loads, to obtain all the state quantities of the whole system. Take the constant E'_q model of a generator as an example. The transient electromotive force of the generator q axis, which is called E'_q, is constant. By eliminating the flux linkage in the Park's equations of the generator model, we can obtain the equations in $dq0$ coordinates as follows:

$$\begin{cases} u_d = -x'_d \rho i_d - r i_d + x_q i_q \\ u_q = E'_q - x'_d i_d - r i_q + x'_q \rho i_q \end{cases} \tag{2.30}$$

where d and q represent the direct axis and quadrature axis, respectively, which rotate at the synchronous speed of the generator; X'_d and X'_q are the direct and quadrature axis transient reactances of the generator, respectively; and p is the differential operator. To interface with the electrical network, the group of equations must be projected into $\alpha\beta0$ coordinates. After differentiating the equations in the $\alpha\beta0$ coordinate system, we can obtain the following equation:

$$i_{\alpha\beta0}(t) = A_{G_{\alpha\beta0}} u_{\alpha\beta0}(t) + B_{G_{\alpha\beta0}} \tag{2.31}$$

where $A_{G\alpha\beta0}$ is a matrix composed of the coefficients in the generator difference equation and $B_{G\alpha\beta0}$ is a column vector that reflects historical records. By substituting formula (2.31) into formula (2.29), we can obtain the following equation:

$$(G_{\alpha\beta0} - A_{G_{\alpha\beta0}}) u_{\alpha\beta0}(t) = B_{G_{\alpha\beta0}} - I_{\alpha\beta0} \tag{2.32}$$

The voltage of all network nodes can be obtained according to this equation, and then the current of each branch and the internal variables of the generator can be obtained according to the voltage.

Additionally, some important points are discussed. First, for an infinite source, the current injected into the network cannot be expressed by terminal voltage, so the nodal voltage is treated as a known boundary. Second, to simplify the model when studying a subsystem of the electrical power network in electromagnetic transient simulation, we need to dynamically simplify the external power grids. Third, to solve the electrical network equation, we need to initialize all electrical power element models.

2.2.3.2 Thermodynamic System

The simulation model of the thermal network, which is closely related to the network topology, is used to describe the thermal dynamic coupling relationship of the thermal equipment. To solve the problems of boundary conditions, modular modelling, modular generality, portability, etc., nodal pressure equations and branch flow equations are proposed. Nodal pressure equations neglecting the inertia of the branches can generate a large sparse matrix with symmetrical and diagonally dominant features. Computational stability can be improved by using the implicit Euler integral method and efficient sparse matrix techniques.

- **Mathematical Model**

The thermodynamic system is a large fluid network system that includes nodes and branches with the remaining network topology. The nodes are connected by vector branches and have a certain volume. Suppose that the fluid in a node reaches equilibrium, has mass and exchanges heat with the environment. A branch has a certain length and area and has no mass or heat exchange with the environment.

Nodal mass equation:

$$V_i \frac{d\rho_i}{dt} = \sum_{j=1}^{k} INP_{ij} G_m \tag{2.33}$$

where $\frac{d\rho_i}{dt} = \frac{\partial \rho_i}{\partial P_i} * \frac{dP_i}{dt_i} + \frac{\partial \rho_i}{\partial h_i} * \frac{dh_i}{dt_i}$. $\frac{dh_i}{dt_i} \ll \frac{dP_i}{dt_i}$. By ignoring the enthalpy terms, the nodal mass equation is shown as follows:

$$C_i \frac{dP_i}{dt} = \sum_{j=1}^{k} INP_{ij} G_m \tag{2.34}$$

where $C_i = V_i \frac{d\rho_i}{dP_i}$ is the compressibility of the fluid.

Branch momentum equation:

$$L_a \frac{dG_m}{dt} = INP_{ij}(P_j - P_i + INP_{ij} H_m) - R_m G_m |G_m| \tag{2.35}$$

Nodal energy equation:

$$M_i \frac{dh_i}{dt} = \sum_{j=1}^{k} \text{MAX}(INP_{ij} G_m, O)h_j - h_i \sum_{j=1}^{k} \text{MAX}(INP_{ij} G_m, O) + V_i \frac{dP_i}{dt} + Q_i \tag{2.36}$$

where $M_i = V_i * \rho_i$ is the node quality. In the case under small perturbations, $V_i \frac{dP_i}{dt}$ can be ignored.

Therefore, the base equations of the simulation model of the thermal fluid network can be shown as follows:

Mass equation:

$$C_i \frac{dP_i}{dt} = \sum_{j=1}^{k} INP_{ij}G_m + C_i(P_a - P_i)/T_c \tag{2.37}$$

Momentum equation:

$$L_a \frac{dG_m}{dt} = INP_{ij}(P_j - P_i + INP_{ij}H_m) - R_mG_m|G_m| \tag{2.38}$$

Energy equation:

$$M_i \frac{dh_i}{dt} = \sum_{j=1}^{k} \text{MAX}(INP_{ij}G_m, O)h_j - h_i \sum_{j=1}^{k} \text{MAX}(INP_{ij}G_m, O) + V_i \frac{dP_i}{dt}$$
$$+ Q_i + M_i(h_a - h_i)/T_m \tag{2.39}$$

The term $M_i(h_a - h_i)/T_m$ represents the heat dissipation to the environment, and the inertia time constant is T_m; the term $C_i(P_a - P_i)/T_c$ represents the leak to the environment, and the inertia time constant is T_c.

- **Solution Method**

For the sake of convenience, ignoring the pipeline inertia $L_a = 0$, the result obtained from formula (2.35) is shown as follows:

$$G_m = INP_{ij}(P_j - P_i + INP_{ij}H_m)/(R_m * |G_m|) + C_i(P_i - P_a)/T_c \tag{2.40}$$

Then, the mass equation can be expressed as follows:

$$C_i \frac{dP_i}{dt} = \sum_{j=1}^{K} (P_j - P_i + INP_{ij}H_n)/(R_m * |G_m|) + C_i(P_a - P_i)/T_C \tag{2.41}$$

By integrating differential Eqs. (2.39) and (2.41), the algebraic forms of the mass equation and energy equation are as follows:

$$C_i^t \frac{P_i^{t+1} - P_i^t}{\Delta t} = \sum_{j=1}^{K} (P_j^{t+1} - P_i^{t+1} + INP_{ij}H_m^t)/(R_m^t * |G_m^{t+1}|)$$
$$+ C_i^t(P_a^t - P_i^{t+1})/T_c \tag{2.42}$$

$$M_i^t \frac{h_i^{t+1} - h_i^t}{\Delta t} = \sum_{j=1}^{K} \text{MAX}(INP_{ij} * G_m^{t+1}, O) h_j^{t+1}$$

$$- h_i^{t+1} \sum_{j=1}^{K} \text{MAX}(INP_{ij} * G_m^{t+1}, O)$$

$$+ Q_i^t + M_i^t (h_a - h_i^t) / T_m \tag{2.43}$$

The matrix forms of formulas (2.42)–(2.43) are as follows:
Nodal pressure equation:

$$A^t [P^{t+1}] = [B^t] \tag{2.44}$$

Energy equation:

$$D^t [H^{t+1}] = [E^t] \tag{2.45}$$

The coefficient matrixes A and D are diagonally dominant and symmetric. Due to stability and convergence-related problems, the key to the solution is double factorization. A framework diagram of a computing program is shown in Fig. 2.10.

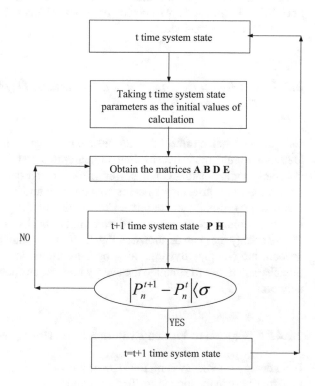

Fig. 2.10 Framework diagram of a thermal system simulation program

2.2.3.3 Fuel Pipeline Grid

According to the law of mass conservation, the input mass flow and output mass flow should be equal at any node of the fuel grid at all times, which yields the node equivalence function

$$Q_n + \sum_{k \in C_n} \alpha_{nk} M_{nk} = 0 \tag{2.46}$$

where

Q_n is the incoming flow from outside to node n,
C_n is the set of components that are directly connected with node n,
M_{nk} is the absolute value of the gas flow from component k to node n and

$$\alpha_{nk} = \begin{cases} 1, & \text{flow is incoming from } k \text{ to } n \\ -1, & \text{flow is outgoing from } n \text{ to } k \end{cases} \tag{2.47}$$

By combining the pipe dynamic functions (2.12)–(2.14) and the node equivalence function (2.47), a dynamic model of the fuel grid is obtained. As mentioned in Sect. 2.2.1.3, this model can be solved by the characteristic method or the centred implicit difference method.

2.2.4 Multi-mode and Phased Evolution Hybrid Simulation Framework

Enlightened by electromechanical/transient multi-mode hybrid simulation and physical/digital multi-mode hybrid simulation architectures and inspired by the theory of interface methods and AC/DC power grids, we can extend the multi-mode hybrid simulation to multi-energy systems containing interactions among different energy forms and multi-network coupling. Combined with the phase change of the working medium, a multi-mode and phased evolution hybrid simulation method is obtained for the energy internet, as shown in Fig. 2.11. The simulation architecture takes into account the complex dynamic long process simulation of energy internet interaction characteristics to simulate the energy interactions between devices and adjacent networks.

2.2.4.1 Solution of Phased Evolution Long Dynamics

In an energy internet system, the state of the working medium in an energy component may vary in the long term. To ensure modelling accuracy, mathematical models

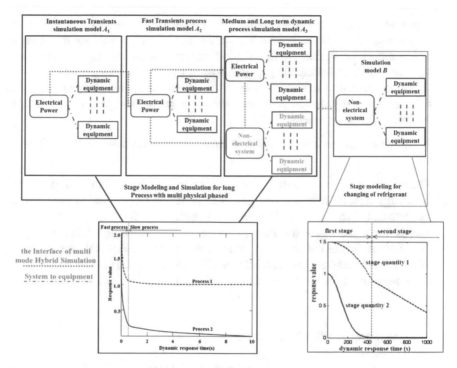

Fig. 2.11 Basic structure of the multi-mode and phased evolution hybrid simulation method

have to be formed separately for the working medium at different stages. The phased evolution hybrid simulation method introduces phases/stages into the dynamic process, where the model and the algorithm are determined by the stage features. The principles of stage definition can be diverse. On the one hand, since characteristics are dramatically different between states, the stages can be divided by the points where the state of the working fluid changes. On the other hand, if the dynamic processes last for a long time, the stages can be divided by the decay time of the dominant dynamic processes.

The simulation modelling method for the phased long process is as follows:

- The dominant dynamic processes of equipment are abstracted and refined. On this basis, the boundary points and boundary conditions for the transition of the energy carrier/working medium state and physical properties are determined. According to the dominant characteristics of the critical state variables before and after a disturbance, a transition connecting method is formulated.
- Mathematical models of dynamic processes for dynamic components in each stage are established, including differential equations describing energy conversion and transmission and algebraic equations describing the physical properties of working fluids and components.

- A physical parameters database should be established for each stage of working fluid transport. By extracting the key state variables and combining the physical constraints of the working fluid, a smooth transition between adjacent stages can be achieved by modelling the physical properties of the working fluid.
- Dynamic simulation models are established for some specific staged deduction of long process dynamic equipment, such as thermal storage and cold storage devices of thermal systems.

A dynamic process can be divided into stages according to the following conditions:

- The state of the energy transfer medium changes drastically. For example, when the energy transfer medium changes from liquid into a gas–liquid mixed state, the state and characteristics of the medium change significantly, and the medium state before and after the qualitative change can be divided into different stages.
- The dynamic process of an energy component includes a physical process with a wide range of response times and can be divided into various stages according to the time required for the decay of the dominant process in each component.

2.2.4.2   Solution of Wide Time Scale Long Dynamics

The large time scale characteristics of the energy internet and the differences in the operating mechanisms of different energy forms have brought substantial difficulties to energy internet simulation. To decompose and overcome these difficulties, a hybrid simulation method is adopted to spatially divide the entire energy system into multiple subsystems and solve each subsystem by a suitable solution method. In the process of solving a subsystem, other parts are simplified and modelled on the dividing boundaries, and components and parameters are exchanged at a fixed interval on the dividing boundaries. A typical example is the electromagnetic/electromechanical transient hybrid simulation of an electrical power grid. The spatial segmentation of the energy system is based on the following conditions:

Devices and local systems with similar response times are arranged into the same subsystem. Large-capacity power electronics devices, as HVDC links in the power system, and the local power grid that needs to be tested and analysed in detail are modelled and simulated by the electromagnetic transient mode; the conventional power grid, generator, load and other links are modelled and simulated by the electromechanical transient mode.

Different energy systems and equipment that are used by different modelling and solution methods should be modelled separately and connected with each other through an established interface at the boundaries.

The equivalent model of the hybrid simulation interface is studied. First, based on the experience of modelling the hybrid simulation mapping interface of the grid, a hybrid simulation interface model of the energy internet system is established. The equivalent modelling of the hybrid simulation interface between different energy subsystems or between different simulation modes should represent the physical process. The physical process of the interface interaction is mapped to a specific time scale.

2.2.4.3 Three-Layer Multi-mode and Phased Evolution Algorithm: Key Techniques

- **Define the boundaries of subnets and sub-blocks**

According to the multi-stage and multi-mode division in the system, it is necessary to clarify the interaction form and the amount of interaction of cross-energy interfaces or different modes.

- **Establishment of equivalent interface model**

The equivalent model of the external network is significant. In multi-mode hybrid simulation, the rest of each subsystem is modelled on the boundary between subsystems. The equivalent model of an interface should reflect the physical process and interaction mechanism of energy dynamic exchange in an energy system and take into account the differences in model equation type, calculation algorithm and step size.

- **Interactive timing design**

According to the different characteristics of the system during normal operation and fault status, a reasonable and efficient interactive time sequence of the hybrid simulation interface is designed to ensure that the speed and accuracy of hybrid simulation meet the best requirements.

In this paper, a *parallel hybrid real-time simulation* interactive computing sequence based on prediction and correction mechanisms is designed. It is summarized as follows.

In a traditional parallel interactive computing time series of a hybrid simulation, prediction and correction mechanisms are added, including (1) when computing resources are allowed, the variable interaction step strategy is adopted to increase the interaction frequency of the hybrid simulation interface in a short time after a fault/large disturbance; (2) the bilateral boundary conditions are predicted at the time of a fault/large disturbance t_{0+}; (3) the waveform prediction of bilateral interface boundary conditions is extrapolated for a time step with no-fault/large disturbance occurrence; and (4) post-correction based on the energy compensation mechanism.

A cooperative interactive calculation mechanism of the bidirectional interface algorithm structure is employed, which uses the **trunk-branch-leaf** three-layer form. The minimum cost of iteration is used to eliminate the interface errors, and multiple sub-network blocks and a multi-layer network cooperative computing system are designed.

In terms of network structure, the energy internet system can be divided into three layers: a leaf layer, branch layer and trunk layer. Among them, the leaf layer represents the dynamic elements; the branch layer represents the energy network of each attribute; and the trunk layer represents the energy internet. From the view of simulation, modelling solves problems of the leaf layer and part of the branch layer, i.e. problems of dynamic component modelling and numerical solution, and the establishment of the interaction relationship between energy equipment and a network interface with the same attributes; the multi-mode hybrid simulation method yields the joint solution of energy networks with the same attributes and the joint solution of the interaction of different energy networks with the energy internet, namely, the problem of solving the branch layer model and the backbone layer model. The key technologies are the equivalent method and model of the cross-attribute energy interface and the interface interaction calculation mechanism of multi-layer coupling.

The equivalent modelling problem of the hybrid simulation interface can be adopted in the fields of non-power system simulation and multi-energy system simulation. The existing research foundation of simulation technology in power systems can be used as a reference, which includes the hybrid simulation interface model of power grid mapping intersection and a distribution network and the electromechanical and electromagnetic transient and medium–long-term process simulation model of equipment in system simulation. The physical process of the interface interaction is mapped to a specific response time scale and modelled based on the coupling mechanism of the energy subnet, dynamic response time scale of the subnet and numerical solution of the model.

Based on the iterative computing mechanism between the equipment and the system in traditional power grid simulation and the iterative computing mechanism between digital simulation neutron networks in a hybrid power grid, the interactive computing mechanism of the hybrid simulation interface is improved. It absorbs the advantages of mature algorithms and extends to multi-layer network coupling of an energy internet system. The "trunk-branch-leaf" three-layer model of the energy internet system is solved, and a flexible interactive sequence of two-way interactions between the trunk layer and the branch layer and between the branch layer and the leaf layer is formed.

2.2.4.4 Three-Layer Multi-mode and Phased Evolution Algorithm: Design and Implementation

As mentioned above, the dynamic response times of the power system, thermodynamic system and chemical fuel pipeline system are quite different; therefore, they can be divided into three simulation stages. The dynamic response time scale of the power system is small, and the simulation step size is set to Δte_{i+1}. Considering that the density of the transmission medium (mainly water) in the thermal system is higher than the density of the medium (mainly gas) in the chemical fuel network, the dynamic response time is relatively faster, so $\Delta tt_{j+1} < \Delta tf_{k+1}$ is set, and $\Delta tf_{k+1}/\Delta tt_{j+1}$ and $\Delta tt_{j+1}/\Delta te_{i+1}$ are integers.

The calculation process of the solution is as follows (Figs. 2.12, 2.13, 2.14 and 2.15):

1. Set the maximum number of iterations for each system, n_{max}, m_{max} and p_{max} in turn.
2. When the algorithm is first started, the program reads the initial data of the three kinds of energy system configurations and exchanges the required variable data at the interface. In subsequent calculations, the variable data are exchanged at the interface. Based on the interface variable data, the boundary conditions calculated by the energy subsystem during simulation are obtained.
3. Determine whether there are faults and/or disturbances in the energy subsystem.

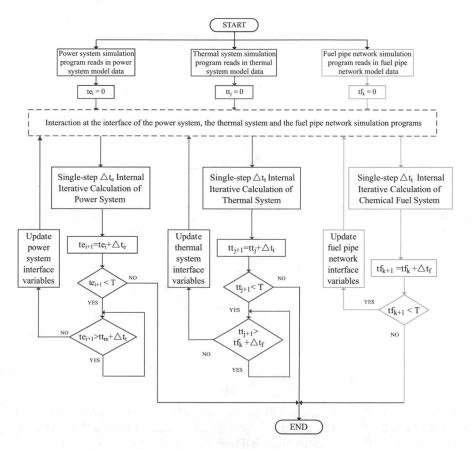

Fig. 2.12 Trunk-branch-leaf three-tier computing architecture

Fig. 2.13 Iterative
calculation of the power
system

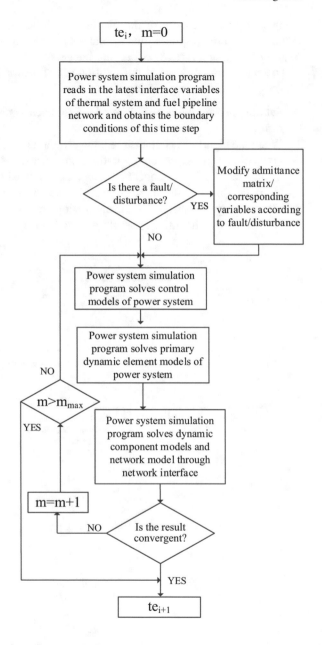

In the power energy subsystem, if there is a fault, the admittance matrix is modified according to the fault, and/or in the first energy subsystem, if there is a disturbance, the corresponding variables are modified according to the disturbance.

Fig. 2.14 Iterative
calculation of the thermal
system

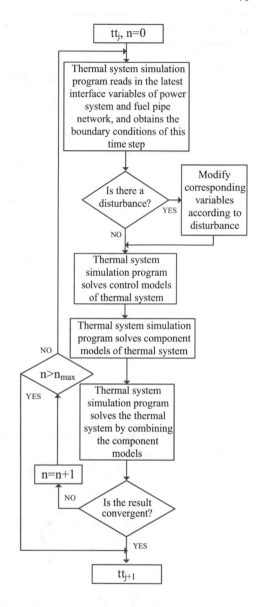

If there is a disturbance in the thermal energy subsystem or the chemical fuel pipeline system, the corresponding variables are modified according to the disturbance.

4. The energy subsystems are modelled, and the results are obtained.

In the above three systems, the control model, dynamic components and network interface of each subsystem are solved in turn. Finally, the dynamic components are solved in conjunction with the system.

Fig. 2.15 Iterative
calculation of the chemical
fuel pipeline system

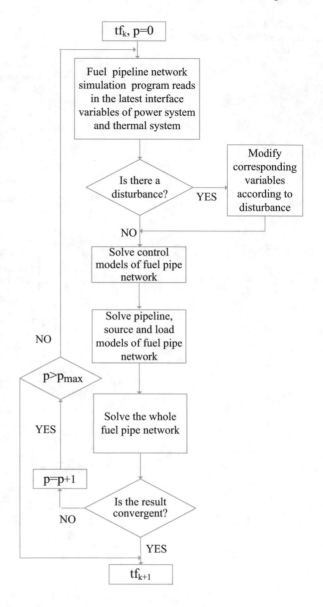

5. The convergence of the energy subsystem solution is judged separately:

If the number of iterations of the power energy subsystem is less than or equal to the maximum number of iterations of the power energy subsystem, step 3 is executed.

   If the number of iterations of the thermal energy subsystem is less than or equal to the maximum number of iterations of the thermal energy subsystem, step 2 is executed, and the interface variables of the power energy subsystem are re-read for calculation.

If the number of iterations of the chemical fuel pipeline subsystem is less than or equal to the maximum number of iterations of the chemical fuel pipeline subsystem, step 2 is executed, and the interface variables of the power energy subsystem and the thermal energy subsystem are re-read for calculation.

In the calculation process, the simulation calculation times of the three kinds of energy systems are judged. If the cumulative simulation time of the small-step simulation subsystem is greater than that of the large-step simulation subsystem, the small-step simulation subsystem will continue to perform the judgement repeatedly until the simulation time of the small-step simulation subsystem is less than or equal to the cumulative simulation time of the large-step simulation subsystem. If the cumulative simulation time of the small-step simulation subsystem is less than or equal to the cumulative simulation time of the large-step simulation subsystem, the small-step simulation subsystem updates the transmitted interface variables and executes step 2.

2.3 Software Suites/Tools for Energy System Simulation

2.3.1 Overview of Energy System Dynamic Simulation Software

Power system dynamic simulation is one of the important means to study and analyze how to build the mathematical model suitable for the actual circuit and power grid. With the continuous expansion and adoption of various new techniques, modern power grids are becoming increasingly complex, and efficient simulations are becoming increasingly important for planning, characteristic analysis, control and protection design, dispatching assistance and testing and troubleshooting in power system engineering. The development of simulation technology of energy system including power system is the same. The following is a brief introduction of commercial simulation platform commonly used in energy system and subsystem, as shown in Table 2.4.

Table 2.4 summarizes the commonly used energy simulation software.

2.3.2 Software Suites or Tools for Electrical Energy Simulation

2.3.2.1 EMTP (Electromagnetic Transients Program)/ATP

The EMTP (Electromagnetic Transients Program)/ATP (Alternative Transients Program) [15] is primarily used to calculate electromagnetic transient processes in power systems. The power system analysis program EMTP/ATP, which is characterized by

Table 2.4 A list of simulation software

Name	Type	Software name
Electrical energy simulation	Electromagnetic transient simulation	EMTP/ATP PSCAD/EMTDC NETOMAC
	Electromechanical transient simulation	PSASP BPA NETOMAC
Non-electrical energy simulation	Non-electrical energy simulation	3KEYMASTER ProTRAX RealPipe-Gas
Multi-energy simulation	Mixed energy simulation	MERTS

large-scale simulation and powerful functions, is an internationally accepted standard electromagnetic transient calculation.

1. Structure

This program consists of an electromagnetic transient calculation program, auxiliary support subroutine and drawing function subroutine. The specific software structure is shown in Fig. 2.16.

2. Function

- Calculating overvoltage;
- Calculating the transient recovery voltage of the circuit breaker;
- Calculating the overvoltage of the isolation switching operation;
- Calculating the inrush current of the transformer;
- Simulating the shaft torsional vibration phenomenon (SSR) of the generator;
- Simulating the start process of the induction motor;
- Simulating the interaction between AC and DC systems;
- Simulating a fault in the DC system;
- Calculating high harmonic;

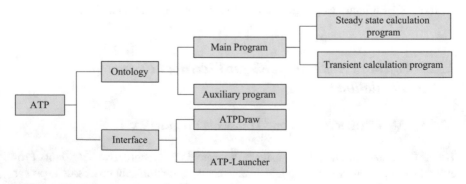

Fig. 2.16 EMTP/ATP software structure

- Calculating the parameters of the overhead line and cable;
- Calculating the ground-return current;
- Calculating the sheath current of the cable;
- Simulating an internal fault of the transformer;
- Simulating and tuning the operating characteristics of the protection and control system.

3. **Application range**

- Applied to the calculation of electromagnetic transient phenomena of any circuit;
- Applied to the steady-state calculation of the power system;
- Applied to the simulation of the generator shaft torsional vibration phenomenon (SSR).

2.3.2.2   PSCAD/EMTDC

PSCAD (PSCAD/EMTDC) [16] (power system computer-aided design/DC electromagnetic transient calculation program) is an electromagnetic transient simulation software suite that is used worldwide. EMTDC is the core of its simulation calculation, and PSCAD is a graphical user interface provided by EMTDC (Electromagnetic Transients including DC).

1. **Structure**

This program consists of the PSCAD graphical user interface and EMTDC simulation calculation program. The main structure is shown in Fig. 2.17.

2. **Function**

- DC electromagnetic transient calculation;
- Computational simulation of the time domain and frequency domain of the power system, which is usually applied when the power system is disturbed and its parameters change with time;
- HVDC transmission, FACTS controller design;
- Harmonic analysis of power systems and simulation calculation in power electronics.

3. **Application range**

- Applied to power system research, including

  - Contingency studies of AC networks consisting of rotating machines, exciters, governors, turbines and transformers;
  - Transmission lines, cables and loads;
  - Insulation coordination of transformers, breakers and arrestors;
  - Sub-synchronous resonance (SSR) studies of networks with machines, transmission lines and an HVDC system;

Fig. 2.17 PSCAD software solution process

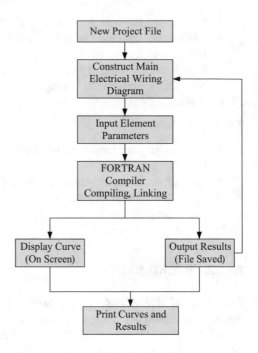

– Lightning strikes, faults or breaker operations, etc.

2.3.2.3 Netomac

NETOMAC (Network Torsion Machine Control) [17] is a mature integrated power system simulation software suite developed by Siemens AG of Germany. It has been widely used for electromechanical, electromagnetic transient and steady-state phenomena by time-frequency domain simulation.

1. **Function**
   - Power flow calculation
     NETOMAC can calculate conventional three-phase and single-phase power flow, as well as multiphase line power flow with inductance and capacitance coupling.
   - Electromagnetic transient and electromechanical transient simulation
     The electrical network in an electromagnetic transient calculation is described by a differential equation and solved by the trapezoidal integral method. An electromechanical transient calculation focuses on the oscillation process of the motor, using complex impedance to describe the grid and differential equations to describe the motor with its excitation system and speed control system; the voltage and frequency characteristics can be considered by a custom load model.
   - Parameter optimization

- One characteristic of the NETOMAC program is parameter identification and optimization. This identification can achieve good performance in both the frequency domain and the time domain. The parameters of the generator voltage regulator can also be optimized to provide better dynamic performance indicators; frequency response.

  NETOMAC can calculate the frequency characteristics of the network and the motor. The frequency response characteristics can provide a basis for the design of the control system for the speed of generator excitation and the parameter design of the diesel generator under the influence of the pulsating torque. This approach can also calculate the harmonic distribution of the passive network.

- Modular simulation language

  NETOMAC uses modular language to simulate the generator excitation system, prime mover governor, steam turbine, etc., which like a block diagram of each link in control theory. The control system consists of the basic modules of proportional, differential, integral, inertia links, etc.

- User openness

  Through BOSL, users can embed their own data expressions, logic expressions and subroutines in the FORTRAN language.

2. **Application range**

- New design verification and new devices tests;
- Applied to power system steady-state and transient simulations;
- Applied to special case simulations, such as transient overvoltage when simulating transformer closing shock, the unloading transient process of saturated and unsaturated transformers, starting and tripping the high-voltage asynchronous motor and the overvoltage caused by arc re-striking.

2.3.2.4 PSASP

The power system analysis integration program PSASP [18] is a highly integrated, open and large software package. PSASP provides convenient interfaces for commonly used software packages, such as Excel, AutoCAD and MATLAB, and can make full use of the resources of other software packages.

1. **Structure**

The structure of PSASP is described by three layers. The first layer is a repository of public data and models, including a power grid base library, integrated model library, user-defined model library and user program. The second layer is a resource library-based application. The package includes steady-state calculation, fault calculation, electromechanical transient calculation and transient stability calculation. The third layer is the calculation result library and analysis tools. After the software performs various analyses and calculations, the generated data results are output or converted to the data forms of Excel, AutoCAD, MATLAB, etc.

2. **Function**

- Power flow calculation and transmission loss analysis;
- Static stability and transient stability analysis;
- Short-circuit current calculation;
- Static and dynamic equivalence;
- Optimal power flow calculation and reactive power optimization;
- Direct method for transient stability and voltage stability determination.

3. **Application range**

- Validation of system planning and design schemes;
- Calculation of system characteristics and control features;
- Determination of system operation mode, analysis of system faults and anti-accident measures;
- Education, training and research in higher education institutions.

4. **Features**

- Calculating large AC-DC hybrid power systems and saving the historical results of various calculations;
- Providing a variety of calculations, common fixed reports and flexible and convenient simulation reports by auto-generation;
- Generating curves and charts automatically;
- Variety of output forms for calculation results;
- Convenient editing report and curve results;
- Provides an interface with Excel, AutoCAD, MATLAB and other software tools.

2.3.2.5 BPA

The BPA [19] program is an offline analysis program for large power systems. This program uses the Newton–Raphson method of sparse matrix techniques and applies the trapezoidal integral method to the calculation of transient stability to generate a more stable numerical solution.

1. **Structure**

This program consists of a platform, core program and auxiliary analysis program. The details are as follows:

- Platform:

PSAP interface; power flow calculation interface; stable computing interface; data file editing interface (TextEdit);

- Core program:

Power flow calculation program of the BPA (Chinses version);
Transient stability calculation program of the BPA (Chinses version);

- Auxiliary analysis program:

One-line diagram format trend graph program (Joy);
 Geographical wiring diagram format trend map program (Clique);
Stability curve mapping tool (CurveMaker);
Power flow calculation interface (PFA.EXE), one-line diagram format trend map program (JOY.EXE) and trend data file.

2. **Function**

- Automatic voltage control;
- Tie line power control;
- System fault and failure analysis;
- Network equivalent;
- Sensitivity analysis;
- Flexible analysis report and load static characteristics model;
- Detailed error detection function and power flow diagram drawing.

3. **Application range**

- Applied to power grid research units;
- Applied to the calculation unit of the grid operation mode;
- Applied to scientific research units in colleges and universities.

4. **Features**

- More than 900 kinds of error detection information are in the program, indicating the cause and nature of an error, which is convenient for users to find and correct errors based on error detection information;
- The calculation results can be repeatedly output, and there are multiple statistical reports in the output.
- The application is simple, convenient and practical.

2.3.3 Software Suites or Tools for Non-electrical Energy Simulation

2.3.3.1 3KEYMASTER

3KEYMASTER is a typical integrated software suite [20–22] that can be used for simulation of cross-energy systems, including power systems, thermal systems and

Table 2.5 A summary of the features of the 3KEYMASTER environment

Environment	Features
Object-oriented design	Yes
Code generation	Not necessary
Humanized interactive interface	Graphical objects, easy modification
Flexibility	Open architecture
DCS systems	Extended with DCS solutions
Integration with other software and hardware	Adopting standard communication protocols, similar to TCP/IP, OPC and others
Embedding engineering codes	RELAP5—NESTLE—MELCOR—MAPP—Simulate 3R are examples of engineering models and codes

fuel pipeline systems. It uses object-oriented graphical modelling and some numerical algorithms. Fluid network modelling tools are mainly oriented to the modelling of thermodynamic and fuel network systems, equipment and related physical processes, such as combustion, heat transfer and multiphase coexistence; power system modelling tools provide a model for the whole process of generation, transmission, distribution and use in the model base; logical control systems can be oriented to the modelling and analysis of signal modulation systems; and relay workers can use the model to describe the phenomena of combustion, heat transfer and multiphase coexistence. This software is used to model logic and control devices. In addition, common models for factories and equipment are provided, and the ability to connect new equipment and factories is provided. Components and model libraries are still being perfected and updated. The features of the 3KEYMASTER environment are summarized in Table 2.5.

Through a Graphical Engineering Station (GES), the 3KEYMASTER simulation platform can access development, operation, testing, integration and deployment of the simulation environment. It has been provided with the ability to develop customized models, reload or connect third-party code or systems, interface with I/O and create new customized modelling tools. It can realize real-time simulation and offline simulation.

The design of the 3KEYMASTER simulation platform can be suited for multiprocessor computing. In the simulation process, simulation tasks can be equally distributed to the processors. This platform is based on a standard off-the-shelf Windows system, and its structure is open and extensible. Several workstations can interact with the main station to realize parallel computing. It provides users with a flexible and efficient application management interface.

2.3.3.2 ProTRAX

ProTRAX is a simulation software suite for dynamic thermal process analysis and is modular [23, 24]. It is suitable for thermal process simulation in power plants, thermal management systems in all-electric ships and operator training. It can fully and dynamically simulate the production process of power plants. The module division in ProTRAX is based on the equipment or equipment components of a power plant, not the thermodynamic nodes (control bodies). In ProTRAX, the common equipment or equipment components, such as pumps, valves, heat transfer surfaces and turbines, are pre-programmed into standard modules and stored in module libraries. Approximately 200 standard process modules in ProTRAX libraries have been widely measured and validated in a large number of training simulators and engineering simulators.

The modules are represented by engineering icons used by engineers, and the icon interfaces are also the imports and exports of the working fluid flow of the actual equipment. Building the system model overlap mainly includes selecting relevant module icons, connecting according to the actual process, and then inputting the physical and operational parameters of the equipment. In addition, ProTRAX allows users to develop new modules in a custom mode. One of the reasons why ProTRAX was selected as an important tool for the simulation of large ship-level thermal management systems is that it has a custom module. Customized models based on ProTRAX should follow the first law of conservation relationship, the second law of thermodynamics and the constitutive relationship between heat transfer and hydrodynamics [23].

2.3.3.3 Real Pipe-Gas

The key technologies and their implementation methods in Real Pipe-Gas simulation software for gas pipelines are discussed in Ref. [25]. Its simulation engine can describe and explain a gas pipeline network with arbitrary topological structure effectively. Linearizing a created system of nonlinear equations and discretizing the system of linear equations in time and space are carried out, and then a large-scale sparse coefficient matrix is formed. An efficient solver has been developed to solve the model, and the "Leapfrog" strategy is used to solve hydraulic systems and thermal systems. Through these methods, the computational efficiency and stability are greatly improved. The computational accuracy, efficiency and stability of Real Pipe-Gas have reached the advanced levels of some international commercial simulation software suites [25].

2.3.4 Multi-energy Real-Time Simulation System (MERTS)

Based on power system real-time simulation, MERTS expands the simulation of thermal systems and fuel pipeline network systems and integrates them through hybrid simulation interfaces to form simulations of multi-energy systems. MERTS fills a gap in energy internet dynamic simulation. MERTS was developed by Tsinghua University.

The platform is mainly applicable to multi-energy micro-grid dynamic and transient simulation including the following elements: non-electrical energy (such as thermal energy and chemical energy pipeline networks) and electrical energy (such as a large amount of distributed new energy generation feeding into the grid) complement each other, coupling wide time scale and complex staged physical processes.

1. **Structure**

The simulation platform consists of a simulation solver kernel program, a human–machine interface program, a case editing program, and an auxiliary analysis program package, as shown in Fig. 2.18.

2. **Function**

- Real-time dynamic simulation of a system containing 5,000 three-phase nodes, 1,500 conventional power generation and new energy generation nodes, 3000 loads, and micro-grids; in addition, it can simulate 50 dynamic element nodes of thermal systems.
- Power flow calculation, electromechanical transient simulation for power networks, quasi-steady-state simulation of new power electronic devices, simulation of dynamic external characteristics of thermal systems with thermal storage and thermal devices.
- Analogue and digital input and output, real-time synchronous communication between the non-electrical energy network simulation and the power grid simulation platform.
- Real-time simulation of the energy internet, energy internet system design, planning verification and control protection configuration, and hardware-in-the-loop testing.

3. **Features**

- It can simulate the steady-state, dynamic and transient characteristics of energy internet systems in real time.
- It has a friendly user interface as a training system and energy management system.
- Based on a CPU and FPGA heterogeneous platform and parallel computing.

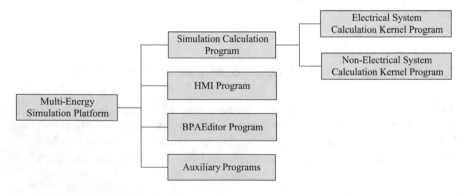

Fig. 2.18 MERTS software structure

2.3.5   Tools for Steady-State and Long-Term Simulation

Tools for modelling and simulating complex energy systems can be basically split into design tools and analysis tools. While design tools are applied in cases when an energy system should be built up and no historical measured data are available, analysis tools are usually applied in cases of existing energy systems with multiple sensors and where a large amount of data collected for a longer time is available (historical operational data). However, it needs to be noted that hybrid analysis systems that combine both types of tools are possible.

Design tools: Design tools, also often called physical tools or white box tools, require detailed knowledge about the system itself, its physical dimension, material, etc. Given the general previous thoughts on the required time resolution, it is necessary to choose tools. It is not always possible to choose only a single tool, as most have strengths in certain areas of modelling and analysis. There are tools that are well suited to model systems and their transient operation as well as control but in turn require knowledge on system characteristics (e.g. Fritzson et al. [26] or its open-source version [27], as shown in Fig. 2.19). These might be applied to power generation systems [28, 29], combined heat and power systems feeding district heating grids [30] and energy systems of buildings [31], to name only a few. These tools usually require the characteristics of all system components as the input, covering the operational conditions to be modelled.

In cases where these characteristics are not available, they need to be generated, either via simulations and modelling or based on measurements. In case they need to be generated via modelling, it is most likely that other tools need to be used. For example, a solar thermal receiver may be simulated and modelled using commercial tools allowing conjugate heat transfer as well as in-stationary simulation. Another example is the transient operation of a steam turbine as one element of a power generation or energy system [32]. Figure 2.20 indicates the model of a steam turbine, with the table giving the number of nodes in the solid domain of [30].

A similar approach applies to the characteristics of chemical processes, etc. In most cases, the direct coupling of tools for modelling energy systems and generating those characteristics is not recommended because of the required modelling effort, level of detail and computing time are quite different for the various tools. However, tools or interfaces used to transfer data between the different solvers may make sense to reduce the necessary manual effort in case multiple simulations are required and to reduce the risk of errors during manual data transfer.

Analysis tools: This type of tool is often referred to as data-driven, self-learning or black box tools. They can usually be applied in cases when a large amount of measured data is available, meaning that the energy systems need to be equipped with many sensors collecting data frequently. They also require a training period that is based on collected and mostly historical data. This automatically means that these tools are not necessarily used in cases of designing a new energy system, as historical data that train the algorithm are not yet available. One well-known type of these tools is ANN (artificial neural network) tools, and their application towards, for example, a micro gas turbine, is described in [33, 34]. These papers highlight the requirements for developing such a tool using only measured data. Even though these

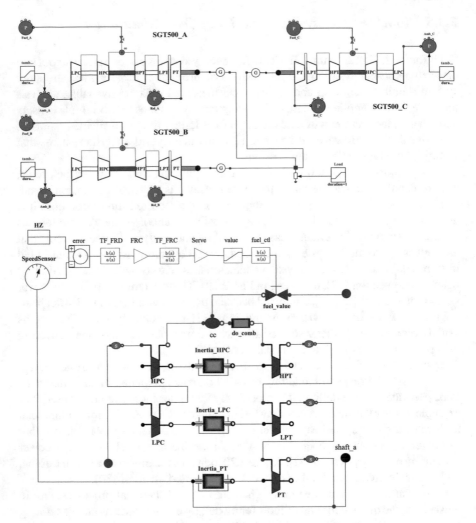

Fig. 2.19 Examples of energy systems for power generation when using Open Modelica as a tool [27]

tools do not require very detailed information on the system to be modelled (in these cases, the systems were gas turbines), they never require less knowledge about the basics of these systems. It is not necessary to know details of the design, the material used, geometrical dimensions, etc., but the basics about the relevant parameters still need to be known. When such a tool exists and is trained, it is characterized by short simulation times, making it well suited for online monitoring and diagnosing energy systems. However, a downside is that for training purposes, a large set of measured data must be available covering at least one full range of operating conditions. For large energy systems, such a typical range of operation covers at least one year.

Hybrid tools are those combining design and analysis tools. The idea is to make use of the advantages of both worlds. Physical-based tools that can be applied from

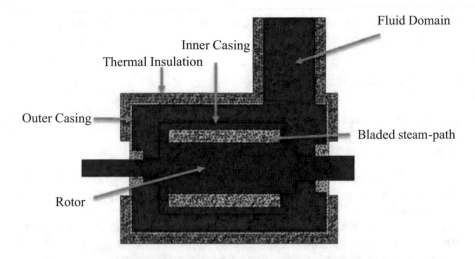

Computational grid in the solid domain	
Blade path	11745
Rotor	12516
Outer casing	22090
Inner casing	46586
Insulation	10464

Fig. 2.20 Example of conjugate heat transfer modelling of a steam turbine for transient simulation [30]

the beginning are combined with learning algorithms that refine the physical model while collecting measured data. These tools are not as established as pure design or analysis tools, and development is still ongoing. However, it can be expected that their application towards complex energy systems will increase.

It needs to be noted that, whichever tool is used, the tool itself as well as the way of modelling should have been validated or should be validated with measured data or cases from the literature.

2.4 Test and Application Examples Based on a Park Energy Internet

2.4.1 Component Model Testing

2.4.1.1 Component Examples

- **Photovoltaic model**

The overall structure of a photovoltaic power generation model is shown in Fig. 2.21. The model of a photovoltaic power generation system mainly includes a photovoltaic array and a VSC grid-connected converter with a control system. This subsection studies the photovoltaic array model. The converter model is shown in the fourth part.

The photovoltaic array is composed of several photovoltaic cells in series or in parallel. Its model includes a photovoltaic cell model and a photovoltaic array integrated model.

The equivalent circuit of photovoltaic cells based on the diode model is shown in Fig. 2.22.

Photovoltaic cells usually involve several parameters: I_{SC} (short-circuit current), V_{OC} (open circuit voltage), I_m (maximum power point load current), V_m (maximum power point load voltage), photovoltaic cell temperature T and light intensity S.

The current of photovoltaic cells can be calculated by the following formula:

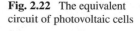

Fig. 2.21 Structure of the photovoltaic power generation model

Fig. 2.22 The equivalent circuit of photovoltaic cells

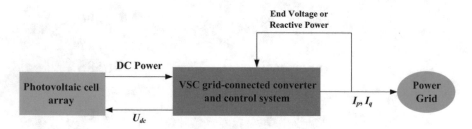

$$I_L = I_{sc}\left[1 - C_1(\exp\frac{V}{C_2 V_{oc}} - 1)\right]$$

$$C_1 = \left(1 - \frac{I_m}{I_{sc}}\right)\exp\left(-\frac{V_m}{C_2 V_{oc}}\right)$$

$$C_2 = \left(\frac{V_m}{V_{oc}} - 1\right)\left[\ln\left(1 - \frac{I_m}{I_{sc}}\right)\right]^{-1} \qquad (2.48)$$

The integrated photovoltaic array model, which is based on a combination of the photovoltaic cell model and the series–parallel relationship, is modified from the photovoltaic cell model as follows:

$$I_{scc\_AR} = N_{se} \cdot I_{scc}$$

$$I_{mm\_AR} = N_{se} \cdot I_{mm}$$

$$U_{occ\_AR} = N_{sh} \cdot U_{occ}$$

$$U_{mm\_AR} = N_{sh} \cdot U_{mm} \qquad (2.49)$$

In this formula, I_{ssc_AR}, U_{osc_AR}, I_{mm_AR} and U_{mm_AR} are the short-circuit current, open-circuit voltage, maximum power point current and maximum power point voltage of the photovoltaic array; N_{se} is the number of photovoltaic cells in series; and N_{sh} is the number of photovoltaic cells in parallel.

The above photovoltaic power generation system is connected to the IEEE 14-bus benchmark system through BUS-5, and a three-phase metal grounding fault is set at the end of the connecting line between BUS-4 and BUS-3 to test the dynamic response characteristics of the photovoltaic power generation system model. The time of fault occurrence is 1.0 s, and the time of fault elimination is 1.1 s.

The voltage, current and output power of the photovoltaic system are observed, and the results are presented as follows (Figs. 2.23, 2.24 and 2.25). The results show that when the system fails at t = 1.0 s, the voltage of the photovoltaic port falls from 11 to 8.8 kV. When the fault is eliminated, the amplitude of the voltage of the

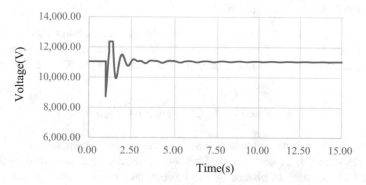

Fig. 2.23 Output voltage of the photovoltaic system

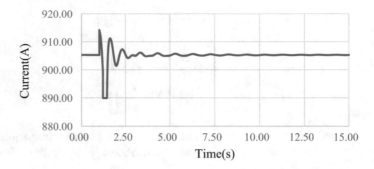

Fig. 2.24 Output current of the photovoltaic system

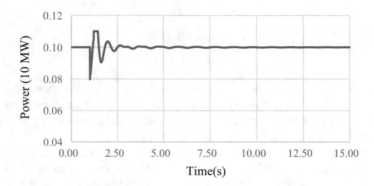

Fig. 2.25 Output power of the photovoltaic system

photovoltaic port oscillates, attenuates and tends towards the stable operation. The output current and power of the photovoltaic system tend to run steadily after the fault, and the amplitude of the photovoltaic system is attenuated by the dynamic process oscillation. The results show that the modelling method is correct.

- **Electrical energy storage model**

An equivalent circuit diagram of an electromechanical transient energy storage battery is shown in Fig. 2.26, and the inner voltage of the battery is related to the temperature and SOC state and is close to the static voltage. The expression is as follows:

$$E = E_0 - K_E(273 + \theta)(1 + S_{soc}) \tag{2.50}$$

where S_{soc} is the unit state of charge based on the battery capacity.

The resistance R is composed of ohmic resistance and polarized resistance, is related to the state of charge of the battery and can describe all polarization characteristics.

Fig. 2.26 Equivalent circuit diagram of an electromechanical transient energy storage battery

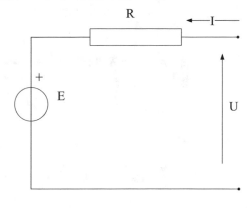

$$R = R_0 + R_1/S_{\text{soc}}^k \qquad (2.51)$$

In this formula, R_0 is the equivalent ohmic resistance of the energy storage battery, which is constant; R_1 is the equivalent polarization resistance of the energy storage battery when S_{soc} is 1; and K is a coefficient that can be determined by a chart provided by the battery manufacturer.

Tests give the results of storage system voltage, current and output power, which are shown in Figs. 2.27, 2.28 and 2.29, respectively. The results show that when the system fails at $t = 1.0$ s, the voltage variations are very small. The output current and power of the storage system quickly tend to operate steadily. The results reflect the dynamic process of the energy storage model and prove the validity of the model.

- **Generator**

Ordinary models used in power system dynamic simulation are compared with those of a commercial software suite. Some of the output results are listed in Table 2.6.

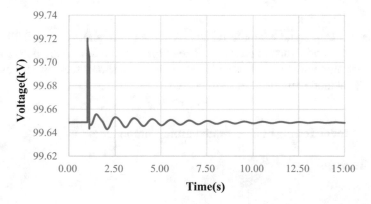

Fig. 2.27 Terminal voltage of the energy storage system

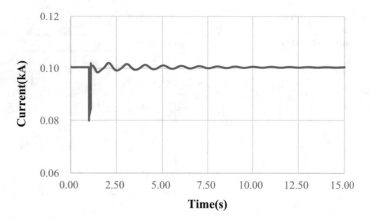

Fig. 2.28 Output current of the energy storage and transport system

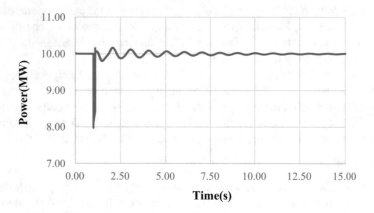

Fig. 2.29 Output power of the energy storage system

Table 2.6 lists some example results. A three-phase short-circuit fault excites the power grid and dynamic elements. The outputs of the developed power models are in good agreement with those of BPA. This result shows the correctness of the modelling method mentioned in subsection 4.

2.4.1.2 Model Testing of Thermodynamic Components

- **Gas Turbine**

Table 2.6 Transient response test

Conditions	Results
The first three-phase short circuit occurs in the ANSHUN line, and the fault is eliminated by jumping off all three-phase switches	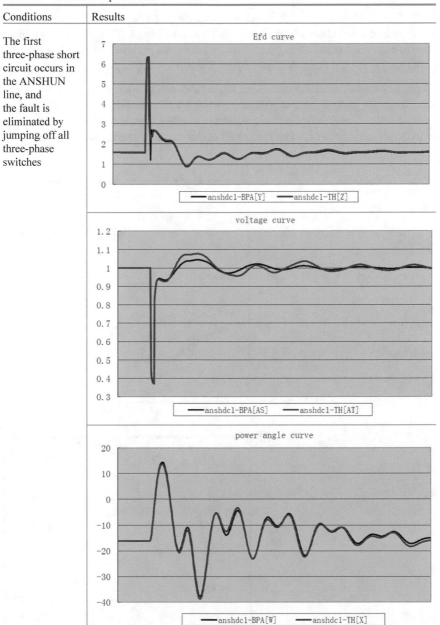

To describe the part-load performance of a turbine and compressor, the converted parameters (for example, for a compressor, $G_c\sqrt{T}\big/P, n_c\big/\sqrt{T}, \pi_c, \eta_c$) are generally used and often drawn into curves, as shown in Fig. 2.30. The relative expressions of converted parameters (the ratio of a converted parameter to its design value, for example, $\dot{G}_c = \left(G_c\sqrt{T}\big/P\right)\Big/\left(G_c\sqrt{T}\big/P\right)_0$) are used. Figure 2.30 shows the general characteristic curve of an axial-flow compressor in a gas turbine [35, 36]. The upper part of the curve cluster shows the relationship between the relative pressure ratio $\dot{\pi}_c$ and the relative converted flow rate $\dot{G}_c$ under different similar converted rotational speeds $\dot{n}_c$. The lower part of the curve cluster shows the relationship

Fig. 2.30 The general characteristic curve of an axial-flow compressor

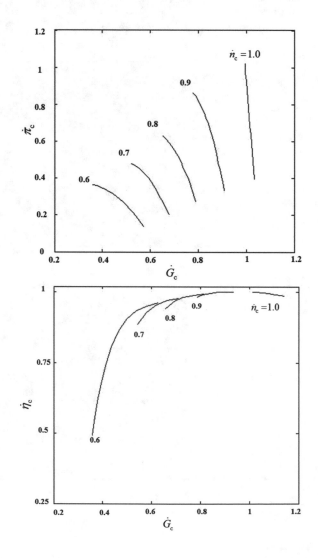

between the relative efficiency $\dot{n}_c$ and the similar converted flow rate $\dot{G}_c$ under different relative converted rotational speeds $\dot{n}_c$. Characteristic curve-based modelling for compressors has been widely used.

By fitting the compressor characteristic curve shown in Fig. 2.30, the characteristic equations are obtained as follows, and the results are shown in Fig. 2.31:

$$\dot{\pi}_c = c_1(\dot{n}_c)\dot{G}_c^2 + c_2(\dot{n}_c)\dot{G}_c + c_3(\dot{n}_c)$$
$$\dot{n}_c = \left[1 - c_4(1 - \dot{n}_c)^2\right](\dot{n}_c/\dot{G}_c)(2 - \dot{n}_c/\dot{G}_c) \qquad (2.52)$$

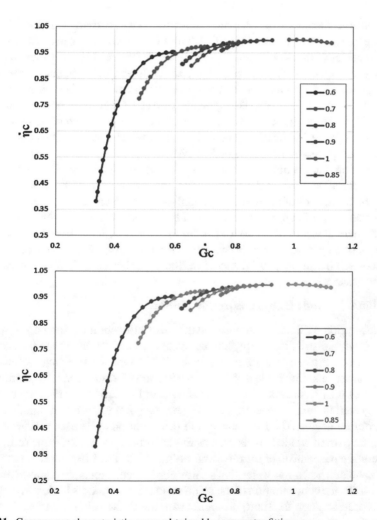

Fig. 2.31 Compressor characteristic curve obtained by parameter fitting

$$\begin{cases} c_1 = \dot{n}_c / \left[p(1 - m/\dot{n}_c) + \dot{n}_c(\dot{n}_c - m)^2 \right] \\ c_2 = p - 2m\,\dot{n}_c^2 / \left[p(1 - m/\dot{n}_c) + \dot{n}_c(\dot{n}_c - m)^2 \right] \\ c_3 = \left(pm\,\dot{n}_c - m^2\dot{n}_c^2 \right) / \left[p(1 - m/\dot{n}_c) + \dot{n}_c(\dot{n}_c - m)^2 \right] \end{cases} \qquad (2.53)$$

where c_1, c_2, c_3 and c_4 are the coefficients of the fitted characteristic equation and m and p are the fitting factors.

To ensure that all points on the constant speed line fall on the right half of the curve, the values of m and p should also satisfy the following relationship:

$$\sqrt[3]{p} \geq 2m/3 \qquad (2.54)$$

A dynamic simulation of a PG6541B gas turbine was carried out in Ref. [37], and the correctness of the model was verified. Therefore, this model can be used to verify the correctness of the 50 MW gas turbine model established in this paper. When the load step decreases to 50 and 75% of the rated power, the dynamic curves of the rotational speed of the model in the reference and the model in this paper are shown in Fig. 2.32a, b, respectively. The dynamic curves of fuel mass flow are shown in Fig. 2.33a, b. These figures show that the durations of the two models returning to the stationary state are the same, approximately 40 s. The reason is that both models have a power generation capacity of 50 MW.

In addition, the maximum deviation rate of the rotational speed, the change rate of the inlet temperature of the turbine, and the change rate of the fuel mass flow during the transitions of the reference model and building model are compared. Due to the differences of equipment capacity and initial state, there are some differences in the numerical values of the above two figures, but the change trend is the same, as shown in Table 2.7. As shown in this table, there is only a slight error between the two models. This shows that the model in this paper has considerable accuracy and effectiveness.

- **Lithium bromide (LiBr) absorption chiller**

LiBr absorption chillers, as common waste heat utilization equipment, have been widely used in distributed energy supply systems, in which the waste heat of a gas turbine is converted into cold energy in a CCHP system. The single-effect LiBr absorption refrigeration cycle is the most basic form of absorption refrigeration. The structure of a LiBr absorption chiller is shown in Fig. 2.34. The main components of the system include a generator, condenser, evaporator, absorber, solution pump and throttling device. The LiBr absorption refrigeration cycle uses LiBr-water as a refrigerant pair, in which LiBr is used as the absorbent and water as the refrigerant.

Operating mechanism of the refrigeration machine: The LiBr solution in the generator is heated when hot water flows through the heat exchange equipment of the generator (as the concentration of the solution in the generator is high, it is called the concentrated solution for short, different from the dilute solution in the absorber); then, water vapour is produced and separated. Next, the water vapour enters the condenser and is condensed in the condenser, releasing heat and then is converted

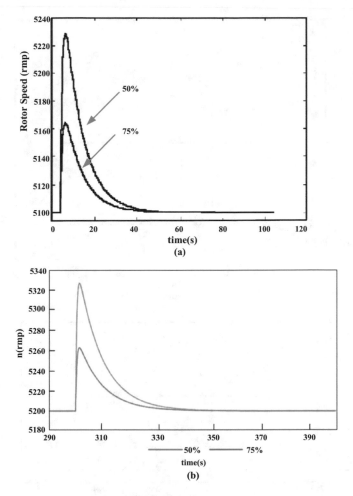

Fig. 2.32 Dynamic curves of the rotational speed

into the liquid state. The condensate water enters the evaporator through a throttling valve. The condensed water absorbs the heat of the refrigerant water through the heat exchange equipment in the evaporator so that the temperature of the refrigerant water is lowered to achieve cooling, and the condensed water absorbs heat and evaporates. The water vapour from the evaporator enters the absorber and is absorbed by the concentrated solution from the generator to form a dilute solution. Finally, the dilute solution is transferred to the generator by the solution pump, thus ensuring that the concentration of the solution in the generator maintains a certain level.

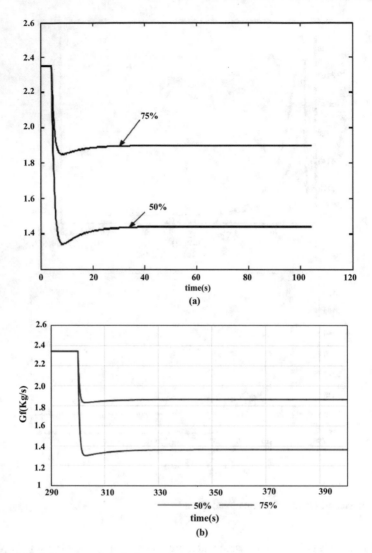

Fig. 2.33 Dynamic curves of the fuel mass flow

Table 2.7 Output results comparison between the two models

Load step	50%		75%	
Model	Reference (%)	Building (%)	Reference (%)	Building (%)
Rotational speed maximum deviation rate	2.55	2.48	1.27	1.23
Turbine inlet temperature change rate	73.86	71.86	87.5	86.37
Fuel mass flow change rate	61.28	58.06	80	79.45

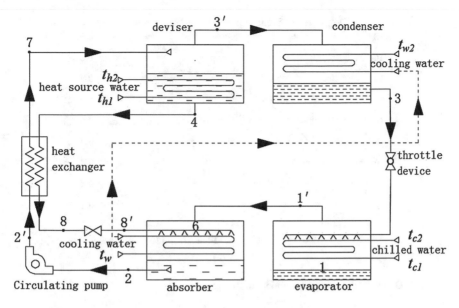

Fig. 2.34 The structure of a lithium bromide absorption chiller

In this process, there are complex processes such as heat transfer, phase change and flow and flash vaporization. The process satisfies the following laws: the law of conservation of mass, the law of conservation of momentum and the law of conservation of energy. Therefore, the dynamic process of each component in the system can be described by using the three laws of conservation.

The above two graphs show the curves provided in the reference (Fig. 2.35) and the curves of the results calculated by the model established according to the methods mentioned in this paper (Fig. 2.36) [38]. As shown in the figures, when the temperature of the heat source increases by 10 degrees, the legends from left to right indicate the heat dissipation amount in the cooling process, the heat absorption in the generating process, the heat absorption in the vaporizing process and the refrigeration coefficient of performance. The variation characteristics of the parameters of the two models are consistent. The accuracy and reliability of the model are verified.

The cooling capacity of the model in this paper is 150 kW, which is greater than the value of 6 kW for the refrigerator model in Ref. [38]. Therefore, the transition time of parameter changes in this model is longer than that of the model in the reference.

2.4.2 Examples of Dynamics of Energy Internet Systems

As shown in Fig. 2.37, this case is abstracted and designed on the basis of a planned industrial park project. The case includes a 110 kV power transmission system as the backbone network and low-voltage distribution network, a fuel network system

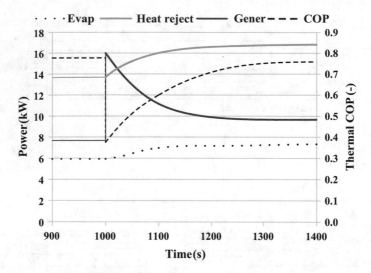

Fig. 2.35 Output of the refrigerator model in Ref. [38]

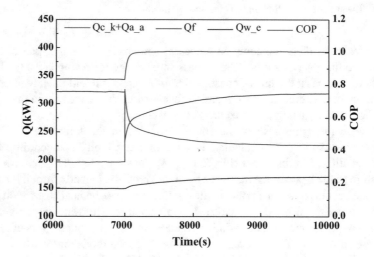

Fig. 2.36 The output of the refrigerator model established by the method described in this chapter

with natural gas as the main part and a multi-node thermal network system. It also includes three 100 MW gas turbine-driven generators, a 10 MW photovoltaic power generation system, a 10 MW electric energy storage system and a 10 MW thermal energy storage system. The system is a typical energy internet system.

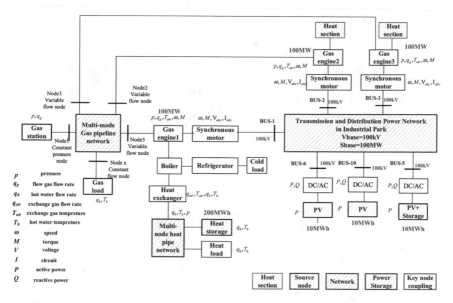

Fig. 2.37 Planned system case structure

2.4.2.1 Steady State

When the system is in steady-state operation, according to the photovoltaic interface voltage curve, the energy storage interface voltage curve and the gas turbine operation state curve (for the same speed of the generator and gas turbine), which are shown in Figs. 2.38, 2.39 and 2.40, respectively, both the photovoltaic interface voltage and the gas turbine speed remain at a constant value, and the voltage at the energy storage interface changes slightly due to the existence of energy storage equipment and AC/DC power conversion equipment.

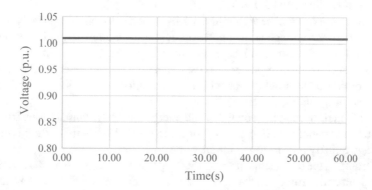

Fig. 2.38 Photovoltaic interface voltage

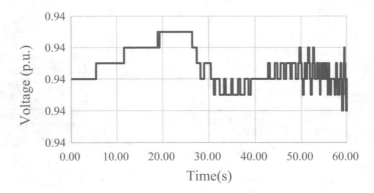

Fig. 2.39 Electrical energy storage interface voltage

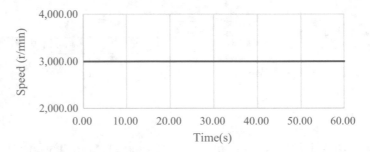

Fig. 2.40 Speed of gas turbine

2.4.2.2 Dynamic Process Analysis

A three-phase short-circuit fault occurs on the line between bus 5 and bus 1 and is located at 0.3 times the length of the line from bus 5. The fault occurs at $t = 1.0$ s, and the fault is eliminated at $t = 1.1$ s.

When the fault occurs, the voltage decreases at the photovoltaic interface. After the fault is cleared, the voltage at the photovoltaic interface tends to gradually stabilize after several periods of oscillation, as shown in Fig. 2.41 in detail. At this time, the output photovoltaic electro-power also oscillates to a certain extent and then enters a region of constant output power with the oscillation attenuation of the interface voltage, as shown in Fig. 2.42.

Once the fault occurs, the voltage of the energy storage terminal drops from 0.94 p.u. in normal operation to 0.33 p.u., as shown in Fig. 2.43. When the fault is cleared, the voltage quickly returns to the initial value.

According to the test results of gas turbine operation (coaxiality of the gas turbine rotor and generator rotor), shown in Fig. 2.44, the three-phase short-circuit fault on the bus 5-bus 1 line has little effect on the speed of the gas turbine, the adjustment process takes a long time and the recovery to normal operation is slow.

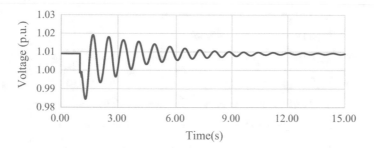

Fig. 2.41   Photovoltaic interface voltage

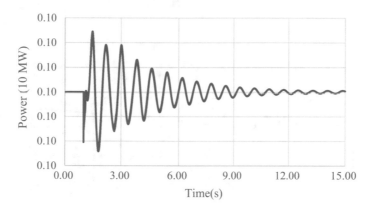

Fig. 2.42   Photovoltaic power output

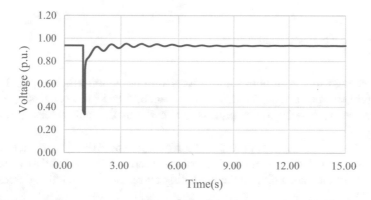

Fig. 2.43   Electrical energy storage interface voltage

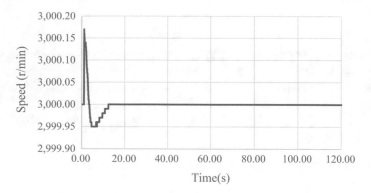

Fig. 2.44 Speed of the gas turbine

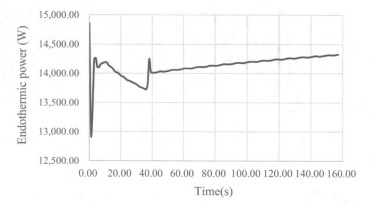

Fig. 2.45 Power of heat absorption Q_f

The change in gas turbine speed affects its conversion efficiency and output power (electric power and thermal power). According to the conversion process of refrigerators, its refrigeration process is affected by the fault process (Figs. 2.45 and 2.46).

According to the response process of each component in the system after the fault in this case, it is effective to establish an energy internet system model to analyse the system behaviours according to the modelling and simulation method described in this section.

2.5 Conclusion

This chapter briefly gives the principles, basic methods and fundamental algorithms for dynamic simulation of the energy internet and further proposes an entire practical solution for simulation of the main body of the energy internet. In addition, simulation

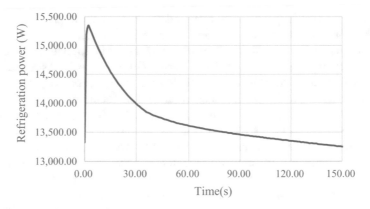

Fig. 2.46 Refrigeration power Q_{w_e}

cases are presented. Finally, some relevant software suites and tools for dynamic simulation are summarized and listed.

Quasi-steady-state and long-term dynamics are always related to system planning and design, calculation of operation modes, energy balance after variations in load and generation, and long-term processes. Quasi-steady-state and long-term dynamic modelling, simulation and analysis of the energy internet are usually closely connected to modelling, algorithms and the underlying numerical schemes. It is essential to evaluate the purpose of evaluation as well as the characteristics of the system and its main components.

Depending on a case-by-case evaluation, it is necessary to clearly define the system boundaries and the target of evaluation. This approach will pave the way to identify the required input parameters and characteristics, the level of detail to which the information needs to be available and which tools should be used.

Especially when measured and/or historical data are incorporated into the model, simulating and analysing energy systems involves knowledge about sensors, the data acquisition system and component internal control mechanisms of interest. Background information and knowledge of system components, especially connected to their operational behaviour, is helpful for well-targeted modelling to cover the important parameters. This is even more important when evaluating fast transients and/or safety-relevant operating conditions while at the same time aiming to evaluate the safe and reliable energy supply.

Dynamics and transients are always related to system control, faults and disturbances, performance tests and so on. Modelling, solution steps and underlying numerical issues are also discussed.

First, modelling of dynamic and transient energy elements includes two parts: traditional elements and elements containing complex long dynamics. Generally, most dynamic and transient element models can directly employ models of conventional electrical power grids, thermodynamic systems and fuel pipes. For new elements containing complex long dynamics, two modelling methods are presented for wide time scale long dynamics and phased evolution long dynamics.

Second, network models also play very important roles in simulation, and they link all the element models together, so that all the models can be solved simultaneously. Node equations in matrix form are employed for electrical power grids. Thermal grids use the nodal pressure equation and branch flow equation.

Third, established models of traditional power grids, micro-grids, and other forms of energy systems, including thermal systems, have been widely applied in corresponding fields as well as solution methods. This chapter proposes a three-layer phased multi-mode hybrid simulation method to solve the dynamic and transient coupling and interactions between devices and adjacent networks of different energy types in the energy internet. This method involves several key techniques: transition methods between stages, interface techniques for multi-mode simulation and simulation timing at the interface.

For completeness, this chapter also gives a simple list of available tools/software for simulation of the energy internet and tests and application examples.

References

1. Source: Danish Energy Agency: Technology Data for Energy Plants for Electricity and District heating generation, August 2016 (Update 2017 and July 2018), [EB/OL], https://ens.dk/en/our-services/projections-and-models/technology-data
2. IRENA, renewable energy statistics, The International Renewable Energy Agency, Abu Dhabi, [EB/OL] (2018). ISBN 978–92-9260-077-8, https://www.irena.org/publicationsearch?irena_topic=440b8182ba8941d291-b9aefac97fef47
3. T. Jamal, T. Urmee, M. Calais et al., Technical challenges of PV deployment into remote Australian electricity networks: a review. Renew. Sustain. Energy Rev. **77**, 1309–1325 (2017)
4. X Xu, Y Huang, G He, et.al. (2009) Modeling of large grid-integrated PV station and analysis its impact on grid voltage [C]. Paper presented at 2009 International Conference on Sustainable Power Generation and Supply. Nanjing, China
5. M.J.E Alam, K.M. Muttaqi, D. Sutanto et al., A performance analysis of distribution networks under high penetration of solar PV, in *Proceedings of 44th International Conference on Large High Voltage Electric Systems* (2012)
6. K. Fekete, Z. Klaic, L. Majdandzic, Expansion of the residential photovoltaic systems and its harmonic impact on the distribution grid. Renew. Energy **43**, 140–148 (2012)
7. S. Kosai, Dynamic vulnerability in standalone hybrid renewable energy system. Energy Convers. Manag. **180**, 258–268 (2019)
8. H. Lund, *Renewable Energy Systems – A Smart Energy Systems Approach to the Choice and Modelling of 100% Renewable Solutions*, 2nd edn. (Elsevier). ISBN 978-0-12-401423-5, http://www.energyplan.eu/
9. M. Rezkalla, A. Zecchino, S. Martinenas et al., Comparison between synthetic inertia and fast frequency containment control based on single phase EVs in a micro grid. Appl. Energy **210**, 764–775 (2018)
10. D. Yoganathan, S. Kondepudi, B. Kalluri et al., Optimal sensor placement strategy for office buildings using clustering algorithms. Energy Build. **158**, 1206–1225 (2018)
11. For example see: [EB/OL], http://scanivalve.com/products/pressure-measurement/

12. N.A. Cumpsty, J.H. Horlock, Averaging non-uniform flow for a purpose, in *Proceedings of GT2005 ASME Turbo Expo 2005*, Power for Land, Sea and Air, 6–9 June 2005. Reno-Tahoe, Nevada, USA, GT2005-68081

13. B.W. Abegaz, T. Datta, S.M. Mahajan, Sensor technology for the energy-water nexus – a review. Appl. Energy **210**, 451–466 (2018)

14. S. Yoon, Y. Yu, Hidden factors and handing strategy for accuracy of virtual in-situ sensor calibration in building energy systems: sensitivity effect and reviving calibration. Energy Build. **170**, 217–228 (2018)

15. [EB/OL], https://en.wikipedia.org/wiki/Emtp

16. [EB/OL], https://baike.baidu.com/item/pscad/1633247

17. Z. Sun, P. Wei, Y. Zhang, Dynamic simulation of voltage collapse in power systems with NETOMAC. J. Shandong Univ. Technol. (Nat. Sci.) 47–50 (2004)

18. School of Electrical Engineering, Wuhan University, PSASP Software Manual (2004)

19. China Electric Power Academy, BPA User Manual V4.10

20. [EB/OL], https://www.ws-corp.com/

21. M.A. Volman, V.K. Semenov, System of training programs for simulation of reactor measurements, in *2016 2nd International Conference on Industrial Engineering, Applications and Manufacturing (ICIEAM)*. IEEE (2016)

22. P. Luo, H. Liu, X. Rong et al., Research on the simulation of 6.6 kV system in nuclear power plant based on 3KEYMASTER software. Comput. Era **4**, 6–8 (2016)

23. J. Wojcik, J. Wang, (2017). Technical feasibility study of thermal energy storage integration into the conventional power plant cycle. Energies, 10(2): 205

24. [EB/OL], https://energy.traxintl.com/training/protrax-overview/

25. J. Zheng, F. Song, G. Chen et al., Development of RealPipe-Gas simulation software for gas pipeline network. Oil & Gas Storage Transp. **30**(9), 652–659 (2011)

26. P. Fritzson, B. Bachmann, K. Moudgalya et al., Introduction to Modelica with Examples in Modeling, Technology, and Applications, [EB/OL], http://omwebbook.openmodelica.org/

27. See also: [EB/OL], https://www.openmodelica.org/

28. F. Casella, P. Parini, Optimal control of power generation systems using realistic object-oriented modelica models. IFAC PapersOnLine 50:11100–11106 (2017)

29. L. Pierobon, K. Iyengar, P. Breuhaus et al., Dynamic performance of power generation systems for off-shore oil and gas platforms, in *Proceedings of ASME Turbo Expo 2014: Turbine Technical Conference and Exposition – GT2014*, Düsseldorf, Germany

30. R. Sangi, P. Jahangiri, A. Thamm et al., Dynamic exergy analysis – modelica-based tool development: a case study of CHP district heating in Bottrop, Germany. Therm. Sci. Eng. Prog, S2451904917302214

31. F. Bünning, R. Sangi, D. Müller, A Modelica library for the agent-based control of building energy systems. Appl. Energy **193**, 52–59 (2017)

32. M. Fadl, P. Stein, L. He, Full conjugate heat transfer modelling for steam turbines in transient operations. Int. J. Therm. Sci. **124**, 240–250 (2018)

33. H. Nikpey, M. Assadi, P. Breuhaus, Development of an optimized artificial neural network model for combined heat and power micro gas turbines [J]. Appl. Energy **108**, 137–148 (2013)

34. T. Palmé, P. Breuhaus, M. Assadi, et al. (2011) Early Warning of Gas Turbine Failure by Nonlinear Feature Extraction Using an Auto-Associative Neural Network Approach[C]// Asme Turbo Expo: Turbine Technical Conference & Exposition. American Society of Mechanical Engineers

35. Z. Na, C, Ruixian, Explicit analytical part-load performance solution of constant speed single shaft gas turbine and its cogeneration. J. Eng. Therophys. **19**(2), 141–144 (1998)

36. Z. Na, C. Ruixian, Analytical solutions and typical characteristics of part-load performances of single shaft gas turbine and its cogeneration. Energy Convers. Manag. **43**(2002), 1323–1337 (2002)

37. L. Xichao Single-axis gas turbine dynamic simulation. College of Power Engineering, Chongqing University (2006)

38. G. Evola, N. Le Pierrès, F. Boudehenn et al., Proposal and validation of a model for the dynamic simulation of a solar-assisted single-stage LiBr/water absorption chiller. Int. J. Refrig. **36**(3), 1015–1028 (2013)
39. T.V. Nguyen, L. Pierobon, B. Elmegaard et al., Exergetic assessment of energy systems on North Sea oil and gas platforms. Energy **62**, 23–36 (2013)

Chapter 3
Cyber-Physical System Security

Heping Jia, Yi Ding, Yishuang Hu, and Yonghua Song

Abstract The rapid development of advanced information technologies, for example, the Internet of Things and Big Data techniques, has made the energy internet achieve a deep integration of physical systems and cyber systems and realize an effective combination of energy flow and information flow among various networks. However, with increasing automation of the energy internet, the scale of physical networks, the size of cyber networks and the numbers of smart sensors and decision-making units have greatly increased, resulting in complex external or internal factors directly or indirectly impacting the control and decisions of networks through various approaches. The interaction mechanisms between cyber networks and physical networks are becoming increasingly complex in the energy internet, resulting in the security and reliability analysis of cyber-physical systems becoming more complicated. In this chapter, the security of components in cyber-physical systems is first introduced. Multiple uncertainties in cyber-physical system operation are also developed, including different types of cyber attacks and corresponding mitigation strategies as well as the volatility of energy sources and stochastic energy consumption. Moreover, the correlation and cascading failures in cyber-physical systems are analysed to demonstrate the coupling between cyber systems and physical systems. Furthermore, challenges in the security of cyber-physical systems are provided. This chapter mainly analyses cyber-physical system security in the energy internet considering various uncertainties, which can provide technical support for the planning and operation of the energy internet.

H. Jia
School of Economics & Management, North China Electric Power University, Beijing, China
e-mail: jiaheping@ncepu.edu.cn

Y. Ding (✉) · Y. Hu
College of Electrical Engineering, Zhejiang University, Hangzhou, China
e-mail: yiding@zju.edu.cn

Y. Hu
e-mail: huyishuang@zju.edu.cn

Y. Song
State Key Laboratory of Internet of Things for Smart City, University of Macau, Macau, China
e-mail: yhsong@um.edu.mo

© Springer Nature Switzerland AG 2020
A. F. Zobaa and J. Cao (eds.), *Energy Internet*,
https://doi.org/10.1007/978-3-030-45453-1_3

3.1 Introduction

The innovative concept of the energy internet has provided a vision for enhancing the capability of multiple energy sources and various types of energy consumption in the form of the internet. Moreover, improvements in advanced information and communication technologies have accomplished close correlations between cyberspace and physical infrastructure in the energy internet.

However, uncertainties from both cyber attacks and physical devices may have negative impacts on the secure and reliable operation of cyber-physical systems, which will definitely influence the reliable implementation of the Energy internet. For example, energy routers, as some of the most critical equipment used to achieve interconnections among multiple local area networks of energy in infrastructure, are vital to fulfil the effective implementation of the energy internet [1]. Figure 3.1 illustrates the framework of a cyber-physical system for the energy internet.

A well-functioning energy internet depends on the reliable operation of cyber networks, which can transfer information or commands from one node to another. Therefore, a critical requirement of cyber-physical systems is that reliable and secure operation must be guaranteed [2]. The security and reliability of cyber networks are dependent on the credibility of physical equipment. When physical equipment is affected by the natural environment, operating conditions or human factors, the reduction in reliability will seriously affect the security of the entire cyber-physical system [3]. For example, under some extreme weather conditions or by incorrect human operation, some communication equipment might be in failure and not usable for signal transmission, which results in unreliable system operation. In addition, in

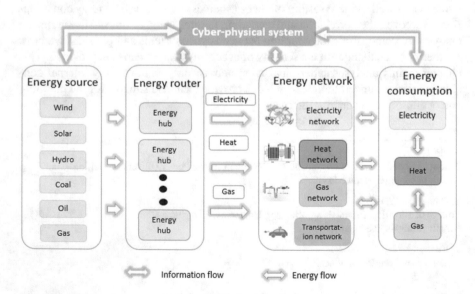

Fig. 3.1 The framework of a cyber-physical system for the energy internet

energy generation, energy transmission, energy distribution, and energy consumption, cyber-physical systems of the energy internet can be influenced by cyber attacks as well. Compared with physical failures in energy systems, cyber attacks have the characteristics of low cost, high concealment, and large scope. Once successful, they can lead to serious consequences.

The security of cyber-physical systems for the energy internet includes analysing cyber networks, energy networks and all the interactions between them and can be described as the ability to secure the information and operational capabilities of energy systems from unauthorized access [4]. A vulnerability analysis of cyber security in power systems was conducted in [5], which mainly focused on cyber networks. Cyber attacks have also been analysed to identify weaknesses in substations and control centres for power systems utilizing probabilistic methods [6]. A security evaluation for cyber-physical systems in power systems was proposed in [7]. A mathematical framework for failure identification and security assessment for water networks was conducted in [8]. References [9, 10] proposed reliability analysis for linear networks with interconnected nodes.

Nevertheless, security analysis containing both physical and cyber aspects is insufficient. Physical devices, cyber components, and their interdependences are supposed to be embedded for security evaluation of them in the framework of the energy internet [11].

3.2 Security of Components in Cyber-Physical Systems

Components in cyber-physical systems under the framework of the energy internet can be classified into physical energy components and cyber components.

Physical energy components in the energy internet can be divided into components with different time characteristics, including energy generation components, energy consumption components, energy storage components, and energy conversion components.

Among these various components, an energy generation component refers to a unit in the energy internet that can convert primary energy into available secondary energy or a unit that can directly utilize primary energy and convert it into usable resources, such as different kinds of power plants, including thermal power plants, gas-fired power plants, hydropower plants, wind farms, and photovoltaic power generation bases. An energy consumption component corresponds to a terminal unit of energy transmission and utilization in the energy internet and exists with industrial customers, commercial customers and residential customers, for instance, heat pumps, air conditioners, and electric vehicles. An energy storage component can achieve long-term, large-scale storage of energy and realize energy transfer or shifting on a time scale, including pumped-storage hydropower stations, battery storage, hydrogen storage, and compressed natural gas/liquefied storage.

Moreover, there is a critical type of component for energy conversion. An energy conversion component is a coupling component between different energy systems, including electrical-gas conversion devices, heat pumps, and so on. In particular, electrical-gas conversion devices and gas-fired power plants can realize couplings and interconnections between electric power systems and natural gas systems; a heat pump is a realization of the mutual coupling of electric and thermal networks; a combined heat and power unit realizes the coupling of electricity, gas and heat; and electric vehicle charging piles can achieve the coupling of a power system and transportation system.

Furthermore, cyber components can provide computational and control support to cyber-physical systems of the Energy internet and have the ability to gather and analyse data or information received from diverse sources and make overall decisions to control the energy internet.

The security of these components will definitely affect the reliable operation of cyber-physical systems for the energy internet. Analysing the reliabilities of these components can ensure the systems' secure operation and provide approaches to enhance system reliability. Risk models for different kinds of components can be established. For energy generation components, reliability modelling can be performed by utilizing failure rates and considering the characteristics of long start-up times for conventional coal-fired units and short start-up times for natural gas generators. For energy consumption components, it is possible to take advantage of their physical characteristics and consider the corresponding time characteristics of fast response for air conditioners and the time characteristics of long duration for electric vehicles, for instance, thermodynamic models for air conditioners and charge and discharge models for electric vehicles. For energy storage components, as a medium for realizing the transfer of the time domain for energy, risk models are supposed to consider the failure rates and characteristics of long-term storage where energy is stored during valley hours and released at the peak load to satisfy the system demand.

3.3 Multiple Uncertainties in Cyber-Physical System Operation

The operation of cyber-physical systems faces multiple uncertainties, including both external and internal factors.

For external factors, extreme weather, such as hurricanes, snowstorms and earthquakes, definitely influences the operation of physical components. Moreover, customers' behaviours can lead their energy consumption patterns to satisfy their requirements, which will impact energy flows. The volatility of certain energy sources, for example, solar power and wind power, will also affect the security of the energy internet. Furthermore, cyber attacks such as cyber theft and cyber corruption are another type of external factor. For internal factors, the ageing, degradation and failure of both cyber components and physical components are inevitable. The multiple uncertainties in cyber-physical systems are illustrated in Fig. 3.2.

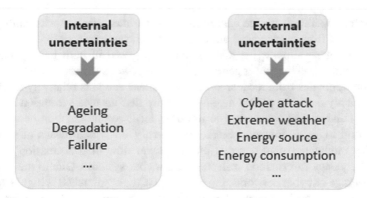

Fig. 3.2 Multiple uncertainties in cyber-physical systems

3.3.1 Cyber Attacks and Corresponding Mitigation Strategies

Vulnerability analysis of cyber-physical systems subject to cyber attacks has recently become a growing concern. A typical technique is to analyse the effects of particular attacks against specific systems [12].

Deception attacks against control systems are defined where deception attacks correspond to an interaction between two sides, an attacker and a target, where an attacker effectively causes a target to receive particular faulty information as true, with the intention of causing the target to perform in a manner benefiting the attacker [13]. False data injection attacks can be considered a kind of deception attack and are devised with restricted resources for the attacker [14]. Stealthy deception attacks against data acquisition and control systems, for example, supervisory control and data acquisition (SCADA) systems, were studied in [15].

Replay attacks (also known as playback attacks) refer to a form of network attack, where effective information communication is spitefully or deceitfully repeated or delayed [16]. Covert attacks consisting of a combined operation of both a system identification attack and data injection attack for service degradation were proposed in [17], which were based on conventional intelligence operations and performed to gather information. Denial-of-service (DoS) attacks are focused on destroying or prohibiting authorized users from obtaining access to websites, applications, or other resources [18]. Malware attacks generally occur through delivering malicious attachments in a phishing email or by downloading from dubious websites [19]. This kind of infection can take place the moment a user opens an attachment or website. In rarer instances, malware can be downloaded on a computer without approval, while spyware can remain as concealed as possible to silently gather useful data in the background [20].

Although specific cyber attacks in cyber-physical systems have received considerable attention, corresponding security analysis, attack detection technologies, and precautionary methods should be further studied to ensure the reliable operation of cyber-physical systems in the energy internet.

In system operation for the energy internet, various cyber attacks can occur in different loops from energy generation to energy consumption.

In energy generation, cyber attacks may take place in both local control and wide-area control on the power generation side of power systems. Local control data collection originates from local areas, and the attack range is relatively small. This control mainly needs to protect against malware that invades substation local area networks (LANs) and attackers who infiltrate local communication networks and alter control logic. Automatic generation control (AGC) is a means of secondary frequency modulation that depends on the power flow of transmission lines and system frequency information collected from cyber systems. Due to the relatively large coverage of cyber networks, the risk of AGC being attacked is high. Common cyber attacks can be implemented by tampering with the control parameters of the system.

In energy transmission, SCADA, energy management systems (EMSs), phasor measurement units (PMUs) and relay protection equipment are mainly utilized to observe, check, analyse and control energy transmission networks. The methods of attacking a SCADA system include the illegal operation of data exchanged in the network, such as intercepting and tampering with data transmitted by the SCADA system to the control centre through fibre-optic eavesdropping technology [21]. Operations for attacking an EMS include illegal modifications of state estimation information, load prediction values, operation commands, and control actions. Moreover, a wide-area measurement system (WAMS) provides simultaneous acquisition and time-stamped voltage and current values for the protection, control and monitoring of power systems. A critical factor in reliable system operation is that the sampling clock of the entire system needs to be synchronized with the same time reference. Common attacks include DoS attacks, loss of synchronization attacks (time-based attacks), and false data injection attacks. Furthermore, spoofing of PMUs is another threat in various cyber attacks. In a spoofing attack process, a spoofer first obtains real GPS signals and then disguises them as false signals [22].

In energy distribution, possible scenarios for cyber attacks include advanced metering infrastructure (AMI) systems, distribution automation equipment, distributed power management, etc. An AMI system involves communication between end users, control systems, and third-party organizations, ensuring peer-to-peer information security. Malicious cyber attacks will destroy the confidentiality of information from smart metres and the integrity and availability of switch control instructions, which might seriously affect the security, stability and economic operation of energy systems. Distribution automation control is definitely significant for the secure operation of distribution networks and economical power supply for customers. Malicious attacks can monitor equipment on a feeder, which may lead to errors in remote operation, such as remote switch opening and closing, and emergency response. Malicious attacks in load management or relay protection communication facilities may cause trips of distribution network feeders, leading to power interruption. Furthermore, since distributed generation can be utilized as a primary or emergency backup power source in the power grid to improve power quality and system operation stability, malicious cyber attacks can destroy the confidentiality of user information in distributed generation and the accuracy and timeliness of measurement information for networks.

In energy consumption, demand response refers to customers responding according to current or future electricity price information adjustment requirements, mainly involving real-time electricity prices, access to energy data and electric vehicles. For small users, real-time electricity price signals are supposed to be transferred by AMI systems, the Internet, or other information channels, which increases the compromise likelihood of pricing information integrity, availability, and credibility by malicious attacks. Moreover, as transferable and mobile energy storage batteries, electric vehicles can use electricity price information to determine when to charge. A distributed energy management program allows electric vehicles to feed the grid based on system status. The privacy of customer information and the accurateness of electricity price data involved in this process may become targets of damage by network attackers.

By analysing the impact of cyber attacks on each part of energy generation, transmission, distribution and consumption, cyber attack models under the background of physical and cyber integration of the energy internet can be established.

It should also be noted that approaches to alleviating these cyber attacks have been developed in recent years. In reference [23], a mitigation strategy to eliminate potential cyber attacks was proposed based on dynamic state estimation. Distributed control-based frameworks are also utilized to mitigate cyber attacks [24]. A hybrid control scheme utilizing various controllers matched to different types of cyber attacks was introduced in [25] to improve system robustness. To mitigate cyber attacks for distributed energy resources, stealthy worst-case strategies were demonstrated based on differential games to construct reliable boundaries for cyber-physical smart grids [26]. In [27], false data injection attacks were investigated for state estimation in power systems, and a corresponding optimal protection strategy for obtaining the smallest number of measurements to be protected was proposed. A mitigation scheme for DoS attacks in autonomous microgrids with energy storage systems was constructed based on rule fallback control to enhance the system resiliency [28]. In [29], an abstract model for distributed DoS attacks and an inference algorithm for estimating possible botnets hidden in a network were developed. False data injection attacks in shipboard power systems can be mitigated by a twofold strategy based on intelligent agents that are utilized to detect malicious data [30]. A platform embedded with cyber attack detection and risk assessment frameworks was presented in [31] for SCADA systems. In [32], security metrics against data integrity attacks were developed to obtain a characteristic SCADA system considering the communication infrastructure of power systems. Detection and localization strategies against data injection attacks in gossip-based networks were introduced in [33]. The influences of wormhole attacks on a networked control system were studied, and a passivity-based control-theoretic framework was developed to demonstrate and alleviate the corresponding wormhole attacks [34].

3.3.2 Volatility of Energy Sources and Stochastic Energy Consumption

With the deep integration of multiple energy sources in the energy internet, uncertainties from the energy supply have greatly increased.

The power outputs of renewable energies, for instance, solar power and wind power, are generally determined by weather conditions, such as wind speed and radiation intensity, which are hard to accurately forecast. Therefore, the power generation from these renewable energy resources has the inherent characteristics of volatility and intermittency, which increase the uncontrollability of system operation.

Moreover, in energy consumption, stochastic behaviours of energy customers also introduce uncertainties for the energy internet. Owing to the diversity and complexity of different kinds of customers as well as the random nature of environmental conditions, customers' energy consumption is stochastic and difficult to precisely predict [35]. It has been shown that varying weather conditions, energy flows, occupancy, and their complicated interactions can result in uncertainties and stochastic behaviours [36]. A physical–statistical approach was adopted in [37] to forecast energy consumption for heterogeneous buildings to avoid uncertainties and heterogeneity from individual customers.

Furthermore, as one approach to enhance network efficiency, demand response technologies have been successfully applied to the implementation of the energy internet. However, uncertainties of demand-side participation might affect the security of energy systems and were integrated into elasticity estimation uncertainty in [38] and multi-state models in [39].

In addition, both energy providers [40] and energy consumers [41] can behave according to uncertainties in the energy market, for example, fluctuating energy prices and changeable incentives or compensation for different market participants [42, 43], which also increase the complexity of secure system operation. Moreover, the investment decisions for different types of energy sources also influence the secure operation of the entire energy internet.

3.4 Correlation and Cascading Failures in Cyber-Physical Systems

In the energy internet, several coupling components or equipment can be utilized to achieve conversion between various kinds of energy sources [44]. For example, combined heat and power (CHP) units have the ability to simultaneously generate electricity and heat. Power-to-gas technologies have made possible the bidirectional interaction of energies in the electricity grid and natural gas systems, which gradually deepens the degree of coupling between power systems and natural gas systems.

Moreover, control centres, such as dispatch centres, represent combinations of energy flows and information flows. It should be noted that a control centre can represent a combination of different types of flows. In a control centre, sensors can collect data to monitor the system operation, while controllers may dominate energy flows by signals. The security of this kind of node in the energy internet is undoubtedly important because its malfunction could not only cause failures of cyber systems but also physical systems [45].

Considering the interdependence in cyber-physical systems, failures of nodes in one network may cause malfunctions in other networks, which in turn result in cascading failures of whole cyber-physical systems [46]. For example, in the energy internet, the malfunction of information transmission in cyber networks might result in faulty energy transmission networks and thereby cause cascading failures of the whole system.

The mutuality between physical infrastructures and cyber networks makes the energy internet more complicated for studying cascading failures of cyber-physical systems [47]. Node failures in a single network might not only lead to a malfunction in that network but also in other networks. For example, the Italy blackout that happened in 2003 was a consequence of cascading failures [48], where the original failure of a power station caused the interruption of communication networks and further affected the operation of the whole power system.

Common assumptions when analysing nodes in interdependent networks are (1) a node is necessarily the property of an equipment as the largest linked sub-network and (2) interdependence relations between different nodes [49]. For cyber-physical networks, a unique modelization seems to be impracticable owing to precise details of propulsion and responses of physical systems [50]. The challenges lie in the recognition and modelling of typical elements presented in various scenarios, as well as combining and applying them to corresponding concrete systems. Most existing studies have focused on extracting characteristics from physical systems and making assumptions related to corresponding cyber systems. For example, the modelling of cyber-physical systems considering intrinsic heterogeneity, concurrency, and sensitivity was proposed utilizing domain-specific ontologies for improving the modularity and functionality of the implementation [51].

The interaction relationships between physical devices and cyber components are critical to analyse embedded computers and communication networks. The interdependent characteristics of components in a battery-supported cyber-physical system were studied to achieve robustness analysis in [52]. The robustness of n interdependent networks considering a limited support-dependence correlation was analysed in [53]. Cascading failures occurring in interdependent complex networks were investigated and the corresponding robustness was analysed by considering random attacks or failures in [3].

3.5   Challenges in Cyber-Physical System Security

The demand for availability of cyber-physical systems, real-time decisions, and resource limitations have brought challenges to the security protection of cyber-physical systems for the energy internet.

The challenge of time sensitivity: Protective measures must not affect the response time of the system. In traditional information systems, access control can easily be implemented without influencing information flows. However, in a cyber-physical system, the response time is important. Therefore, any protective measures should be carefully tested and discussed. For example, password authentication and authorization on a human–machine interface might hinder or interfere with emergency behaviours of control systems.

Resource constraints: Protective measures such as anti-virus software are almost impossible to install. The industrial control systems in the energy internet and their operating systems are usually resource-constrained and cannot increase the security of traditional IT systems. In addition, third-party security solutions are not allowed due to vendor licenses and service agreements. Service technical support may be lost due to the lack of supplier approval or consent for the installation of third-party applications. Furthermore, adding security measures to control systems in the energy internet may introduce new operation risks, such as new interference to original system operation.

Management constraints: Patches and updates are not safe in real time. In an industrial control system, a software update must be fully tested by suppliers and users to install an industrial application. Moreover, an interruption of the industrial control system must be prepared in a few days or weeks because the update operation of the patch may affect the normal operation of the system. For energy systems in the energy internet, service support is usually through a single vendor, which may not allow different but compatible support solutions from other vendors.

The challenge of physical system recovery increases the recovery difficulty and recovery time for cyber-physical systems. In traditional IT systems, if a system failure or attack occurs, most problems can be solved by a simple system restart. However, in a cyber-physical system of the energy internet, the failure of a cyber system will cause the physical system to be damaged to some extent. Damage to the physical system is generally irreversible. For example, an interruption caused by a short circuit in a smart grid requires manpower, material resources and time to replace the physical object.

3.6   Conclusions

Cyber-physical systems will further change the way people interact with the physical world and will achieve efficient coordination and deep integration between cyber and physical resources. At present, the development of cyber-physical systems faces various difficulties and obstacles, especially security issues. The human, machine and material fusion capabilities of cyber-physical systems can achieve ubiquitous information monitoring and precise control in the energy internet and truly realize comprehensive management of complex systems.

Multiple uncertainties, including external and internal factors, might affect the secure and reliable operation of cyber-physical systems, especially cyber attacks, and the corresponding volatility of energy sources and consumption. Moreover, the correlation and cascading failures of cyber-physical systems in the energy internet were introduced in this chapter. Several challenges in cyber-physical systems for the energy internet were also concluded.

References

1. A. Humayed, J. Lin, F. Li, B. Luo, Cyber-physical systems security—a survey. IEEE Internet of Things J. **4**(6), 1802–1831 (2017)
2. J. Madden, Security analysis of a cyber physical system: a car example. Missouri University of Science and Technology (2012)
3. Z. Zhang, W. An, F. Shao, Cascading failures on reliability in cyber-physical system. IEEE Trans. Reliab. **65**(4), 1745–1754 (2016)
4. A. Banerjee, K.K. Venkatasubramanian, T. Mukherjee, S.K.S. Gupta, Ensuring safety, security, and sustainability of mission-critical cyber-physical systems. Proc. IEEE **100**, 283–299 (2012)
5. C.W. Ten, C.C. Liu, G. Manimaran, Vulnerability assessment of cybersecurity for SCADA systems. IEEE Trans. Power Syst. **23**(4), 1836–1846 (2008)
6. C.W. Ten, G. Manimaran, C.C. Liu, Cybersecurity for critical infrastructures: Attack and defense modeling. IEEE Trans. Syst. Man Cybern.-Part A: Syst. Hum. **40**(4), 853–865 (2010)
7. S. Sridhar, A. Hahn, M. Govindarasu, Cyber-physical system security for the electric power grid. Proc. IEEE **99**(1), 1–15 (2012)
8. D.G. Eliades, M.M. Polycarpou, A fault diagnosis and security framework for water systems. IEEE Trans. Control. Syst. Technol. **18**(6), 1254–1265 (2010)
9. S. Sundaram, C. Hadjicostis, Distributed function calculation via linear iterative strategies in the presence of malicious agents. IEEE Trans. Autom. Control. **56**(7), 1495–1508 (2011)
10. F. Pasqualetti, A. Bicchi, F. Bullo, Consensus computation in unreliable networks: A system theoretic approach. IEEE Trans. Autom. Control. **57**(1), 90–104 (2012)
11. R. Akella, H. Tang, B.M. McMillin, Analysis of information flow security in cyber-physical systems. Int. J. Crit. Infrastruct. Prot. **3**, 157–173 (2010)
12. F. Pasqualetti, F. Dörfler, F. Bullo, Attack detection and identification in cyber-physical systems. IEEE Trans. Autom. Control. **58**(11), 2715–2729 (2013)
13. S. Amin, A. Cárdenas, S. Sastry, Safe and secure networked control systems under denial-of-service attacks. Hybrid Syst.: Comput. Control. **5469**, 31–45 (2009)
14. Y. Liu, M.K. Reiter, P. Ning, False data injection attacks against state estimation in electric power grids, in *Proceedings of ACM Conference Computation Communication Security*, Chicago, IL, USA, November 2009 (2009), pp. 21–32
15. A. Teixeira, S. Amin, H. Sandberg, K.H. Johansson, S. Sastry, Cyber security analysis of state estimators in electric power systems, in *Proceedings of IEEE Conference Decision Control*, Atlanta, GA, USA, December 2010 (2010), pp. 5991–5998
16. Y. Mo, B. Sinopoli, Secure control against replay attacks, in *Proceedings Allerton Conference Communication, Control, Computation*, Monticello, IL, USA, September 2010 (2010), pp. 911–918

17. A.O. de Sá, L.F.R. Carmo da Costa, R.C. Machado, Covert attacks in cyber-physical control systems. IEEE Trans. Ind. Inf. **13**(4), 1641–1651 (2017)
18. Q. Yan, F.R. Yu, Distributed denial of service attacks in software-defined networking with cloud computing. IEEE Commun. Mag. **53**(4), 52–59 (2015)
19. T.J. Holt, G.W. Burruss, A.M. Bossler, Assessing the macro-level correlates of malware infections using a routine activities framework. Int. J. Offender Ther. Comp. Criminol. **62**(6), 1720–1741 (2018)
20. H. Abualola, H. Alhawai, M. Kadadha, H. Otrok, A. Mourad, An android-based Trojan spyware to study the notification listener service vulnerability. Proc. Comput. Sci. **83**, 465–471 (2016)
21. B. Zhu, A. Joseph, S. Sastry, A taxonomy of cyber attacks on SCADA systems, in *2011 IEEE International Conferences on Internet of Things, and Cyber, Physical and Social Computing*, October 2011 (2011), pp. 380–388
22. M. Yasinzadeh, M. Akhbari, Detection of PMU spoofing in power grid based on phasor measurement analysis. IET Gener. Transm. Distrib. **12**(9), 1980–1987 (2018)
23. A.F. Taha, J. Qi, J. Wang, J.H. Panchal, Risk mitigation for dynamic state estimation against cyber attacks and unknown inputs. IEEE Trans. Smart Grid **9**(2), 886–899 (2018)
24. A. Farraj, E. Hammad, A.A. Daoud, D. Kundur, A game-theoretic analysis of cyber switching attacks and mitigation in smart grid systems. IEEE Trans. Smart Grid **7**(4), 1846–1855 (2016)
25. C. Kwon, I. Hwang, Cyber attack mitigation for cyber–physical systems: hybrid system approach to controller design. IET Control. Theory Appl. **10**(7), 731–741 (2016)
26. P. Srikantha, D. Kundur, A DER attack-mitigation differential game for smart grid security analysis. IEEE Trans. Smart Grid **7**(3), 1476–1485 (2016)
27. X. Liu, Z. Li, Z. Li, Optimal protection strategy against false data injection attacks in power systems. IEEE Trans. Smart Grid **8**(4), 1802–1810 (2017)
28. M. Chlela, D. Mascarella, G. Joós, M. Kassouf, Fallback control for isochronous energy storage systems in autonomous microgrids under denial-of-service cyber-attacks. IEEE Trans. Smart Grid **9**(5), 4702–4711 (2018)
29. V. Matta, M. Di Mauro, M. Longo, DDoS attacks with randomized traffic innovation: botnet identification challenges and strategies. IEEE Trans. Inf. Forensics Secur. **12**(8), 1844–1859 (2017)
30. T.R.B. Kushal, K. Lai, M.S. Illindala, Risk-based mitigation of load curtailment cyber attack using intelligent agents in a shipboard power system. IEEE Trans. Smart Grid. https://doi.org/10.1109/tsg.2018.2867809
31. C. Foglietta et al., From detecting cyber-attacks to mitigating risk within a hybrid environment. IEEE Syst. J. https://doi.org/10.1109/jsyst.2018.2824252
32. O. Vukovic, K.C. Sou, G. Dan, H. Sandberg, Network-aware mitigation of data integrity attacks on power system state estimation. IEEE J. Sel. Areas Commun. **30**(6), 1108–1118 (2012)
33. R. Gentz, S.X. Wu, H. Wai, A. Scaglione, A. Leshem, Data injection attacks in randomized gossiping. IEEE Trans. Signal Inf. Process. Over Netw. **2**(4), 523–538 (2016)
34. P. Lee, A. Clark, L. Bushnell, R. Poovendran, A passivity framework for modeling and mitigating wormhole attacks on networked control systems. IEEE Trans. Autom. Control. **59**(12), 3224–3237 (2014)
35. F.F. Wu, P.P. Varaiya, R.S. Hui, Smart grids with intelligent periphery: an architecture for the energy internet. Engineering **1**(4), 436–446 (2015)
36. L. Ju, Z. Tan, J. Yuan, Q. Tan, H. Li, F. Dong, A bi-level stochastic scheduling optimization model for a virtual power plant connected to a wind–photovoltaic–energy storage system considering the uncertainty and demand response. Appl. Energy **171**, 184–199 (2016)
37. X. Lü, T. Lu, C.J. Kibert, M. Viljanen, Modeling and forecasting energy consumption for heterogeneous buildings using a physical–statistical approach. Appl. Energy **144**, 261–275 (2015)
38. A. Moshari, A. Ebrahimi, M. Fotuhi-Firuzabad, Short-term impacts of DR programs on reliability of wind integrated power systems considering demand-side uncertainties. IEEE Trans. Power Syst. **31**(3), 2481–2490 (2016)

39. H. Jia, Y. Ding, Y. Song, C. Singh, M. Li, Operating reliability evaluation of power systems considering flexible reserve provider in demand side. IEEE Trans. Smart Grid (2018)
40. Q. Tang, K. Yang, D. Zhou, Y. Luo, F. Yu, A real-time dynamic pricing algorithm for smart grid with unstable energy providers and malicious users. IEEE Internet of Things J. **3**(4), 554–562 (2016)
41. H.A. Aalami, S. Nojavan, Energy storage system and demand response program effects on stochastic energy procurement of large consumers considering renewable generation. IET Gener. Transm. Distrib. **10**(1), 107–114 (2016)
42. S. Nojavan, B. Mohammadi-Ivatloo, K. Zare, Optimal bidding strategy of electricity retailers using robust optimisation approach considering time-of-use rate demand response programs under market price uncertainties. IET Gener. Transm. Distrib. **9**(4), 328–338 (2015)
43. M. Marzband, M. Javadi, J.L. Domínguez-García, M.M. Moghaddam, Non-cooperative game theory based energy management systems for energy district in the retail market considering DER uncertainties. IET Gener. Transm. Distrib. **10**(12), 2999–3009 (2016)
44. S. Ali, T.A. Balushi, Z. Nadir, O.K. Hussain, in *Cyber Security for Cyber Physical Systems* (Springer, Switzerland, 2018)
45. S. Sridhar, A. Hahn, M. Govindarasu, Cyber-physical system security for the electric power grid. Proc. IEEE **100**(1), 210–224 (2012)
46. S.C. Suh, U.J. Tanik, J.N. Carbone, A. Eroglu, *Applied Cyber-Physical Systems* (Springer, New York, 2014)
47. S.V. Buldyrev, R. Parshani, G. Paul, H.E. Stanley, S. Havlin, Catastrophic cascade of failures in interdependent networks. Nature **464**, 1025–1028 (2010)
48. S. Corsi, C. Sabelli, General blackout in Italy sunday september 28, 2003, h. 03: 28: 00, in *IEEE Power Engineering Society General Meeting*, June 2004 (2004), pp. 1691–1702
49. M. Newman, *Networks: An Introduction* (Oxford University Press, London, 2010)
50. Z. Huang, C. Wang, M. Stojmenovic, A. Nayak, Characterization of cascading failures in interdependent cyber-physical systems. IEEE Trans. Comput. **64**(8), 2158–2168 (2015)
51. P. Derler, E.A. Lee, A.S. Vincentelli, Modeling cyber–physical systems. Proc. IEEE **100**(1), 13–28 (2012)
52. F. Zhang, Z. Shi, S. Mukhopadhyay, Robustness analysis for battery-supported cyber-physical systems. ACM Trans. Embed. Comput. Syst. **12**(3), 69 (2013)
53. G. Dong, L. Tian, D. Zhou, R. Du, J. Xiao, H.E. Stanley, Robustness of n interdependent networks with partial support-dependence relationship. Europhys. Lett. **105**(4), 68004 (2013)

Chapter 4
Early Experience of the Energy Internet: A Review of Demonstrations and Pilot Applications in Europe

Shi You and Hanmin Cai

Abstract A demonstration is an essential process that ensures a successful transition from an innovative idea to a successful product. As an energy ecosystem, the energy internet (EI) is built on a wide variety of solutions that are designed and developed to achieve advanced connectivity, intelligent management and seamless integration of billions of smart devices, machines, and systems from multiple energy sectors. This chapter presents a brief overview of some demonstrations and pilot applications of EI-related solutions developed in Europe. The technology readiness levels (TRLs) of these solutions are between TRL 3 (experimental proof of concept) and TRL 9 (successful system operation). The main effect of these demonstrations and pilot applications is creating learning effects and showcasing opportunities for various stakeholders that are involved in developing and realizing EI solutions.

4.1 The Development of the Energy Internet

Motivated by the success of internet technologies, the energy internet (EI), also known as a web-based smart grid (SG) or SG 2.0, has become the most recognized visionary concept for future energy systems. The origin of the EI can be traced back to 2007, when many initiatives were launched worldwide [1–5]. These initiatives investigated how to integrate maturing information and communication technologies (ICTs) into the development of electrical power systems. In [1], a future renewable electric energy delivery and management (FREEDM) system built on automated and flexible electric power distribution resembled the seed of the EI in the USA. In [2], researchers from the UK proposed an overarching structure of a virtual power plant (VPP) that integrates some distributed energy resources (DERs) into one operation portfolio. Because the key enabling technology of this VPP is plug-and-play, which

S. You (✉) · H. Cai
Department of Electrical Engineering, Technical University of Denmark,
Elektrovej 325, Kgs, Lyngby, DK 2800, USA
e-mail: sy@elektro.dtu.dk

H. Cai
e-mail: hacai@elektro.dtu.dk

© Springer Nature Switzerland AG 2020
A. F. Zobaa and J. Cao (eds.), *Energy Internet*,
https://doi.org/10.1007/978-3-030-45453-1_4

is also the primary accelerator of the Internet of Things (IoT), the VPP is often taken as a preliminary application of the EI. In Germany, the E-Energy flagship project supported by the German Federal Ministry of Economics and Technology for the first time presented a systematic solution for achieving the 2020 energy plan of Germany, i.e. the EI. The solution aimed to integrate advanced ICT into efficient management of the power system with growing penetration of DERs and intermittent renewables [3, 4]. Similarly, the iPower platform operated between 2010 and 2015 exhibited the Danish perspective of the EI. Through iPower, universities and industrial partners in Denmark showed the feasibility of using market-based and service-oriented solutions to manage a large number of flexible consumption units [5].

In 2013, a book written by Jeremy Rifkin explored how the two defining technologies of the twenty-first century—the Internet and renewable energy—are merging to create a powerful "third industrial revolution" [6]. As a result, hundreds of millions of people will produce green energy anywhere and will share it with anyone through an EI, just like how information is created and exchanged online. Rifkin's vision of the EI has quickly gained traction in international communities in Europe, Asia, Africa, and the Americas who have started working on initiatives to transform the vision into new technical and economic paradigms.

Today, the EI has evolved into an energy ecosystem that builds upon advanced connectivity, intelligent management and seamless integration of billions of smart devices, machines, and systems from multiple energy sectors. Although such an ecosystem lacks a standard definition, there is a consensus that the EI shall provide an open and equal platform to all energy stakeholders and solutions from different energy sectors while possessing a spirit of sharing and collaboration. As a result, the ecological thinking behind the EI will create a sustainable energy future where the synergies among electricity, heat, gas, transportation, and water sectors are maximized as much as possible to address both individual challenges (such as wind power balancing) within each sector and global challenges (such as energy scarcity and sustainable development) with a holistic view.

This chapter presents an overview of several selected demonstrations and pilot applications of EI solutions in Europe. The selected solutions cover diverse technological, commercial, regulatory and social aspects of the EI; on the other hand, the technology readiness levels (TRLs) of these solutions vary from case to case. Most of these EI solutions are still in the early development phase but have acquired a significant amount of interest. Section 4.2 presents a short overview of the status of EI-related demonstrations and pilot applications in the EU, as well as their categorization. Section 4.3 gives a case-by-case overview of several representative pilot activities, with a short list of projects selected for further reading. Section 4.4 concludes this chapter.

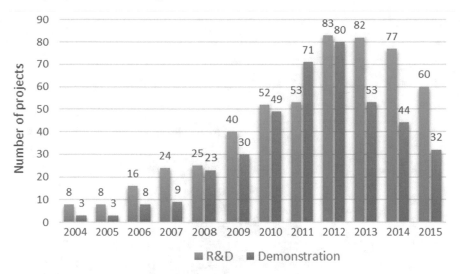

Fig. 4.1   Time distribution of R&D and demonstration projects in the EU between 2004 and 2015

4.2   An Overview of EI Demonstrations in Europe: Status and Categorization

In Europe, the need to meet the EU's climate change and energy policy objectives for 2020 and beyond has led to strong momentum for transforming energy infrastructure by investigating smart grid/EI solutions. Thanks to this growing interest, many stakeholders within the energy sector have participated in a variety of R&D and demonstration projects in recent years. R&D often constitutes the first stage of the development of new methods, technologies, products, and services. A demonstration is a proof-of-concept prototype of a conceivable smart grid/EI solution for continued learning and showcasing opportunities. Figures 4.1 and 4.2 show the trends of smart grid R&D and demonstration projects and investment between 2004 and 2015 in the EU based on a survey [7].[1] Within this observed period, €3.36 billion was invested in 410 demonstration projects by various national and European funding schemes, while 540 R&D projects were funded with a total investment of 1.61 billion. Approximately 800 implementation sites and 36 countries were involved in these demonstrations. These figures point to the fact that demonstration is an irreplaceable element in the process of realizing the envisaged benefits of the smart grid/EI. The rationale behind this is that the smart grid/EI is an extraordinarily complex and practical systematic engineering solution. When designing such a solution, the advancement of energy and ICT infrastructures, new business models and practices, policy/reregulation revisions, and understanding and acceptance of energy consumers all have to be taken into account from a practical perspective.

---

[1]More information about smart grid project facts, figures and trends in the EU and its member states is available in [7].

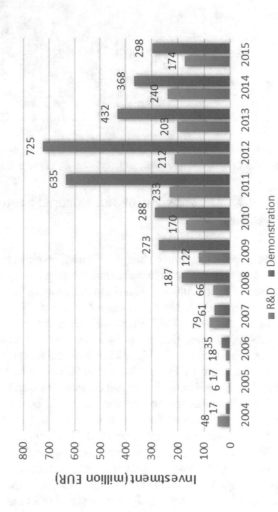

Fig. 4.2 Time distribution of R&D and demonstration investment in the EU between 2004 and 2015

These solutions of the EI can be classified into the following six categories:

(Ca. 1)   Intelligent and efficient energy component technologies: that cover intelligent and efficient control, operation and management solutions related to energy production, energy conservation, energy storage, energy consumption, measurement, etc., at the component level.

(Ca. 2)   Safe, speedy, and reliable ICTs: that store, retrieve, manipulate, transmit, or receive mostly digital information for achieving high-level connectivity and management of energy components through the internet. Representative technologies include big data, artificial intelligence, cloud computing, and blockchain.

(Ca. 3)   Cyber-physical technologies: that explore different aspects between physical energy systems and ICT (including the software) systems that are deeply intertwined with each other, such as how cybersecurity influences energy security and how to integrate advanced ICTs into energy systems.

(Ca. 4)   Integrated energy systems (IESs): are often related to smart, reliable, and cost-effective energy planning, operation, management, and control solutions for energy systems that include a set of technologies associated with more than one energy carrier/sector. The primary purpose of IESs is to explore synergies between different energy components/systems/sectors to create innovative value streams.

(Ca. 5)   Innovative business and market solutions: aim to support the transition towards the EI from a market-oriented business perspective. These solutions are either new ones that are developed under the existing market and regulatory frameworks, such as the real-time energy market, or disruptive solutions that are developed out of the box, such as peer-to-peer energy trading.

(Ca. 6)   Fully resembling the EI: is often large-scale solutions such as smart cities/smart communities that intend to show a good integration of various EI technologies and functions into one live place and to prove the envisaged benefits of the EI.

Within each category, the TRLs defined by the European Commission (EC) [8] are applied to clarify the difference between a theory-focused investigation and a market-oriented implementation. As shown in Table 4.1, solutions rated between TRL 6 and TRL 7 require live field-based organizing demonstrations based on private and public partnerships to address relevant environmental challenges, while laboratory-based demonstrations are performed for solutions that are rated between TRL 3 and TRL 5. When a solution is rated between TRL 8 and TRL 9, the demonstrations or pilot applications become more market-oriented, meaning the demonstration fields are either live marketplaces or resemble real marketplaces as much as possible. Usually, these live fields comprise both relevant infrastructure and a group of stakeholders (e.g., research institutes, industrial partners, technology end-users, and policymakers), while the lab fields mainly involve their operators.

The TRL-based categorization of R&D projects has been widely used and proven relatively successful for supporting technology development by creating learning

Table 4.1 TRLs defined by the EC [8]

TRL	Description	Interpretation
TRL 1	Observation of fundamental principles	R&D (often theory-based)
TRL 2	Formulation of concepts	
TRL 3	Proof of concept by experiments	R&D with lab-based prototyping demo
TRL 4	Lab-based technology validation	
TRL 5	Validation of technology in relevant environment	
TRL 6	Demonstration of technology in a relevant environment	R&D with live field-based organizing demo
TRL 7	Demonstration of system prototype in the operational environment	
TRL 8	Completion of qualification of system	Market-oriented Demos and application
TRL 9	Successful operation in the operational environment	

effects at different stages. A research article [9] gives a further explanation of demonstrations developed at each stage, including their cooperative forms and the corresponding learning effects. For instance, prototyping demonstrations can generate a strong learning effect to boost technology development, while market demonstrations support the development of market niches where policymakers and technology end-users learn to understand and use the new technology products.

Today, a number of approaches have been developed and used in Europe to increase the value and impact of demonstration infrastructure, such as collaborative labs and living labs. For instance, lab-based associations such as DERlab bring independent laboratories together to offer standardized services for testing, demonstration and pre-standardization of energy component/system technologies [10]. Living lab platforms, such as PowerLabDK [11] and the Smart Mobility Living Lab [12], have also been strategically developed by technology developers and end-users based on long-term public–private partnerships and therefore allow for the use of real energy systems as laboratory test fields. This collaborative approach enables the testing and demonstration of a single technology in multiple locations with multiple methods and therefore can speed up the R&D process. The living lab approach to a large degree allows for reusing invested experimental facilities and long-term observations of end-users' responses.

4.3   Demonstrations and Pilot Applications of the EI in Europe: A Review of Selected EI Projects

Today, there are tens or hundreds of EI-relevant demonstrations and applications that are completed, ongoing or under planning. This section presents several projects of EI solutions with different technological and commercial aspects. A list of selected projects is attached for readers who are interested in learning more about demonstrations and applications of the EI in Europe.

4.3.1   The E-Energy Programme in Germany

As one of the earliest EI initiatives, during 2008–2013, the German E-Energy programme successfully demonstrated several EI solutions in a variety of locations, as indicated in Fig. 4.3 [13].

Among the six performed demonstrations, eTelligence is an electricity market-based VPP with wind and PV power plants, a combined heat and power station, two industrial cooling plants and 450 residential customers. E-DeMa is a demonstration of end-user-involved virtual electricity trading platform solutions within local communities. MeRegio demonstrated optimal electricity network management through price-based demand response. Moma demonstrated a hierarchical cell-based energy system structure with self-management within each cell and market-based coordination between cells. With the participation of 250 households, Smart Watts demonstrated the value of ICTs in price-based demand response and EEBus technology

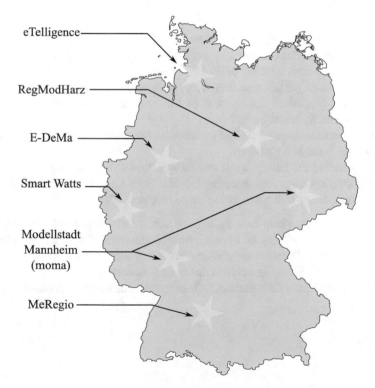

Fig. 4.3 The demonstration projects conducted within the E-Energy programme

for addressing interoperability issues between different communication protocols that are used by different stakeholder groups within the electricity sector. RegMod-Haz is a synthesized smart electricity system with various kinds of active elements (including multiple electricity production technologies, pumped hydrogen storage, and plug and play-based home energy management systems) integrated to realize cost-effective and reliable system operation.

These solutions to a large degree prove the value and feasibility of EI solutions for the electricity sector and pave the way for further developments that aim to integrate multiple energy sectors with the intelligent use of digital solutions.

4.3.2 EnergyLab Nordhavn in Denmark

EnergyLab Nordhavn is a Danish lighthouse project to demonstrate the new urban integrated energy infrastructure and solutions in central Copenhagen [14]. With a total investment of DKK 143 million, a number of innovative technologies and business solutions were developed and demonstrated over five years from 2015 to 2019.

Among these developments and demonstrations, several particularly interesting solutions distinguish the EnergyLab Nordhavn project from other smart city projects, such as

1. Including a data warehouse solution that provides safe and efficient big data management services to different stakeholder groups, particularly researchers. It enables real-time tracking of all kinds of data collected from different experiments and third parties and facilitates the development of data-based EI solutions such as pattern recognition of end-user behaviour, intelligent energy management, forecasting, and so on.
2. Using a service-oriented design to integrate the operation and management of electricity, heat, and transport.
3. Demonstrating the feasibility of low-temperature district heating.
4. Demonstrating integrated management of various kinds of flexibility offered by smart low-energy buildings, utility-scale batteries, EVs, etc.

This project shows a best-case example of how stakeholders from different industries and sectors can collaborate to develop and demonstrate future energy solutions. In addition, it provides an open platform for *small and medium*-sized enterprises (*SMEs*) to become actively involved in different demonstrations to show their innovative ideas related to data visualization, CO_2 measurement, intelligent communication, etc.

4.3.3   Decentralized Energy Control, Trading and Management Platforms

One of the focuses of EI solutions is to facilitate the operation and management of millions of units through distributed/decentralized solutions [15]. Peer-to-peer (P2P) energy trading, as illustrated in Fig. 4.4, describes flexible energy trades between energy peers, therefore enabling market-oriented optimal energy resource management [16]. Compared to a pool-based energy market organization, the main benefit may be societal. This is primarily due to the involvement rate of actors in the system at all levels being significantly high. Notably, residential actors are motivated to improve their awareness of various opportunities for energy trading and energy services, which contributes to energy transitions such as the adoption of new green technologies.

In principle, a peer can be interpreted as anything that constitutes an energy system, such as an energy producer, a consumer, an agent or even a system. Today, most existing P2P applications in the EU are developed for the electricity sector, such as the Vandebron platform [17] and the PeerEnergyCloud project [18], which allow renewable energy producers such as PV owners to sell their excess electricity production to anyone, such as the PV owners' neighbours, who might need electricity.

Instead of trading energy, developing a marketplace for trading different kinds of energy services is another mission of EI solutions because energy infrastructure must be continuously reliable and stable. The flexibility clearing house (FLEX) launched by the iPower Consortium presents a platform for realizing flexibility trading, wherein grid operators can acquire flexibility services (e.g., congestion management, frequency support, and voltage control) directly from energy end-users or their representatives, such as an aggregator [19]. Such market-based solutions have also gained strong support from the European Research Council. The H2020 project Magnitude will extend electricity-focused flexibility trading mechanisms towards multi-energy sector-oriented solutions [20]. The new business, market mechanisms, and supporting coordination tools will support and demonstrate the integrated use of flexibility options from electricity, heating/cooling, and gas systems with different sizes and technological features in seven European countries.

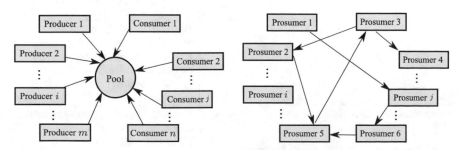

Fig. 4.4 Different organizational structures for electricity markets: from pool (left) to peer-to-peer (right)

4.3.4   The Digitization of District Heating in Aarhus

Following the large-scale deployment of smart metres in the electricity sector, there has also been an accelerated rollout of digital solutions in other energy sectors such as heat. In 2017, AffaldVarme Aarhus (AVA), the heating utility of Denmark's second largest city, completed the implementation of 56,000 remotely and hourly read heat metres [21]. One of the most substantial benefits brought by this digitalization process is to provide the heating utility a real-time overview of its network status and energy consumption, as well as speedy identification of fault locations. In addition, the collected big data are used to improve customer relations and to optimize network investment.

4.3.5   A List of EI Demonstrations and Applications in Europe

Among the hundreds of demonstrations and pilot applications related to different solution categories of the EI, Table 4.2 gives a list of selected activities for inspiring reading.

Table 4.2 Selected EI demonstrations and applications in Europe

Title/acronym	Demo location	Description
Hybalance [22]	Denmark	Demonstrate the use of a 1 MW PEM electrolyzer to produce hydrogen from excess wind power and to provide balancing services to the electrical network operator
Parker [23]	Denmark	Demonstrate using EVs to provide electrical grid services
Cell [24]	Denmark	Demonstrate a decentralized control solution "cell" for intelligent electrical network management in the presence of a high share of DERs
EStorage [25]	France	Demonstrate the feasibility and benefits of converting a fixed-speed pumped hydro storage into a variable-speed solution while providing ancillary services to grid operators
SonnenCommunity [26]	Germany	An application of P2P energy trading developed for SonneBatterie PV-battery system owners

(continued)

Table 4.2 (continued)

Title/acronym	Demo location	Description
The Stockholm Royal seaport [27]	Sweden	Demonstrate a smart energy city with optimal management of integrated energy resources and active involvement of energy users
Piclo [28]	UK	An application supporting P2P energy trading
Electron [29]	UK	An application of an energy meter and billing platform using blockchain
CPSBuDj [30]	UK	A toolset for the multidisciplinary design of cyber-physical systems for smart energy control in buildings and districts
E-HUB [31]	Multiple sites in Europe	Demonstrate the feasibility of 100% renewable energy on-site within an "Energy Hub District" with heating, cooling and electricity needs
ARTEMIS [32]	Multiple sites in Europe	Demonstrate e-mobility and real-time EV integration into a smart grid through the IoT
IDE4L [33]	Multiple sites in Europe	Demonstrate an active distribution network solutions
STORE&GO [34]	Multiple sites in Europe	Demonstrate the integration of power-to-gas storage in energy systems with a high share of renewable energy
SENSIBLE [35]	Multiple sites in Europe	Demonstrate the integration of multiple types of energy storage technologies and micro-generation (CHP, heat pumps) and renewable energy sources (PV) into power and energy networks as well as homes and buildings
SmartEnCity [36]	Multiple sites in Europe	Demonstrate the concept of a carbon-neutral smart city, wherein coordinated stakeholders from public and private sectors apply integrated green energy supply and demand response technologies

4.4   Conclusion and Outlook

The EI is a promising set of solutions that will enable a paradigm shift in energy towards an energy ecosystem with advanced connectivity, intelligent management and seamless integration of billions of smart devices, machines, and systems from multiple energy sectors. Tremendous effort has been given to developing EI demonstrations and pilot applications that are massively implemented in Europe to prove feasibility, demonstrate values, and create various kinds of learning effects.

This chapter presents a brief overview of demonstrations and pilot applications of EI-related solutions in Europe. Although the electricity sector is still the major field where EI solutions are built or developed, there is an apparent trend of demonstrating EI solutions at the district/city level with advanced integration with other energy sectors and relevant stakeholders. Through participating demonstrations that demonstrate solutions at various TRLs, participants are offered learning opportunities to improve their technical, organizational, and market insights for the EI. Direct outcomes are often clear understandings of the further needs of supporting technologies, public policies, market designs, and regulation improvements. This learning effect naturally increases the capability of participants to contribute to further development and realization of the EI.

References

1. A. Huang, M. Crow, G. Heydt et al., The future renewable electric energy delivery and management (FREEDM) system: The energy internet. Proc. IEEE **99**, 133–148 (2011). https://doi.org/10.1109/jproc.2010.2081330
2. D. Pudjianto, C. Ramsay, G. Strbac, Virtual power plant and system integration of distributed energy resources. IET Renew. Power Gener. **1**, 10 (2007). https://doi.org/10.1049/iet-rpg:20060023
3. C. Block, F. Briegel, *Internet of EnergyICT for Energy Markets of the Future* (Federation of German Industries (BDI e.V.), Berlin, 2010)
4. O. Vermesan et al., Internet of energy—Connecting energy anywhere anytime, in *Advanced Microsystems for Automotive Applications 2011*, ed. by G. Meyer, J. Valldorf (VDI-Buch, Springer, Berlin, Heidelberg, 2011)
5. S. You, J. Lin, J. Hu et al., The Danish perspective of energy internet: from service-oriented flexibility trading to integrated design, planning and operation of multiple cross-sectoral energy systems. Proc. CSEE **35**(14), 3470–3481 (2015)
6. J. Rifkin, *The Third Industrial Revolution* (Palgrave Macmillan, Basingstoke, 2013), pp. 31–46
7. F. Gangale, J. Vasiljevska, C. Covrig, et al., Smart grid projects outlook 2017: facts, figures and trends in Europe, pp. 13–14 (2017)
8. D. Tawil-Jamault, M. Buna, L. Peeters, et al., Technology readiness level: guidance principles for renewable energy technologies: annexes. Pulication office of the European Union (2017)
9. B. Bossink, Demonstrating sustainable energy: a review based model of sustainable energy demonstration projects. Renew. Sustain. Energy Rev. **77**, 1349–1362 (2017). https://doi.org/10.1016/j.rser.2017.02.002
10. DERlab—DERlab: European Distributed Energy Resources Laboratories (DERlab) e.V (2019). https://der-lab.net/. Accessed 20 Feb 2019
11. (2019). http://www.powerlab.dk/. Accessed 20 Feb 2019
12. Smart mobility living lab. in *Smart Mobility Living Lab* (2018). https://www.smartmobility.london/. Accessed 20 Feb 2019
13. E-Energy—Smart grids made in Germany—Power plus communications AG. in *Power Plus Communications AG* (2014). https://www.ppc-ag.com/projekte/e-energy/. Accessed 20 Feb 2019
14. (2015). http://energylabnordhavn.weebly.com/

15. X. Han, K. Heussen, O. Gehrke et al., Taxonomy for evaluation of distributed control strategies for distributed energy resources. IEEE Trans. Smart Grid **9**, 5185–5195 (2018). https://doi.org/10.1109/tsg.2017.2682924
16. F. Moret, P. Pinson Energy collectives: a community and fairness based approach to future electricity markets. IEEE Trans. Power Syst. 1–1 (2018). https://doi.org/10.1109/tpwrs.2018.2808961
17. Vandebron—Duurzame energie van Nederlandse bodem. in *Vandebron* (2016). https://vandebron.nl/. Accessed 20 Feb 2019
18. C. Zhang, J. Wu, Y. Zhou et al., Peer-to-Peer energy trading in a Microgrid. Appl. Energy **220**, 1–12 (2018). https://doi.org/10.1016/j.apenergy.2018.03.010
19. K. Heussen, D. Bondy, J. Hu, et al., A clearinghouse concept for distribution-level flexibility services. IEEE PES ISGT Europe 2013 (2013). https://doi.org/10.1109/isgteurope.2013.6695483
20. MAGNITUDE. in *Magnitude* (2018). http://www.magnitude-project.eu/. Accessed 20 Feb 2019
21. S. Tougaard, Up close with the world's biggest wireless district heating network. Euroheat Power (English Edition) **12**(2), 46–47 (2015)
22. HyBalance—Green energy project denmark (2016). http://hybalance.eu/. Accessed 20 Feb 2019
23. Parker | Parker-project (2017). http://parker-project.com/. Accessed 20 Feb 2019
24. S. Cherian, V. Knazkins The Danish cell project—Part 2: verification of control approach via modeling and laboratory tests. in *2007 IEEE Power Engineering Society General Meeting* (2007). https://doi.org/10.1109/pes.2007.386224
25. eStorage project (2014). http://www.estorage-project.eu/. Accessed 20 Feb 2019
26. sonnenCommunity. in Sonnenbatterie.de (2015). https://www.sonnenbatterie.de/en/sonnenCommunity. Accessed 20 Feb 2019
27. H. Shahrokni, L. Årman, D. Lazarevic et al., Implementing smart urban metabolism in the stockholm royal seaport: smart city SRS. J. Ind. Ecol. **19**, 917–929 (2015). https://doi.org/10.1111/jiec.12308
28. A glimpse into the future of Britain's energy economy (Open Utility Ltd, 2016)
29. Electron | Blockchain systems for the energy sector. in Electron.org.uk (2016). http://www.electron.org.uk. Accessed 20 Feb 2019
30. L. Etxeberria, F. Larrinaga, U. Markiegi, et al., Enabling co-simulation of smart energy control systems for buildings and districts. in *2017 22nd IEEE International Conference on Emerging Technologies and Factory Automation (ETFA)* (2017). https://doi.org/10.1109/etfa.2017.8247746
31. Energy Hub | Energy Hub District | District Heating, Cooling & Power| on-site renewable energy | Energy Hub | Heat Sharing | Seventh Framework Programme FP7 Smartgrid Technology| Zero energy districts. in E-hub.org (2013). http://www.e-hub.org/. Accessed 20 Feb 2019
32. IoE—Project. in Artemis-ioe.eu (2013). http://www.artemis-ioe.eu/ioe_project.htm. Accessed 20 Feb 2019
33. IDE4L—Ideal grid for all (2016). http://ide4l.eu/. Accessed 20 Feb 2019
34. STORE&GO (2017). https://www.storeandgo.info/. Accessed 20 Feb 2019
35. Project Sensible (2017). https://www.projectsensible.eu/. Accessed 20 Feb 2019
36. SmartEnCity.eu (2016). http://smartencity.eu/. Accessed 20 Feb 2019

Part II
Energy Switching and Routing for Energy Internet

Chapter 5
Modified P&O Approach Based Detection of the Optimal Power-Speed Curve for MPPT of Wind Turbines

Liuying Li, Yaxing Ren, Jian Chen, Kai Shi, and Lin Jiang

Abstract Improving operation efficiency has gained attention for wind turbine operation and maintenance. This chapter proposes a method to dynamically calibrate the optimal power-speed curve (OPSC) based on a modified perturbation and observation (P&O) approach to improve power signal feedback (PSF) based maximum power point tracking (MPPT). The detection takes advantage of the P&O approach without detailed modelling. By controlling the wind turbine rotational speed and recording operating data, the maximum power points (MPPs) can be calculated. The OPSC is obtained from the calculated MPPs and applied in PSF control to improve the MPPT efficiency. Simulation and experimental results show that this method can effectively obtain the OPSC for wind turbines to achieve MPPT.

5.1 Introduction

With the rapid development of wind turbines (WTs) in the past decade, improving the energy conversion efficiency has become one of the main concerns for wind turbine operation and maintenance [1]. Most modern turbines operate to achieve better energy conversion efficiency and MPPT below the rated power region [2]. There are three methods to achieve MPPT, i.e. the power signal feedback (PSF) method, perturbation and observation (P&O) method and tip-speed ratio (TSR) method [2–4].

The TSR method has the advantage of fast response due to direct control of the rotation speed. However, it requires wind speed measurement and the optimal TSR value as a control reference. The P&O method disturbs a control variable and observes changes in output until the operating point is close to a maximum power point (MPP). P&O-based strategies are mainly applied for small-scale wind turbines because they cause disturbance to the WT and do not need a model of the WT. However, the tracking ability may be interrupted under rapidly changing wind speeds. The turbine rotational speed is controlled by tracking the optimal power-speed curve (OPSC) [3, 5]. The accuracy of the OPSC will affect the efficiency of MPPT.

L. Li · Y. Ren · J. Chen · K. Shi · L. Jiang (✉)
Department of Electrical Engineering and Electronics, University of Liverpool, Liverpool, UK
e-mail: l.jiang@liverpool.ac.uk

© Springer Nature Switzerland AG 2020
A. F. Zobaa and J. Cao (eds.), *Energy Internet*,
https://doi.org/10.1007/978-3-030-45453-1_5

For a particular type of wind turbine, the OPSC of the WT can be obtained accurately via an aerodynamic experimental test when WTs are manufactured. However, the accuracy of the OPSC will be reduced during the lifelong operation period because of ageing, structural deformation and erosion of the blade, and the external environment, e.g. icing, dirt, insects [5]. In some cases, the power loss in a year is typically in the range of 20–50% due to ice accretion on blades [6]. Thus, it is necessary to calibrate wind turbines' characteristics, e.g. the OPSC, during the wind turbines' life cycles to reduce the losses caused by an inaccurate OPSC. Furthermore, the preconfigured OPSC should be ideally tuned to adapt to the local conditions of different installation sites and updated during lifetime operation.

In this chapter, a strategy to calibrate the OPSC with variable wind speed to achieve MPPT by the PSF method is proposed. The proposed method will operate the wind turbine at a constant rotational speed and under variable wind to record the operating data. The process will continue running until enough data are recorded. The rotational speed reference will then change to a different rotational speed reference value, and the process will continue recording data. Based on the recorded data, the MPP at each wind speed and the OPSC can be obtained. This method is a new type of P&O approach with the advantage of not requiring a detailed model of the WT. Furthermore, this method will not cause an additional disturbance in the control loop, as it is only used for data collection and detection of the OPSC. Although the data recording process will lose MPPT operation during the test, it can still be used for on-site wind turbine characteristics or OPSC detection and calibration.

Moreover, the proposed method is verified by simulation and experimental tests. The experiment tests are based on a wind power generation system rig with a PMSG and a DC motor coupled to imitate the wind turbine profile.

5.2 Model

A wind power generation system (WPGS) is composed of a drive train, generator-side converter and grid-side converter. The converters adjust the rotation and extend the active power or maintain it at the rated value and also transfer active power to the grid [7]. In a WPGS, MPPT is used to adjust the wind turbine rotation speed through the generator-side converter to maximize the power captured.

In a WPGS, the mechanical power P_m is given by

$$P_m = \frac{1}{2}\rho\pi R_b^2 V_w^3 C_p(\lambda, \beta) \tag{5.1}$$

where ρ is the density; C_p describes the efficiency power extraction, which is calculated from λ to β; and λ is defined by the wind speed V_w and rotational speed ω as given in (5.2).

$$\lambda = \frac{R_b \omega}{V_w} \tag{5.2}$$

MPPT is used to control the wind turbine continuous operation at this point. In the region in which the turbine control objective is MPPT, the turbine blade pitch β has a constant value to enable the generation of the highest power. C_p in this paper is defined as a function of λ [8]:

$$C_p(\lambda) = 0.22(116\lambda_t - 0.4\beta - 5)e^{-12.5\lambda_t} \tag{5.3}$$

where

$$\lambda_t = \frac{1}{\lambda + 0.08\beta} - \frac{0.035}{\beta^3 + 1} \tag{5.4}$$

With a fixed β, λ_{opt} is the optimal point where C_p has its highest value (C_{p_max}). Thus, P_{opt} can be described by

$$P_{\text{opt}} = \frac{1}{2}\rho\pi R_b^5 \frac{C_{p\_\text{max}}}{\lambda_{\text{opt}}^3} \omega^3$$

For a direct drive wind turbine-based WPGS, the rotor speed is equal to the generator speed. With wind, the optimal rotation speed is that at which the turbine captures maximum power from the wind, and their relation can be given by [9]:

$$P_{\text{opt}} = k_{\text{opt}}\omega^3 \tag{5.5}$$

where k_{opt} is the parameter used for defining the OPSC. The curve of P_{opt} defines the optimal power generated by the turbine tracking on this curve during variable wind.

The power-speed curves of a WPGS at different wind speeds are shown in Fig. 5.1. There is an optimal power-speed curve in the figure. This curve is the combination of all the MPPs at different wind speeds. The MPPs give information on the optimal rotation speed to maximize the generated power. The objective of MPPT is to control the extraction of optimal power continuously in variable wind. Therefore, MPPT controls the rotational speed to follow the optimal power-speed curve, as shown in Fig. 5.1 [10].

5.3 Proposed Optimal Power-Speed Curve Detection Method

The proposed optimal power-speed curve detection method contains the following five steps:

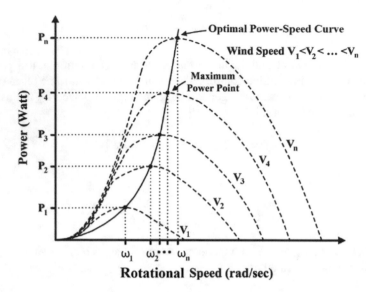

Fig. 5.1 Power-speed relationship with rotational speed at different wind speeds [10]

1. Give the rotational speed reference ω_{ref} a specified constant value, and control the wind turbine operation at the reference rotational speed.
2. Measure the output power, wind speed and rotational speed. Record data when the measured speed is near the pre-set values.
3. When all pre-set data are recorded for the current rotational speed reference, change to a new rotational speed reference value, and repeat the first two steps.
4. When all pre-set rotational speed reference values have been called and all operating data have been recorded, stop recording the operating data. Compare all recorded power values under each wind speed, and calculate the maximum power points.
5. Calculate k_{opt} as the optimal power-speed curve parameter.

These five steps can be classified as two main operation processes: recording operating data and detection of the OPSC, which are described in detail as follows.

5.3.1 Recording Operating Data

Figure 5.2 shows operating data recorded under three different wind speeds. ω_1^* is the rotational speed reference ω_{ref} at the beginning. By controlling the turbine's rotational speed, when ω is near ω_{ref}, it will satisfy Eq. 5.6.

$$|\omega_{ref} - \omega| < \varepsilon_\omega \tag{5.6}$$

Fig. 5.2 Recording operating data on a power-speed curve

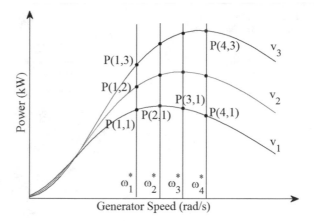

where ε_ω is a small value used to define the threshold of controlling the rotational speed at ω_{ref}.

In Fig. 5.2, $P(1, 1)$ is the measured power in the situation that under v_1, the corresponding rotation speed is ω_1. When the wind speed changes from v_1 to v_2, $P(1, 2)$ will be measured and recorded under the same rotational speed reference ω_1^*. Thus, n operating data are recorded as the wind speed varies between v_1 and v_n under the same rotational speed reference. Then, the rotational speed reference changes from ω_1^* to ω_2^*, where $\omega_2^* = \omega_1^* + \Delta\omega$. $\Delta\omega$ is a constant step size between two reference rotational speeds. The operating data from $P(2, 1)$ to $P(2, n)$ under ω_2^* will be recorded in the same way as for ω_1^*. By controlling the wind turbine rotational speed varying from ω_1^* to ω_m^*, there will be $m \times n$ recorded operating points. Comparing the recorded power values for different wind speeds, the recorded maximum power points are shown in red in Fig. 5.2.

Figure 5.3 presents the flowchart of recording operating data. Equation 5.7 is used to decide whether the operating data should be recorded.

$$|v - v_i| < \varepsilon_v \tag{5.7}$$

where ε_v is a constant parameter used to decide this at the pre-set value. The rotational speed changes from ω_1^* to ω_n^*. When all the rotational speed reference ω_n^* values are recorded, the recording operating data process ends. Then, the controller will move on to the detection of the optimal power curve and calculation of k_{opt}.

In the flowchart of recording operating data (Fig. 5.3), there are two parts marked in blue and green. These two added parts in the flowchart are for generation system operation and operation time reduction.

Fig. 5.3 Flowchart of
recording operating data

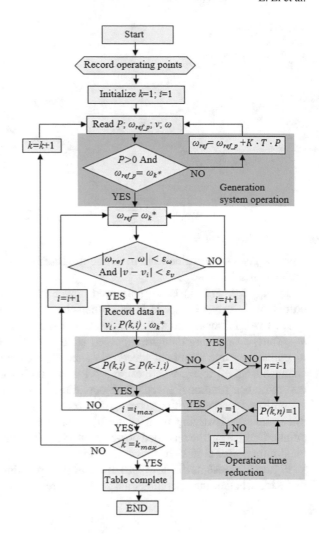

5.3.1.1 Generation System Operation

The wind turbine rotational speed is controlled at specified constant values during
the recording operating data process. In the situation that the wind speed decreases to
a low value, this approach may cause the generator to not be able to generate power
at the pre-set constant reference value. This will decrease the generation efficiency
and should be avoided.

To avert the above situation, we need to decrease the rotational speed reference
ω_{ref} to acclimate to the decrease in wind speed and avoid the case of the wind turbine
not generating the pre-set reference power. Until the wind speed increases to a value
that the WPGS is able to generate positive power, the rotational speed reference ω_{ref}
will switch back to the pre-set value and continue recording the operating data. The

blue area in Fig. 5.3 shows this algorithm in the method flowchart. When the power is negative or the rotational speed reference of the previous step $\omega_{\text{ref\_}p}$ is not the pre-set value ω_k^*, then the rotational speed reference is given by

$$\omega_{\text{ref}} = \omega_{\text{ref\_}p} + K \cdot T \cdot P \tag{5.8}$$

where K is a constant parameter and T is the system period. The setting of this function is used to control the rotational speed value in the case that power P is negative.

5.3.1.2 Operation Time Reduction

In the flowchart of recording operating data (Fig. 5.3), there is a green area presenting the logic of the method for operation time reduction. During the control of the wind turbine rotational speed at specified constant values, if the measured power $P(k, i)$ is less than $P(k - 1, i)$, then the recording of further operating data under wind speed v_i will be stopped. For example, as shown in Fig. 5.2, $P(3, 1) < P(2, 1)$ signifies that the maximum power point of this wind speed has already been recorded. Then, there is no need to record $P(4, 1)$ or additional data under wind speed v_1. Thus, in Fig. 5.3, the green area recodes as 1 to represent skipping these unnecessary data. This process can skip unnecessary data to reduce the time of recording the operating data.

5.3.2 Detection of the OPSC

Based on the pre-set values of wind speed v_i and reference rotational speed ω_i^*, it is probable that the maximum power points may not be detected directly. Thus, a remedy detection algorithm is designed to solve this problem. The implementation is shown in Fig. 5.4. In Fig. 5.4a, the operating points are recorded at the rotational speeds ω_k and ω_{k+1}. However, the actual MPP is between ω_k and ω_{k+1}, shown by the red point. Figure 5.4b shows the zoomed-in top part of the curve in Fig. 5.4a. Under wind speed v_i, point A is the recorded MPP; C is the second-highest point; and B is the third-highest point. Then, the real MPP must be between points A and C. The real MPP can be estimated if we assume the curve is symmetrical about the axis passing through the real MPP in a small region. Then, the approximate MPP can be estimated as follows. Draw a line connecting points A and B. It will create an angle $\angle b$ with the x-axis. Then, make a line pass through point C with an angle $\angle c$ equal to $\angle b$. The extensions of these two lines meet at point D. Point D would be at the central axis of this symmetric curve from a geometric principle. As it was assumed that the curve is symmetrical about the axis passing through the real MPP in a small region, point D will be recorded as the MPP under wind speed v_i instead of point A. Thus, the following formulas are deduced.

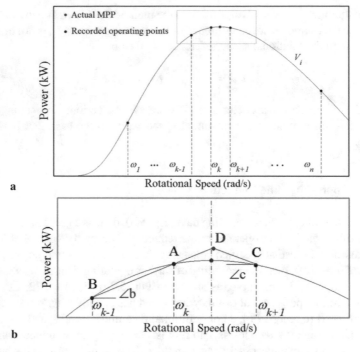

Fig. 5.4 a The overall recorded operating points at a constant wind speed, **b** the zoomed-in top part of a to derive the maximum power point

The maximum power point calculated when $P(k-1, i) > P(k+1, i)$ is

$$\begin{cases} \omega_{i\_opt} = \frac{1}{2}\left[\omega_k + \omega_{k-1} + \frac{P_{(k,i)} - P_{(k-1,i)}}{P_{(k,i)} - P_{(k+1,i)}}(\omega_{k+1} - \omega_k)\right] \\ P_{i\_max} = P_{(k,i)} + \frac{P_{(k,i)} - P_{(k+1,i)}}{\omega_k - \omega_{k+1}}(\omega_{i\_opt} - \omega_k) \end{cases} \tag{5.9}$$

The maximum power point calculated when $P(k-1, i) < P(k+1, i)$ is

$$\begin{cases} \omega_{i\_opt} = \frac{1}{2}\left[\omega_k + \omega_{k+1} + \frac{P_{(k,i)} - P_{(k+1,i)}}{P_{(k,i)} - P_{(k-1,i)}}(\omega_k - \omega_{k-1})\right] \\ P_{i\_max} = P_{(k,i)} + \frac{P_{(k,i)} - P_{(k-1,i)}}{\omega_k - \omega_{k-1}}(\omega_{i\_opt} - \omega_k) \end{cases} \tag{5.10}$$

Equations 5.9 and 5.10 can obtain a more accurate MPP when the recorded operating points are not located accurately at the real MPP. Thus, there are no restrictions on the pre-set operating points if small enough step sizes are used.

Figure 5.5 shows the flowchart of the detection of the optimal power-speed curve and k_{opt}. After the recording operating data process, the maximum power points under each wind speed can be derived. Then, the OPSC and k_{opt} for each wind speed can be calculated by a polynomial interpolation method. For example, for v_i, the obtained MPP is P_{i_max}. When P_{i_max} is achieved, k_i is the calculated k_{opt} under wind

Fig. 5.5 Flowchart of
detection of the optimal
power-speed curve and k_{opt}

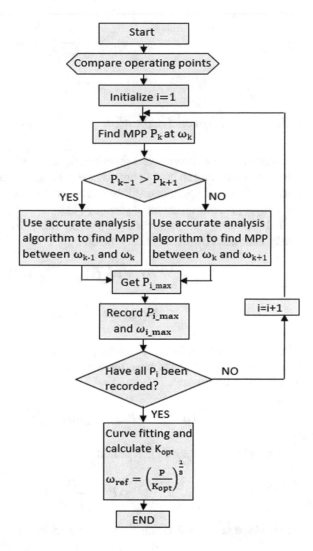

speed v_i. ω_k and $P_{(k,\, i)}$ are the rotational speed and power when the wind speed is v_i,
respectively. After k_i is recorded, the optimal curve parameter k_{opt} can be estimated
by calculating the average value of all obtained k_i.

$$k_{\mathrm{opt}} = \frac{1}{n}\sum_{i=1}^{n}k_i \tag{5.11}$$

After k_{opt} is calculated, the wind turbine will operate in PSF control with the
detected OPSC with the parameter k_{opt} to achieve MPPT. The PSF control will be
based on the rotational speed reference ω_{ref} calculated by Eq. 5.12 to achieve MPPT.

$$\omega_{\text{ref}} = \left(\frac{P_e}{k_{\text{opt}}}\right)^{1/3} \tag{5.12}$$

5.3.3 Wind Speed Estimation

The optimal power curve detection in this paper requires wind speed measurement. An anemometer can be used to provide the wind speed value. In the case of a lack of anemometers, this speed can still be estimated by wind speed estimation methods. This paper presents an example given by [11].

This example uses the error of power. By minimizing the error, the wind speed can be estimated from the function below:

$$J(t, V) = (P_r(t) - f_r(V_w))^2 \tag{5.13}$$

$$f_r(V_w) = \frac{1}{2}\pi\rho R^2 V_w^3 C_p(\beta, \lambda) \tag{5.14}$$

where $P_r(t)$ is the real-time measured power and $f_r(V_w)$ is the turbine mechanical power, which is calculated from the wind V_w. Combining Eqs. 5.13 and 5.14, the cost function becomes:

$$I(t, V_w) = P_r(t) - \frac{1}{2}\pi\rho R^2 V_w^3 C_p(\beta, \lambda) = 0 \tag{5.15}$$

From the partial derivative equation

$$\Delta P_r = \frac{\partial P_r}{\partial V_w}\Delta V_w \tag{5.16}$$

the wind speed estimation value can then be derived as

$$\hat{V}_w = \Delta P_r\left(\frac{\partial P_r}{\partial V_w}\right)^{-1} \tag{5.17}$$

where

$$\frac{\partial P_r}{\partial V_w} = -\frac{3}{2}\pi\rho R^2 V_w^2 C_p(\beta, \lambda) - \frac{1}{2}\pi\rho R^2 V_w^2\frac{\partial C_p}{\partial V_w} \tag{5.18}$$

$$\frac{\partial C_p}{\partial V_w} = -\frac{0.22}{\omega_r R}\frac{178.5 - 1450\lambda_t + 5x_4}{(\lambda + 0.08x_4)^2}e^{-12.5\lambda_t} \tag{5.19}$$

Thus, the estimated cost function can be written as:

$$I\left(t, \hat{V}_t\right) = P_r(t) - f_r\left(\hat{V}_t\right) \; < \; \varepsilon \qquad (5.20)$$

where ε is a pre-set constant value and $\hat{V}_t$ is the estimated wind speed.

5.4 Simulation Test

The WPGS includes a PMSG and SIMULINK software. The WPGS parameters are shown in Table 5.1. In the simulation test, the system detects the optimal power-speed curve without any preliminary information of the OPSC and operates the WT in MPPT mode. The power coefficient C_p is given (Figs. 5.6d, 5.8b and 5.11c) to verify the accuracy of the detected OPSC.

The simulation test detects the optimal power-speed curve at the beginning (from 0 to 85 s) and then uses the detected optimal power-speed curve to control the turbine in MPPT mode (after 85 s). The simulation calculates the wind speed v, generator rotational speed ω_g, rotor power P_r and coefficient C_p. Figure 5.6a shows a profile generated by TurbSim [12], with an average wind speed of 8 m/s. Figure 5.6b shows control at a specified constant reference value, as mentioned in Sect. 5.3. Speed control of the WT is achieved by vector control with a PI loop. At time $t = 30$ and 60 s in Fig. 5.6, the wind speed decreases and causes a power drop, and the rotational speed is controlled to decrease to maintain positive output power, which demonstrates the effectiveness of the negative power rejection function mentioned in Sect. 5.3.1.

At $t = 84.2$ s in Fig. 5.6, a sufficient number of operating points are obtained from the method given in Sect. 5.3.2. The calculated maximum power points are shown in Table 5.2, and the curve fitting result is shown in Fig. 5.7. The calculated k_{opt} value is 1.303×10^5, which is near the accurate k_{opt} value of 1.306×10^5 (with 0.2% relative error).

After k_{opt} is detected, the WT is operated. Figure 5.6d shows that after $t = 84.2$ s, C_p reaches C_{p_max}, which shows the WT is tracking accurate points.

Figure 5.8 compares the power coefficient before and after the calibration, and Fig. 5.8a shows that the wind turbine does not achieve the maximum power before calibration. The gap between the red and blue lines means that the MPPT efficiency

Table 5.1 Wind turbine and PMSG parameters in simulation	Items	Specification
	Blade radius R_b	37.6 m
	Equivalent inertia J_r	3×10^5 kg m^2
	PMSG stator resistance R	0.1 Ω
	PMSG stator inductance L_d, L_q	0.005 H
	Pole pairs p	80
	PMSG field flux	10.68 Wb

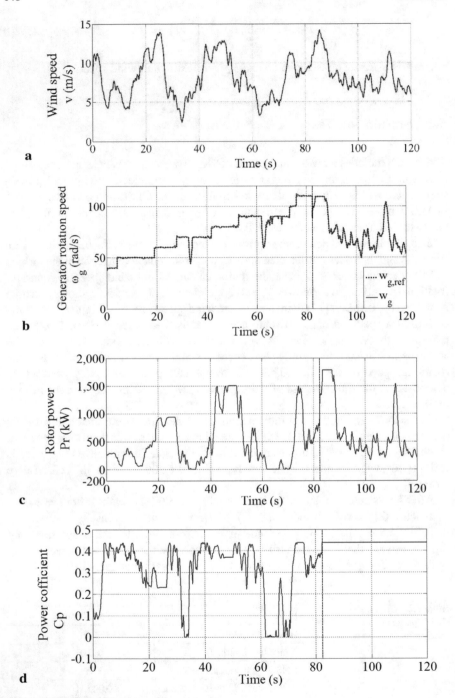

Fig. 5.6 Case 1 initial process simulation results

Table 5.2 Maximum power points recalculated

Items	Calculate	Calculate	Calculate	Calculate	Calculate	Calculate	Calculate
Generator rotation speed ω_g (rad/s)	1	1.295	1.501	1.724	1.921	2.14	2.358
Power P (kW)	1.62×10^5	2.965×10^5	4.664×10^5	6.824×10^5	9.719×10^5	1.315×10^6	1.747×10^6

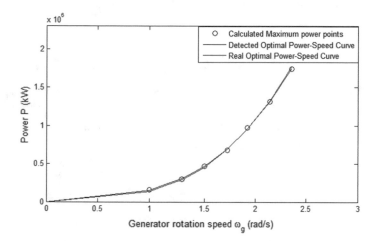

Fig. 5.7 Optimal power-speed curve fitting from the calculated maximum power points

needs to be improved and the turbine should capture more power from the wind. After the calibration process, as shown in Fig. 5.8b, C_p can nearly track its maximum value. The MPPT process is optimized.

To illustrate the effect of the optimal power-speed curve error in energy generation, more simulation studies were performed, as shown in Table 5.3. The results show that a 10% error of k_{opt} will cause a 0.28% loss in energy generation. When the error increases, the loss will increase as well. When k_{opt} has a 40% error, the generated energy will decrease by 5.24%.

5.5 Experimental Test

5.5.1 Experimental Platform

The schematics of the WPGS controller and the experimental test rig are illustrated in Fig. 5.9a, b. The experiment is implemented with dSPACE as a hardware-in-the-loop

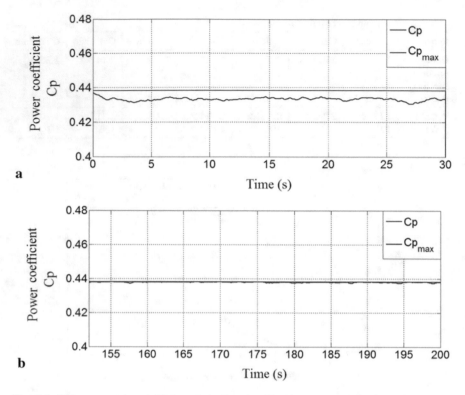

Fig. 5.8 Power conversion coefficient C_p in Case 2 calibration process simulation test. **a** Before the calibration process, **b** After the calibration process

Table 5.3 MPPT energy generation comparison

Error in the optimal power-speed curve	Energy generation loss (%)
10% relative error	0.28
20% relative error	1.10
30% relative error	2.70
40% relative error	5.24

platform. A DC motor is controlled to perform like a wind turbine by generating the torque from the model under variable wind conditions. A PMSG is coupled with the DC motor to output electricity. The dSPACE control panel outputs control signals as PWM signals. The PWM signals drive an IGBT-based converter to control the PMSG side of the WPGS (Table 5.4).

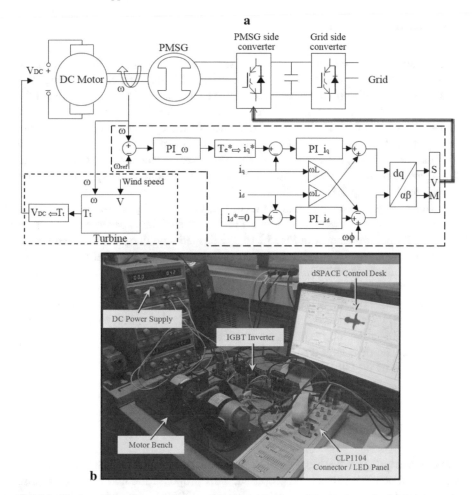

Fig. 5.9 Hardware experimental setup. **a** Schematic diagram of the hardware-in-the-loop experiment, **b** experiment bench photo

Table 5.4 Wind turbine and PMSG parameters in the experiment

Items	Specification
Blade radius R_b	35 m
Equivalent inertia J_r	3×10^6 kg m^2
PMSG stator resistance R	0.19 Ω
PMSG stator inductance L_d, L_q	0.0005 H
Pole pairs p	5
PMSG field flux	0.0127 Wb

5.6   Results

The system was tested for 120 s with a variable wind speed as the input to the system. Figures 5.10 and 5.11 present the results of recording the operating data and detection of the optimal power-speed curve (0 s–60 s) and validation in MPPT mode (60 s–120 s), respectively. Figure 5.10a shows the wind speed and Fig. 5.10b shows the system during the stage of detection of the optimal power-speed curve. In this stage, the power captured by the turbine is shown in Fig. 5.10c. From t = 35 to 50 s in Fig. 5.10b, the wind turbine rotational speed decreased corresponding to the decrease in the wind speed. This result verifies the generation system operation mentioned in Sect. 5.3.1.

After enough operating points are recorded, the optimal power-speed curve and k_{opt} are calculated at t = 60 s. Figure 5.11a, b, c shows the experimental results in the MPPT validation mode. In Fig. 5.11d, the dashed blue line presents the detected OPSC by the proposed method, and the red line presents the pre-set real OPSC of the simulated wind turbine in the experiment. The OPSC detected by the proposed method is very close to the accurate curve. The calculated k_{opt} from the curve in the experiment is 1.787×10^5. This means that the accuracy of the detection is approximately 97%. The detected OPSC is then used in the control process of MPPT. The hardware-in-the-loop experiment verified that the proposed method can obtain the OPSC. The detected OPSC can be used in the control process to improve the performance of WPGS MPPT.

5.7   Conclusion

A method to optimize the MPPT efficiency for wind turbines using PSF under variable wind speed conditions has been proposed. This proposed method does not require system pre-knowledge and can obtain an accurate OPSC under variable wind speed. This makes the method feasible for different wind turbine types and useable for on-site conditions. The method obtains the MPP by applying a modified P&O approach to control the WPGS at a specified constant rotational speed and by recording the power and wind speed. Methods to estimate accurate MPP values and accelerate the detection time were proposed. After recording the operating data and calculating the optimal power-speed curve with quadratic interpolation, the WPGS is controlled to obtain the OPSC and achieve MPPT. The OPSC detected by the proposed method has been verified to improve the MPPT performance and increase the WPGS energy generation. In experimental verification, the accuracy of the detected optimal power-speed curve and k_{opt} is approximately 97% of the actual values.

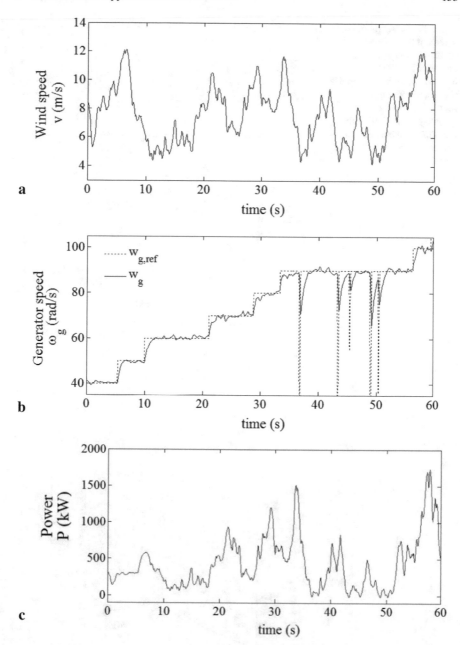

Fig. 5.10 The stage of recording the operating data experimental results

Fig. 5.11 Experimental results of the detection of the optimal power-speed curve and MPPT validation. **a** Wind speed, **b** rotational speed and reference rotational speed, **c** wind turbine power coefficient, **d** comparison between the detected OPSC and the real OPSC

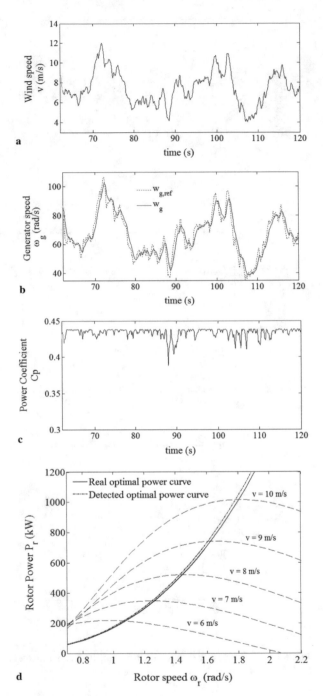

References

1. M.A. Abdullah et al., A review of maximum power point tracking algorithms for wind energy systems. Renew. Sustain. Energy Rev. **16**(5), 3220–3227 (2012)
2. M. Malinowski et al., Optimized energy-conversion systems for small wind turbines: renewable energy sources in modern distributed power generation systems. IEEE Power Electron. Mag. **2**(3), 16–30 (2015)
3. S. Shamshirband, et al., *RETRACTED: wind Turbine Power Coefficient Estimation by Soft Computing Methodologies: comparative Study* (Elsevier, 2014)
4. Y. Xia, K.H. Ahmed, B.W. Williams, A new maximum power point tracking technique for permanent magnet synchronous generator based wind energy conversion system. IEEE Trans. Power Electron. **26**(12), 3609–3620 (2011)
5. I. Staffell, R. Green, How does wind farm performance decline with age? Renew. Energy **66**, 775–786 (2014)
6. N. Dalili, A. Edrisy, R. Carriveau, A review of surface engineering issues critical to wind turbine performance. Renew. Sustain. Energy Rev. **13**(2), 428–438 (2009)
7. S.M.R. Kazmi, et al., Review and critical analysis of the research papers published till date on maximum power point tracking in wind energy conversion system. in *Energy Conversion Congress and Exposition (ECCE), 2010 IEEE* (IEEE, 2010)
8. B. Beltran, T. Ahmed-Ali, M.E.H. Benbouzid, Sliding mode power control of variable-speed wind energy conversion systems. IEEE Trans. Energy Convers. **23**(2), 551–558 (2008)
9. D. Kumar, K. Chatterjee, A review of conventional and advanced MPPT algorithms for wind energy systems. Renew. Sustain. Energy Rev. **55**, 957–970 (2016)
10. M. Yin, et al., Modeling of the wind turbine with a permanent magnet synchronous generator for integration. in *Power Engineering Society General Meeting, 2007. IEEE* (IEEE, 2007)
11. A. Kumar, K. Stol, Simulating feedback linearization control of wind turbines using high-order models. Wind Energy **13**(5), 419–432 (2010)
12. B.L. Jonkman, TurbSim user's guide: version 1.50. in *National Renewable Energy Lab (NREL)* (Golden, CO, United States, 2009)

Chapter 6
Flexible Substation and Its Demonstration Project

Zhanfeng Deng, Jun Ge, Guoliang Zhao, and Chaobo Dai

Abstract Flexible substations were proposed by Chinese scholars in 2015 as a new generation of substations mainly based on power electronic technology and information communication technology and will play an important role in the energy internet because of their flexibility in converting and distributing electrical energy. This chapter details a state-of-the-art flexible substation and its demonstration project. First, the background and the development of substations are described, and the basic information of the flexible substation is given, including comparisons, the system structure and the topology of the power electronic transformer. Then, with the example of the world's first flexible substation, the Xiaoertai 10 kV flexible substation, the flexible substation is further elaborated from various aspects such as the system proposal, control and protection, core equipment, results of the RTDS test and the site test, and operation of the flexible substation.

6.1 Overview

6.1.1 Background

In terms of energy demand, the annual global power demand will reach 73 trillion kWh in 2050 [1]. To meet such very high power demand, clean energy will find a significant growth in its development and utilization, and its ratio in the energy

Z. Deng · G. Zhao · C. Dai (✉)
Global Energy Interconnection Research Institute, Co. Ltd., Beijing, China
e-mail: chaobodai@qq.com

Z. Deng
e-mail: dengzhanfeng@geiri.sgcc.com.cn

G. Zhao
e-mail: fighter_zhao@126.com

J. Ge
State Grid JiBei Electric Power Company, Beijing, China
e-mail: ge.jun@jibei.sgcc.com.cn

© Springer Nature Switzerland AG 2020
A. F. Zobaa and J. Cao (eds.), *Energy Internet*,
https://doi.org/10.1007/978-3-030-45453-1_6

portfolio will dramatically increase. Clean energy will replace fossil energy as the dominant energy in the future with the background of low-carbon development to address global climatic changes. It is estimated that the supply of global non-fossil energy will increase to 23.7 billion tons of standard coal in 2050, 480% higher than that in 2010 [1].

The development of new energies will enter a new era in which new energies from large centralized bases will dominate with distributed ones as indispensable supplements. As shown in Fig. 6.1, it is estimated that in 2050, the global power generated by clean energies will reach 66 trillion kWh, accounting for 90% of total power generation, where the power generated by solar and wind energies will account for 66%; and the global distributed power generation will reach 11 trillion kWh, accounting for 15% of total power generation, becoming an important contributor to energy supplies [1].

As far as future energy and power supply patterns and development of information technology are concerned, Jeremy Rifkin, a scholar from the USA, predicted in his book The Third Industrial Revolution [2] published in 2011 that a new energy utilization system characterized by a deep combination of new energy technology and information technology will emerge. He named the new energy system he envisioned the "energy internet", and the energy internet is a vast network that efficiently supplies electricity to anyone anywhere. The energy internet can be understood in such a way that a great number of new power network nodes (composed of distributed energy generators, distributed energy storage devices and various types of loads) are interconnected by comprehensive utilization of state-of-the-art power electronic

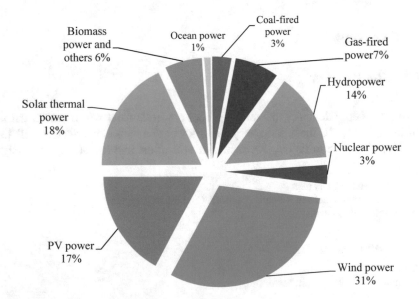

Fig. 6.1 Global power supply portfolio in 2050

technology, information technology and smart management technology to achieve bidirectional energy exchange.

To build the energy internet containing large-scale and distributed connections of clean energies, the grid must be provided with such functions as flexible power flow control, multiform energy connection and transformation and the ability to realize an energy sharing configuration similar to the internet. Currently, the scale of transmission networks has increasingly been growing, and AC/DC hybrid systems have become increasingly complex due to the addition of HVDC transmission and DC distribution. As a consequence, many issues have become prominent, e.g. unbalanced power flow, limited power transmission capacity and excessive short-circuit current in closed loops. In addition, power control technology based on conventional substations has difficulty meeting the strict demand of an energy sharing configuration similar to the internet. There is an urgent need to develop new substation technologies that integrate power flow control, dynamic voltage support and fault current limiting into one to serve as the control nodes of regional transmission networks.

As distributed renewable energies of various types and sensitive AC/DC loads have been connected, widely and extensively, to distribution networks, a key development trend for future distribution networks is to build AC/DC hybrid distribution networks on the basis of AC distribution networks. New challenges have also been identified for substation power flow control, AC/DC connection, power quality isolation and high-quality power supply. There is an urgent need to develop new technologies for grid substations to support power generation, grid connection, transformation and local efficient absorption of renewable energy; flexibly turning AC/DC loads on and off; and energy storage devices to achieve optimal management and flexible control of active distribution networks.

6.1.2   Development of Substations

Having been driven by power technology, automatic technology and information communication technology, substation technology has undergone four stages: the automatic substation, digital substation, smart substation and flexible substation. At present, smart substations have been deeply promoted, while demonstrations of flexible substations have already started to happen.

The stage of automatic substations was from 1970 to 2005. This period can be further subdivided into the conventional automation stage and the comprehensive automation stage. So-called substation automation usually refers to a comprehensive automation system [3] that widely applies micro-computer technology to the measurement and control, protection, monitoring and telecontrol systems of substations; integrates their functions to take full advantage of micro-computers; improves the automation capability of substations and the reliability of the automatic devices; and reduces the amount of wiring used in the secondary systems of the substations.

The stage of digital substations was from 2006 to 2008. During this stage, China began to develop digital substations with great effort [4, 5]. A digital substation refers to a modern one that is based on IEC 61850 and consists of smart equipment (electronic transformers, smart switches, etc.) and network-based secondary equipment (process, bay and substation layers) to realize information sharing and inter-operation among the smart electric equipment in the substation. In this way, operation with few attendants or attendant-free operation is achieved. The development process of substations shows that digital substations can be regarded as transitional from automatic substations to smart substations. Electronic transformers and smart switches were used in these substations, which reduced the layout of secondary cables. However, digital measurement was not fully used.

Stage of smart substations: 2009–present. The first generation of smart substations was promoted in China during this stage. A smart substation based on IEC 61850 is fully designed with advanced, reliable, integrated and environmentally friendly smart primary equipment [6–8]. With digitalization of substation information, networking of communication platforms and standardization of information sharing as the basic requirements, these substations can automatically perform basic functions such as information acquisition, measurement, control, protection, metering and detection and realize attendant-free operation. In addition, a smart substation can take full advantage of the basic data among systems. It is designed with many advanced functions, such as real-time automatic smart control, maintenance, online analysis and decision-making and collaborative interaction.

The above substations all feature automatic, digital and smart secondary equipment; however, their primary equipment is still not power electronics. Consequently, their flexibility and control performance are insufficient for future substations with the background of the energy internet.

Stage of flexible substations: The concept of a solid-state power substation (SSPS) [9] or solid-state substation [10] was proposed by Electric Power Research Institute, USA, in 2006. It is a new substation based on power electronics, where the mechanical ferromagnetic primary equipment in a conventional substation was replaced by power electronic equipment such as power electronic transformers (PETs) [11], solid-state transformers (SSTs) [12–15], and solid-state switches. The University of North Carolina and Future Renewable Electric Energy Delivery and Management (FREEDM) conducted a great deal of research on the relevant technologies of SSPSs. A 13.8 kV/465 V, 855 kVA SSPS was successfully developed in 2011, whose efficiency reached 97%. The testing system for a 13.8 kV/480 V, 100 kVA SSPS was successfully developed in 2014 using silicon carbide (SiC) components, but it was still at the testing stage and not put into commercial operation. The SSPS was a valuable exploration to replace primary equipment with power electronics, but it has not yet developed mainly due to the lack of a comprehensive study on the overall function and system architecture of a substation with power electronic primary equipment. It can be regarded as a rudimentary flexible substation.

Chinese scholars proposed the concept of a "flexible substation" [16, 17] in September 2015. Flexible substations are a new generation of substations based on power electronic technology [18, 19] and information communication technology. Technically, they feature the integration of power electronic technology, control and protection technology, information communication technology and advanced computation. In terms of equipment forms, the functions of several separated installations are integrated into one or two installations. The roles are an energy transmission node, a control node in the grid and a load control node. For operation modes, these substations can be connected to a large grid and operated with or without the grid connection. For information interaction, they can serve as a node to receive information, execute commands and realize deep integration of electrical equipment and information.

A flexible substation can be upgraded from an existing smart substation with the help of power electronic technology. It can be designed with such integrated control functions as energy conversion, power flow control, frequency modulation, system stability control and power quality isolation and control, so that it can improve the absorption capacity of new energies and the flexibility, stability and power quality of the grid. Additionally, the equipment in a flexible substation features "multi-functions in one", greatly reducing the types and quantity of substation equipment, the footprint of the substation and the quantity of spare parts.

The power flow control mode can be changed with the aid of flexible substations from stiff connections because of the lack of effective control means to flexible ones via bidirectional power flow controls among different substation terminals.

Flexible control of flexible substations can relieve adverse effects on the grid due to the fluctuation and intermittence of renewable energies and realize a change from "connection to the grid" to "integration with the grid" of renewable energies, resulting in new energy absorption to the maximum extent.

AC/DC power supply with different qualities: It can make full use of high voltage (HV)/low-voltage (LV) AC/DC multiple electrical terminals of power electronic converter units to realize AC/DC interconnections. It can achieve effective isolation of power quality between the grid and loads to meet power supplies with different power qualities.

Load control: It can adjust the output power frequency and voltage in a dynamic manner to realize the dynamic regulation of load demand and improve the response capacity of the load demand response. It can provide support for rapid grid recovery and improve grid transient stability during post-fault recovery.

Flexible control of reactive power and voltage: It can realize active power flow control in each terminal and provide steady-state and dynamic reactive power according to the grid demand.

The "information flow" and the "power flow" work together, and power and information interact deeply. The collaborative control of grid power flow and information flow shall be built to realize power allocated by demand and collaborative operation among different flexible substations.

6.1.3 Comparisons

Compared with automatic substations, digital substations and smart substations, flexible substations have advantages in such aspects as flexible power flow control, multiple information flow interactions, improvement of power quality, rapid fault removal, efficiency improvement of operation and maintenance and flexible access to other modules such as energy saving and DC loads.

A power electronic converter in a flexible substation has integrated functions such as voltage transformation, circuit breaking and Var compensation. In this way, several equipment, e.g. the traditional HV/LV outgoing circuit breakers of the main transformer and the LV Var compensator in the substation, can be omitted, which simplifies the main circuit and reduces the footprint of the substation.

Based on the real-time power flow of power supplies and loads, a flexible substation can coordinate comprehensively with the grid, distributed power supplies and controllable loads to allocate the power flow by demand in each terminal. The power flow can be optimized through this way to realize mutual complementation among several energies, allocation of power flow by demand and implementation of economic operation at the same time.

A flexible substation can dramatically enhance the management and control function of existing substations. It can effectively organize the power resources and achieve real-time power supply/demand balance in a region via the interaction among the distributed power supplies, the controllable loads and the HV grid.

A power electronic converter can isolate harmonic propagation between the HV and LV sides and effectively suppress the voltage sag, resulting in a voltage decrease duration of less than 10 ms and increasing the grid voltage qualification rate.

The circuit-breaking time of an ordinary AC circuit breaker ranges from ten to less than one hundred milliseconds, while a power electronic converter and a solid-state switch can block fire signals in a range of 3–5 ms. As a consequence, they can dramatically shorten the fault removal time, improve system stability and protect relevant equipment.

A flexible substation employs a modular standard design, where each module can be maintained separately with other modules in normal operation. As a result, it can avoid substation shut down due to module maintenance in the case of substation equipment faults and improve power supply reliability and efficiency of operation and maintenance.

A flexible substation is designed to deal with the development of distributed power supplies and large-scale DC loads. DC terminals, together with AC terminals for ordinary AC lines, are designed for the flexible connection of various AC/DC loads such as electric vehicles (EVs), distributed energy storage and distributed power supplies.

Table 6.1 summarizes comparison results between a flexible substation and a conventional substation concerning the main performance indices.

Table 6.1 Performance comparison of substations

Index		Conventional substations	Flexible substations
Technology advance	Flexible control		√
	Information interaction		√
Operation reliability	Power quality improvement		√
	Rapid fault removal		√
Maintenance convenience	Operation & maintenance improvement		√
Equipment adaptation	Flexible interface		√
Application economy	Lower construction cost	√	
Social benefit	Lower footprint	√	
	Clean energy development		√

6.1.4 System Structure

Figure 6.2 shows a diagram of a flexible substation for a transmission application. It can serve as a network node of various AC/DC transmission lines with different

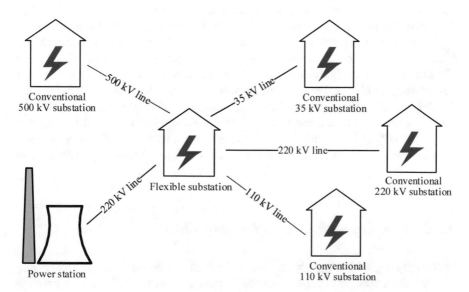

Fig. 6.2 Schematic diagram of a flexible substation system oriented to the transmission grid

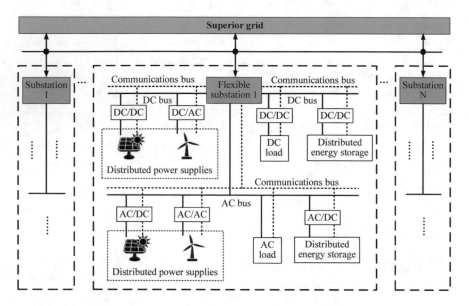

Fig. 6.3 Schematic diagram of a flexible substation for the distribution grid

nominal voltages and a control hub to control system voltage and power flow and to improve system stability. It can improve the dispatching and control ability of the system as a whole and increase system operation efficiency, flexibility and reliability.

Figure 6.3 shows a diagram of a flexible substation for a distribution application. A superior grid means a grid with a higher nominal voltage and/or a large capacity. The flexible substation can serve as an energy router for the AC and DC distribution grids and support DC connection and local consumption of new energies. It can isolate harmonic propagation and directly supply reliable power to the AC/DC loads. In addition, it can simplify the network and reduce conversion stages, improve power quality and provide reactive compensation, etc. Although it is not shown in Fig. 6.3, the flexible substation can coordinate with the controllers of other systems in the energy internet, such as thermal storage systems, combined cooling, heating and power (CCHP) systems with liquefied air, integrated hydrogen refuelling stations, and CCHP systems with a fuel cell, to achieve balanced coordination and optimization of various energy sources.

6.1.5 Topology of the Main Circuit of PETs

Various topologies are available for the main circuit of power electronic transformers (PETs). A scheme of series-connected converters on the HV side and parallel-connected converters on the LV side is usually introduced for a voltage step-down application. Three typical topologies are described below.

The Global Energy Interconnection Research Institute (GEIRI) proposed an H-bridge-based cascaded circuit topology for PETs [20], as shown in Fig. 6.4. This circuit topology can deal with the voltage withstand problem of power electronic devices on the HV input side. The topology design is suitable for other applications, including higher voltage applications. The topology is characterized by multilevel, standard power electronic modules and no reactive power filter with large capacity. As a result, the overall system covers less land, and the overall cost is significantly reduced. It is applicable to AC/AC conversion. However, there is no medium-voltage DC bus.

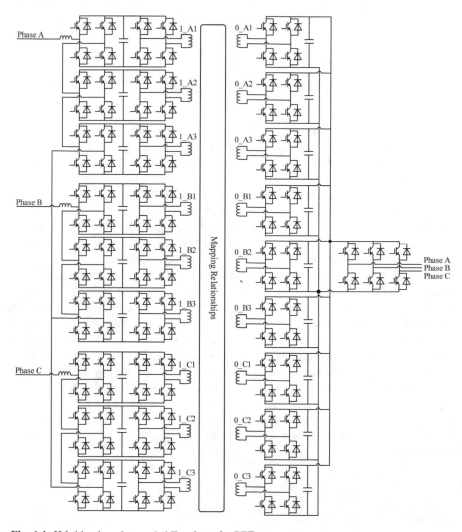

Fig. 6.4 H-bridge-based cascaded Topology for PETs

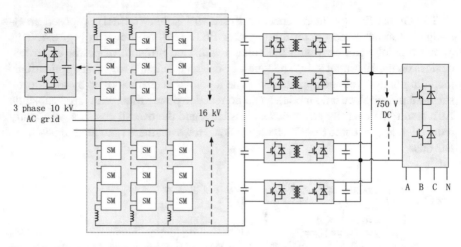

Fig. 6.5 Topology based on MMC and ISOP structure for PETs

To provide the medium-voltage DC bus, the Institute of Electrical Engineering, Chinese Academy of Sciences (IEECAS) proposed a topology based on a modular multilevel converter (MMC) and ISOP (series for inputs and parallel for outputs) topology for PETs [21], as shown in Fig. 6.5. The MMC structure is used on the HV side, and high-frequency DC/DC converters with ISOP topology are used for the isolation. The proposed topology, however, has the following disadvantages: it is difficult to isolate HV AC and DC faults, and the capacitor of the MMC module is large and expensive. In addition, HV DC series capacitors will supply a large short-circuit current, and there are difficulties in voltage balancing.

GEIRI proposed a multi-purpose multi-terminal integrated topology for PETs, as shown in Fig. 6.6 [21]. Four terminals, i.e. 10 kV AC, ±10 kV DC, 750 V DC and 380 V AC, are provided. Bidirectional flow is available among these terminals for power flow to meet the demand for flexible and convenient control over the AC/DC distribution network. Modules with voltage clamping after fire blocking and fast disconnectors are introduced; therefore, HV DC faults can be switched off and isolated, and no DC circuit breaker is required. The technology is employed to balance the power fluctuation among three phases, which can dramatically reduce the capacitance, the volume and the cost of the module capacitor in the MMC module. This approach integrates the functions of several installations, e.g. an HV AC/DC converter, DC/DC transformer, AC/AC transformer, LV DC/AC converter and Var compensator.

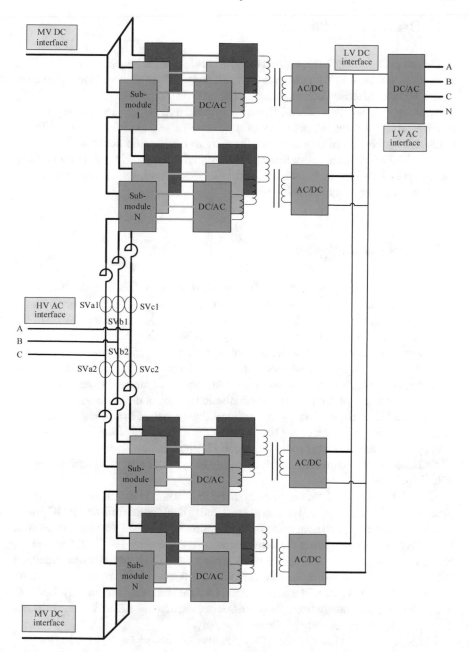

Fig. 6.6 Multi-terminal Topology for PETs

6.2   Overall Introduction

At the end of December 2017, the first flexible substation, i.e. Xiaoertai 10 kV flexible substation, was built in Alibaba Data Port, Zhangbei, Zhangjiakou, Hebei province, China, by Jibei Electric Power Co., Ltd. of State Grid Corporation of China (SGCC) (hereinafter referred to as "Jibei Power"), GEIRI, and NARI Group Corporation. It can effectively meet the demands of a flexible network configuration, different loads and connection of new energies for a smart distribution network. Taking the Xiaoertai 10 kV flexible substation as an example, a detailed introduction is given for flexible substations from such aspects as the background, system scheme, control and protection, key equipment and tests.

6.2.1   Project Background

In August 2015, SGCC prepared the Overall Development Plan Proposal for Smart Grid of Renewable Energy Demonstration Zone, Zhangjiakou, in accordance with the Development Plan of Renewable Energy Demonstration Zone, Zhangjiakou. In the demonstration zone, state-of-the-art technologies shall be applied, including those for power generation, transmission, distribution, storage, consumption and operation control, to build three demonstration zones—the power supply, grid and load zones. By doing so, the aim is to drive technical innovation for energy power and energy transformation so that it can promote economic and social development in Beijing, Tianjin and Hebei. In addition, it can support the themes of the "Green Olympics" and "Low-carbon Olympics" for the Winter Olympics, 2022. Jibei Power issued a white paper on the Development of Innovation Demonstration Zones in Zhangjiakou, Global Energy Interconnection, to promote key technological demonstration projects with full efforts.

Zhangbei is a base of the cloud computing industry. It is estimated that approximately 1.5 million servers will be accommodated in data centres in 2020. If DC power distribution is built, it can save energy consumption by 10 ~ 20% for the data centres. In addition, it can also reduce the investment in equipment by 1% [1, 2] for the data centres. Therefore, there is an urgent need to meet these DC loads. Alibaba Zhangbei Innovation R&D Demonstration Centre, located in Xiaoertai Township Park, a cloud computing base, is a part of the data centres. Its total load is approximately 2.5 MW, where the DC load mostly from the servers is approximately 1.8 MW. Clearly, it has a demand for a high-reliability DC power supply.

Zhangbei 50 MW Photovoltaics (PV) Poverty-Alleviating Demonstration Project, ELION, is 4 km away from the Alibaba Data Centre. The DC power generated by this PV project can be converted into AC power and then connected to the grid through step-up transformers. This energy flow path has several conversion stages, and power losses are large. Moreover, it cannot supply DC power to the DC load in the data centre. At present, the consumption of renewable energy power is less than 20% in the Zhangjiakou District. Clearly, there is a very large capacity to increase renewable energy consumption.

As DC loads, e.g. EVs and LED lights, increase by large amounts and more DC devices for PV, wind power, energy storage, etc., are built, an important trend for future smart distribution networks is to build AC/DC hybrid distribution networks on the basis of AC networks. It is necessary to explore a new pattern for green energy consumption in data centres and promote efficient consumption and absorption of renewable energy.

6.2.2 System Proposal

Power distribution and consumption in a data centre is a typical application for an AC/DC distribution network. To meet the demand for local consumption of large-scale centralized PV power and the DC power supply for the data centre, a flexible substation shall be built in Alibaba Data Port (Xiaoertai Township, Zhangbei County) in the first phase of the proposal, which shall work standby together with a conventional dry-type transformer to meet the demand of the data centre for reliable power supply; a DC step-up substation will be built in the ELION PV station, through which megawatt level PV power shall be stepped up and connected to the flexible substation, realizing DC connection and local consumption of PV power.

Figure 6.7 shows a schematic diagram of the flexible substation proposal in the first phase. The flexible substation is designed with four terminals ±10 kV DC, LV 750 V DC, 10 kV AC and 380 V AC, where the 10 kV AC shall be connected to

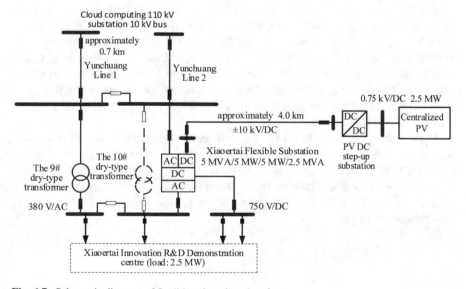

Fig. 6.7 Schematic diagram of flexible substation, 1st phase

the 10 kV bus in the 110 kV substation of the cloud computing base, the ± 10 kV DC shall be connected to the PV DC step-up substation via overhead DC lines, and the 750 V DC and 380 V AC terminals shall supply power to the AC/DC loads in the innovation R&D demonstration centre. Based on engineering experience, equipment manufacturing capability and future computing load demands, the capacities of the four terminals are designed as 5 MW/5 MW/5 MW/2.5 MW for the flexible substation, the capacity of the PV DC step-up substation is 2.5 MW, and the MV is ± 10 kV with a low voltage of 750 V.

The PV power shall be connected to the PV DC step-up substation, and then the DC power shall be supplied to the flexible substation. In this way, the PV power can be locally consumed by the data centre or connected to the distribution network via the 10 kV AC terminal of the flexible substation. The conversion and transmission losses can be reduced by approximately 2% if the DC PV power is locally consumed.

Zhangbei County features centralized loads and convenient transport. The Zhangjiakou Power Supply Company is proposed to build a 10 kV electric vehicle (EV) charging station in the northern region and a roof-mounted PV station in the north for demonstration. Based on the growth of the AC/DC loads in the region, the second flexible substation will be built in the north in the future proposal to connect to the local EVs and the AC loads so that the DC distribution network will fully exhibit such advantages as large transmission capacity, long distance, low losses and flexible power flow control. Additionally, both Xiaoertai Flexible Substation and the flexible substation in the north of the county can transmit power via the DC lines, forming an AC/DC hybrid distribution network with multi-terminals and flexible interconnections. It practices the concept, called "source-network-load" coordination and control, and achieves efficient connection and coordinated control of distributed power supplies and diversified loads.

A schematic diagram of the future flexible substation is shown in Fig. 6.8. The EV charging station (approximately 500 kW) and the roof-mounted PV station (approximately 80 kW peak) shall be connected to the 750 V DC bus of the flexible substation, and the flexible substation will output 380 V AC to supply power to the nearby AC loads (approximately 2 MVA). To ensure reliable consumption of the connected PV power and meet the demand of the AC/DC loads in the Xiaoertai Data Centre and the EV charging station, the flexible substation shall be designed as 5 MW/5 MW/5 MW/2.5 MW.

In the future, two flexible substations will operate together to form the AC/DC hybrid network and connect the distributed power supplies of different nominal voltages. These flexible substations can be used to achieve control over multi-terminal DC operation and to coordinate the interaction of source–network–load. At the same time, it can also show the supporting role of the flexible substations in the flexible interconnection of grid.

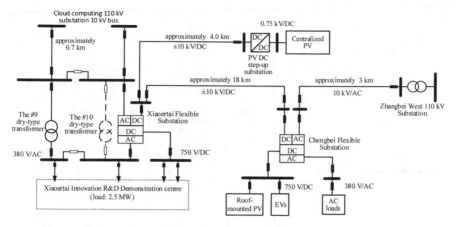

Fig. 6.8 Schematic diagram of flexible substation in the 2nd phase

6.3 Controls and Protections of Flexible Substations

6.3.1 Controls of Flexible Substations

For the Zhangbei AC/DC distribution network and the flexible substation demonstration project, the controls consist of two layers: the system layer and the equipment layer. The system layer includes the flexible substation control system, the PV DC step-up substation control system and the subdomain control system, and it is designed with functions to interact with the automation system of the distribution network, such as equipment startup/shut down control, operation mode management, energy management and power flow optimal control. It is built to realize the optimum operation of the distribution network as a whole. The equipment layer shall receive the commands from the system layer, execute equipment control and monitoring, and realize switch control, energy storage (supercapacitor) control, multi-terminal PET control, and PV step-up substation control, mainly including outer-loop voltage control, inner-loop current control, master/slave control based on voltage dropping, switching of control modes, fault ride-through and disturbance-free grid connection of a single converter station. In this way, the equipment can work in the expected operation status. A structural diagram of the control system is shown in Fig. 6.9.

6.3.1.1 Controls at the System Layer

Based on the actual conditions, the demonstration project shall be designed with the following operation modes: normal operation, shut down of a #9 dry-type transformer as shown in Fig. 6.8, shut down of the flexible substation, shut down of the PV DC

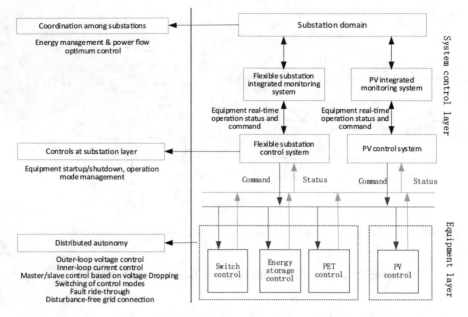

Fig. 6.9 Structural diagram of the control system

step-up substation, and power supply by diesel generators, all of which shall be realized by the controls at the system layer, as shown in Fig. 6.10.

(1) Normal operation

The PET shall run in a split manner with a #9 dry-type transformer, and a #10 dry-type transformer shall be on standby during normal operation.

The PET shall be in constant DC voltage control mode. It shall consume the PV power first to supply PV power for loads. It will feedback the surplus power to the 10 kV AC bus when the PV output is larger than the loads, and it will absorb the power from the 10 kV AC bus to make up the insufficient power when the PV power is lower than the loads.

(2) Shut down of the #9 dry-type transformer

When the #9 dry-type transformer is shut down, the flexible substation shall work in normal operation with the loads of the whole station, and the #10 dry-type transformer shall be on standby.

(3) Shut down of the flexible substation

LV loads shall be supplied by the 10 kV AC system, provided that the flexible substation is shut down. In this case, there are two operation modes:

(a) If the flexible substation is shut down or in maintenance for a long time, the #10 dry-type transformer shall be put into service, and the #9 and #10 dry-type transformers shall be operated in a split manner.

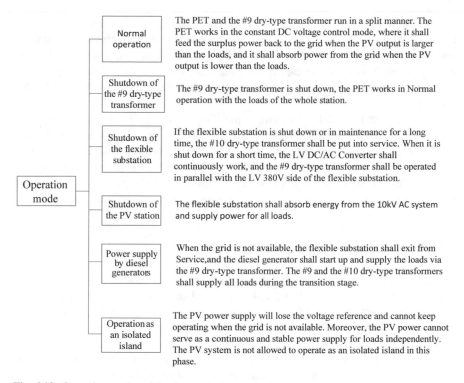

	Normal operation	The PET and the #9 dry-type transformer run in a split manner. The PET works in the constant DC voltage control mode, where it shall feed the surplus power back to the grid when the PV output is larger than the loads, and it shall absorb power from the grid when the PV output is lower than the loads.
	Shutdown of the #9 dry-type transformer	The #9 dry-type transformer is shut down, the PET works in Normal operation with the loads of the whole station.
	Shutdown of the flexible substation	If the flexible substation is shut down or in maintenance for a long time, the #10 dry-type transformer shall be put into service. When it is shut down for a short time, the LV DC/AC Converter shall continuously work, and the #9 dry-type transformer shall be operated in parallel with the LV 380V side of the flexible substation.
Operation mode	Shutdown of the PV station	The flexible substation shall absorb energy from the 10kV AC system and supply power for all loads.
	Power supply by diesel generators	When the grid is not available, the flexible substation shall exit from Service,and the diesel generator shall start up and supply the loads via the #9 dry-type transformer. The #9 and the #10 dry-type transformers shall supply all loads during the transition stage.
	Operation as an isolated island	The PV power supply will lose the voltage reference and cannot keep operating when the grid is not available. Moreover, the PV power cannot serve as a continuous and stable power supply for loads independently. The PV system is not allowed to operate as an isolated island in this phase.

Fig. 6.10 Operation modes of the demonstration project

(b) If the flexible substation is shut down for a short time, the LV DC/AC system will continuously work, and the #9 dry-type transformer shall run in parallel with the LV 380 V terminal of the flexible substation. The LV DC/AC converter of the flexible substation shall be operated in the constant DC voltage control mode to supply power for the DC loads.

(4) Shut down of the PV DC step-up substation

When the PV DC step-up substation is shut down and the flexible substation works normally, the flexible substation will absorb energy from the 10 kV AC system to supply power for all loads.

(5) Power supply by diesel generators

When the grid is not available, the flexible substation shall exit from service, and the diesel generator shall start up and supply the loads via the #9 dry-type transformer. In addition, the #9 and #10 dry-type transformers shall supply all the loads during the transition stage.

(6) Operation as an isolated island

In the proposal in this phase, the design of the control, protection and communications shall meet the demands of operation as an isolated island in the future and provide reserved interfaces for future function extensions. In the future, a new flexible substation will be put into operation and connected with this flexible substation, and more distributed energies and energy storage will also be connected. The new energy supply will meet the capacity requirement for an isolated island operation. Therefore, the isolated island operation mode shall be taken into account in this phase.

In the project proposal in this phase, the PV power supply will lose the voltage reference and cannot keep operating when the grid is not available. Moreover, the PV power cannot serve as a continuous and stable power supply for loads independently. With comprehensive consideration of the capacity ratio of PV power and loads, PV randomness and fluctuation and the actual conditions, the PV system is not allowed to operate as an isolated island in this phase when the 10 kV AC grid is lost.

6.3.1.2 Controls at the Equipment Layer

The controls at the equipment layer mainly include the controls of the multi-terminal PET. The control system is designed in a layered structure, as shown in Fig. 6.11, where three layers are provided: the main controls, the valve-based controller (VBC), and the drive and protection module. The PET function shall be realized by the main controls, which shall acquire the 10 kV AC voltage, AC current and ±10 kV DC voltage as the input signals of constant DC voltage control, reactive power control, startup/shut down control and fault ride-through control. They shall also acquire the 750 V DC voltage, the mean voltage of VBC arm capacitors, which shall serve as the input signals of some control operations. The main controls shall generate and send a modulated wave to the VBC, which shall generate the pulse width modulation (PWM) triggering pulse for each module. The drive and protection module shall execute the VBC commands and realize over-voltage and over-current protection. The DC/AC, DC/DC, PV and switches receive the command from the main control

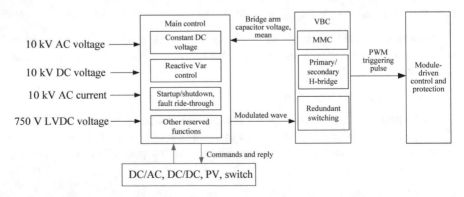

Fig. 6.11 Structural diagram of the control system at the equipment layer

and feedback the corresponding status information. In addition, the main control shall be provided with such functions as reactive power control and AC voltage control.

6.3.2 Protections of Flexible Substations

The protections of flexible substations attempt to provide safety for equipment and power systems and minimize the shut down range and influence upon the grid in the case of faults. In addition, the protection of a flexible substation shall cooperate with the protection of the interfaced equipment and line relay to achieve dead-zone-free protection for the flexible substation and the grid.

For the Zhangbei AC/DC distribution network and the flexible substation demonstration project, the protection zone of the flexible substation can be divided into the following: the AC protection zone, the DC protection zone, the load protection zone, the device protection zone and the PV and step-up substation protection zone, as shown in Fig. 6.12.

The AC protection zone includes the 10 kV AC lines and the relevant equipment, the #9 and #10 dry-type transformers. The AC protection shall respond to the following faults of the AC system: single-phase-to-ground fault, phase-phase short-circuit fault, internal fault of transformer, faults at bushing and lead-out wires of transformer, etc. For the AC single-phase-to-ground fault, the protection shall send out an alarm; for the AC phase-phase short-circuit fault, the protection shall trip the circuit breakers on both sides of the faulted line; and for the internal fault and faults at bushing

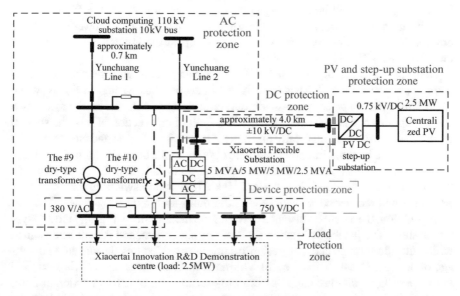

Fig. 6.12 Protection zone diagram of the flexible substation

and lead-out wires of the transformer, the protection shall trip the circuit breaker on each side of the transformer.

The DC protection zone includes ±10 kV DC lines and the relevant equipment, which shall respond to the following faults of the ±10 kV DC system: single-pole ground fault, double-pole short-circuit fault, broken line faults, etc. The protection shall block the multi-terminal PET, turn off the PV system and trip the ±10 kV DC line switch.

The load protection zone includes the 750 V DC bus, the load lines and the relevant equipment and the 380 V AC bus, the load lines and the relevant equipment. The load protection shall respond to the single-pole ground fault and double-pole ground fault in the DC bus and the load lines and the single-phase-to-ground fault and the phase-phase short-circuit fault in the 380 V AC bus and the load lines. For the 750 V DC bus fault, the protection shall trip the 750 V DC outgoing circuit breaker of the multi-terminal PET; for the 750 V DC load line fault, the protection shall trip the DC load circuit breaker; for the 380 V AC bus fault, the protection shall trip the 380 V AC outgoing circuit breaker of the multi-terminal PET; and for the 380 V AC load line fault, the protection shall trip the AC load circuit breaker.

The device protection zone includes the multi-terminal PET equipment. The device protection shall respond to bridge arm reactor faults, ground faults and short-circuit faults of the module, short-circuit faults of several modules, primary/secondary ground and short-circuit faults of high-frequency transformers and faults of power electronics devices. The protection shall block the multi-terminal PET, trip the terminal switches of the multi-terminal PET and turn off the PV system.

The PV and step-up substation protection zone includes the centralized PV and the PV DC step-up substation. These protections shall respond to the centralized PV fault, ground fault and short-circuit fault of the module of the PV DC step-up substation, short-circuit fault of several modules and fault of power electronics devices. The protection shall block the PV step-up substation and trip the switches on both sides of the ±10 kV DC line.

In the project, the multi-terminal PET can exchange loads with the dry-type transformer, and the HV/LV AC bus is designed in section bus wiring. In this way, the important users can be connected with two circuits from different sections, and two power supplies can be used. When there is a fault in one section of bus, the section circuit breaker shall automatically disconnect the faulted section and ensure that the normal section bus supplies power uninterruptedly to the user. For example, if there is a fault in the 10 kV AC line 2, the protection in the AC protection zone shall trip the line circuit breaker and close the AC 10 kV bus section switch so that the 10 kV AC line 1 will recover power supply with the multi-terminal PET.

In addition, the electrical distance between the flexible substation and the AC system is short. The flexible substation shall not be shut down, and it shall be designed with fault ride-through and auto-restart functions provided that the current and voltage of the flexible substation are still within the allowed safe ranges during external faults or voltage ride-through. For example, in the event of a 10 kV external fault, the flexible substation, after detecting under-voltage or over-current, shall block the multi-terminal PET and send commands to block the PV station until the external protection removes the fault. The control and protection of the flexible substation shall automatically unblock the PET when the over-current disappears, the system voltage is recovered after the fault and the delay requirement is met. The PV station will restart when the flexible substation restarts successfully and operates for a preset duration.

In the project, the flexible substation can be connected to and operated with the PV system via the ±10 kV DC line or run independently. In the event there is a fault in the ±10 kV DC line, the whole flexible substation shall not be shut down because of the fault ride-through function. For example, if there is a single-pole ground fault, the ±10 kV DC voltage imbalance protection will block the multi-terminal PET, turn off the PV system, and trip the ±10 kV DC line switch. The PV step-up substation will be out of service after the DC lines are switched off. The flexible substation main controls shall restart the PET when the restart conditions are met.

6.4 Core Equipment of Flexible Substations

As shown in Fig. 6.13, the core equipment in a flexible substation is the multi-terminal PET, which is provided with four different voltage interfaces, i.e. 10 kV AC, ±10 kV DC, 380 V AC and 750 V DC. The multi-terminal PET is designed with a double clamping module that has fault current interruption capacity. It can effectively cut off the ±10 kV DC short-circuit fault current after blocking, which is good for the DC grid. The 10 kV AC terminal can regulate the power flow and realize Var compensation and voltage control. The ±10 kV terminal can support grid connection of the centralized new energy and DC grid. The 380 V AC terminal is designed with such functions as connection of distributed power supplies and control of load power quality. The 750 V DC system can supply power to the data centre, the

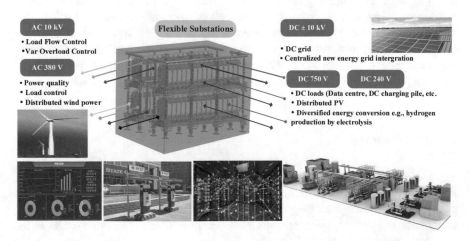

Fig. 6.13 Roles and functions of Flexible Substations

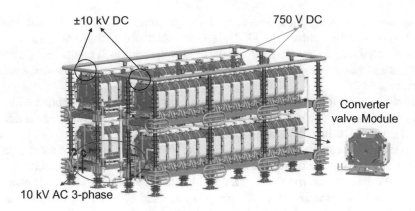

Fig. 6.14 Valve Tower of Flexible Substations

DC charger pile and other relevant DC loads; it is easy to connect with the distributed PV and other DC power supplies and offers power interfaces to the conversion of multiple energies such as hydrogen production by electricity and power generation by fuel cells.

6.4.1 Valve Tower

For the multi-terminal PET, the core is the converter valve, where a horizontal valve tower with two layers and two rows is built. The valve tower is built with a three-interface converter, as shown in Fig. 6.14. According to the topology, there is a 10 kV AC three-phase incoming line interface, ±10 kV HV DC outgoing line interface and 750 V LV DC outgoing line interface. The valve tower consists of two bridge arms—the upper and lower arms, which are symmetrically arranged. Each bridge arm includes 28 valve modules in series in the two layers. Between the layers is supported by the insulator. A total of 56 valve modules are built. The valve tower covers an area of 6.724 m × 3.979 m and is 3.39 m high. The structure is simple and easy for installation, operation and maintenance.

6.4.2 Module

The valve tower of the flexible substation consists of many converter valve modules. As shown in Fig. 6.15, a module consists of the power modules on HV phases A, B and C, high-frequency isolating transformer, LV power module, cooling water pipe, connecting copper bar, epoxy support and communications fibre-optic cable. In the structure design, the 3-phase 4-terminal high-frequency transformer is placed in the

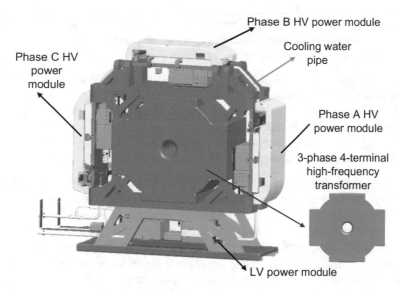

Fig. 6.15 Valve tower module of Flexible Substations

epoxy support, the 3-phase HV power modules are placed in the front/top/rear of the epoxy support, and the LV power module is installed at the bottom of the epoxy support. The power modules on the HV/LV sides shall be connected with the high-frequency transformer via the copper bar. The HV/LV power modules and the epoxy support and the epoxy support and the high-frequency transformer shall be bolted together. The valve module is designed in a ring arrangement. The load support and insulation span shall be reasonably designed to achieve sufficient electrical insulation among the HV power module in different phases and the HV/LV power modules and minimize the volume of the valve module. The epoxy support is made of an epoxy material with good mechanical and electrical performance and can meet the demands of dynamic/static loads of the converter valve.

The cooling water pipe shall flow from the LV power module to the HV power modules of phases A, B and C. The high-frequency transformer shall be designed with water cooling, where the cooling water pipe shall be in parallel to the valve module water circuit. The water loop shall be in series among the HV/LV power modules, which can effectively reduce the quantity of connection points and the risk of water leakage. In this way, the water pipe path can be simplified to an integrated circuit so that it features easy maintenance.

6.4.3 High-Frequency Transformer

The large-capacity high-frequency transformer is the core of the isolated full-bridge DC/DC converter. Since the transformer core volume can be greatly reduced at high frequency, it can effectively increase the power density. The high-frequency transformer features advantages such as small volume, high power density and high efficiency. It mainly has the following three functions:

(1) To realize electrical isolation between the HV 10 kV system and the LV 750 V system and insulation among the three phases of the HV 10 kV system;
(2) To realize voltage matching and conversion based on the output voltage of the primary/secondary power modules;
(3) To exchange active power between the primary and secondary sides and realize reactive power flow among the three phases on the primary side.

To meet the requirements of less land coverage, energy savings, environmental protection, safety, reliability and easy maintenance, the high-frequency transformer in the flexible substation is designed with a "3-phase to one phase" structure, i.e. 3-phase inputs on the primary side shall be connected to the 3-phase DC/AC converter units on the HV side, respectively, to realize coupling of the primary three phases and flowing of reactive power among the phases; single-phase output is used on the secondary side, which is connected to the AC/DC converter unit on the output side to supply the HV 3-phase active power to the LV side.

The core of the high-frequency transformer is made of nanocrystalline material. Compared with ferrite, amorphous alloy and other magnetic materials, the nanocrystalline core features high saturated flux density and low high-frequency losses. The core consists of several small cores to form the required section due to the limitation of the width of the nanocrystalline material. Since the winding conductor has a skin effect and proximity effect at high frequency, the high-frequency transformer winding is made of copper foil to minimize the winding losses at high frequency.

The volume of the high-frequency transformer is mainly determined by the insulation distance between the HV and LV systems and among different phases. To ensure the insulation performance and minimize the volume of the high-frequency transformer, the high-frequency transformer adopts the epoxy solid material as the main insulation material. The epoxy solid insulation material has good mechanical performance, high dielectric performance, resistance to surface power leakage, arc resistance, etc. In addition, it can offer convenience to fire protection design due to the lack of oil in a conventional transformer. The HV/LV windings of the high-frequency transformer can pass the 35 kV/1 min power-frequency voltage withstand test, and the partial discharge is less than 10 pC at 1.3 × rated voltage.

To effectively control the temperature rise of the large-capacity dry-type high-frequency transformer, deionized cooling water, the same as that of the power modules, shall be used to remove the heat of the core and windings, keeping the maximum surface temperature rise to less than 90 K. Figure 6.16 shows the structure of the high-frequency transformer. Table 6.2 presents the associated parameters.

Fig. 6.16 Schematic diagram of the high-frequency transformer Structure

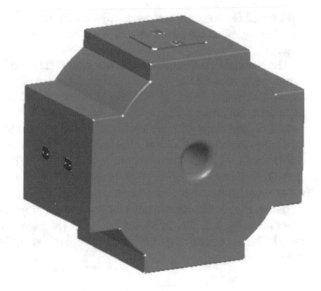

Table 6.2 Main Technical Data of the High-frequency Transformer

Name	Performance index
Primary 3-phase voltage	0.9/0.9/0.9 kV
Secondary voltage	0.75 kV
Working frequency	3 kHz
Primary/secondary capacity	157 * 3/124 kVA
Efficiency	99.2%
Cooling method	Cooled by deionized water
Temperature rise	95 K
Power-frequency voltage withstand	35 kV/1 min
Partial discharge	<10 pC
Volume	$0.28 \times 0.28 \times 0.35$ m

6.5 RTDS Test of Flexible Substation

6.5.1 Building of Testing System

6.5.1.1 Structure of Testing System

Hardware-in-the-loop (HIL) real-time simulation technology [22], represented by real-time digital simulation (RTDS), is very useful for the simulation of power system transients. The RTDS and the control and protection system can be used to build a HIL simulation system, which has advantages such as easy system condition simulation, reproduced results, easy parameter adjustment, high acceptance in the industry and high testing efficiency. This system has been widely used in system performance tests.

The simulation testing platform is based on RTDS and its I/O cards. In this way, the simulation testing platform can be built for the substation control system. Figure 6.17 shows the simulation testing platform of the control system in the flexible substation demonstration project [23].

RTDS shall supply the analogue and digital outputs to the control and protection of the flexible substation and receive the protection commands from the control and protection via digital inputs of the RTDS. The VBC interfaces the control via GTHDLC (Giga-Transceiver High-Level Data Link Control), and the RTDS, i.e. the valve via AURORA (a communications protocol for use in point-to-point serial links). Based on modular verification, the demonstration project can be divided into several subsystems by function, as shown in Fig. 6.18. Each subsystem can be separately tested. The test can be developed from bottom to top until the whole system is finally tested.

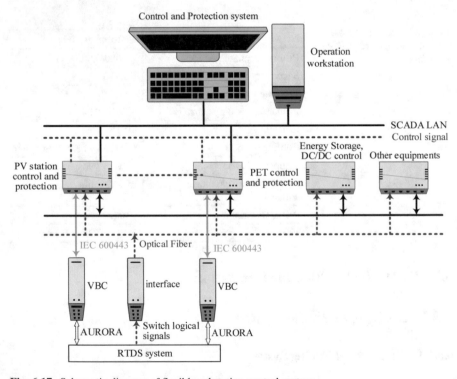

Fig. 6.17 Schematic diagram of flexible substation control system

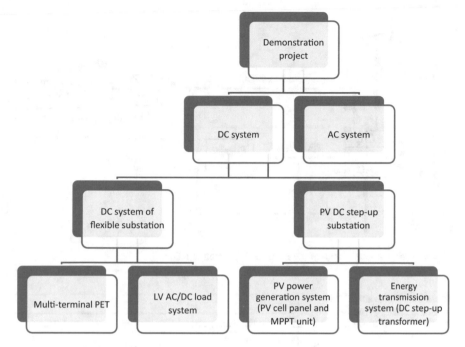

Fig. 6.18 Hierarchical diagram of testing system

6.5.1.2 Testing Equivalent System

To verify the operation performance of the flexible substation, we simulated the local consumption performance of PV power and the operation performance in the case of load variation in RTDS. Appropriate simplifications were made on the basis of the system in Fig. 6.7 with the precondition that no influence shall be exerted on the test and inspection effect. See Fig. 6.19 for the equivalent system of the flexible substation in the RTDS.

6.5.1.3 System Model

The PET valve shall be built in the RTDS small-step subsystem by discrete components. Both the multi-terminal PET converter and PV DC/DC converter are simplified into 4-level structures in RTDS and are able to simulate all the functions of the actual equipment. Table 6.3 presents the main data of the converter models [23].

The equivalent AC system takes the cloud computing 110 kV bus as the boundary, and it mainly includes the 110 kV main transformer, loads and cloud computing in the Xiaoertai line. The 10 kV AC line model shall include the simulation module of the main and backup protection, which shall be used to test the impact of the fault of the DC project on the nearby AC system protection.

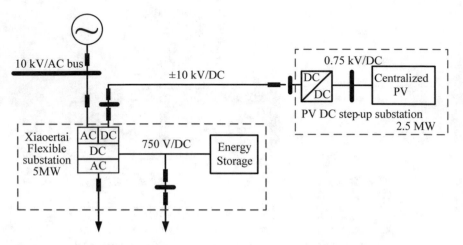

Fig. 6.19 Equivalent system of the Flexible Substation in RTDS

Table 6.3 Main data of the simulation models for the PET and the PV step-up substation

Parameter	Value
AC input line voltage rms/kV	10.5
HV DC voltage/kV	±10.5
PET rated capacity/MVA	5
Number of PET modules	4
PET sub-module capacitor voltage/kV	10.5
PET/high-frequency transformer ratio/kV	10.5/0.76
PET/high-frequency transformer working frequency/Hz	500
PET bridge arm inductance/mH	18
PET sub-module capacitance/μF	2000
Total capacitance of PET 750 LV DC side/mF	30
Rated capacity of PV station/MVA	2.5
HV DC max. output voltage of PV station/kV	26
Number of PV converters in parallel	4
LV DC output voltage of PV station/kV	760

6.5.2   Testing of Control Functions

6.5.2.1   Local Consumption Test of PV Power

To verify the consumption capacity of the flexible substation loads for PV power, the 750 V DC load was input for testing when the system was stable. The actual waveform is shown in Fig. 6.20. The PV output was approximately 2.2 MW before

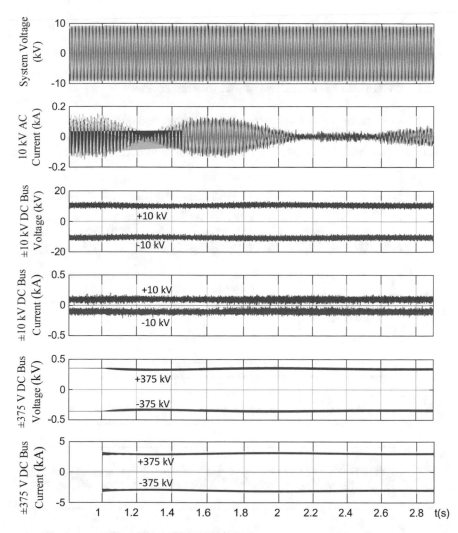

Fig. 6.20 Waveforms before/after LV DC loads were input

1.0 s. During this period, the power generated by the PV station entered the grid due to the absence of load in the flexible substation. The waveform shows that the system voltage, 3-phase incoming current, voltage and current on the HV DC side and voltage and current on the LV DC side were stable. The LV DC load was input at the moment of 1.0 s. Then, the LV DC positive current rose from 0 to 3.1 kA. A 2.35 MW load was introduced to locally consume the PV output power. The waveform after the moment 1.0 s shows that the HV/LV DC voltage only had slight fluctuation; the 3-phase incoming current had current direction adjustment and saw fluctuation at 1.8 s since the current direction changed—originally, it flowed from the PV station to the grid; then, from the PV station and the grid to the DC load—but it had no large impact. Clearly, the transition was smooth.

6.5.2.2 LV Onload Variation Test of Flexible Substation

Since load switching is necessary for the flexible substation during operation, a test to disconnect the LV DC load was conducted when the system was stable. The actual waveform is shown in Fig. 6.21. The PV output was zero before 1.0 s. During this

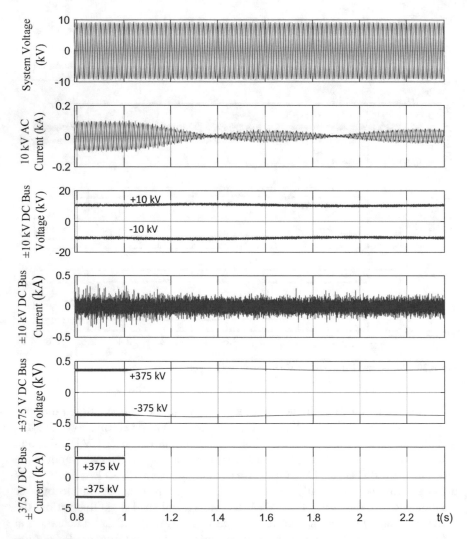

Fig. 6.21 Waveform for disconnection of LV DC loads

period, the load current was supplied by the grid. The waveform shows that the parameters, e.g. system voltage, 3-phase incoming current, voltage and current on the HV DC side and voltage and current on the LV DC side, were all stable. The LV DC load was disconnected at the moment of 1.0 s. The disconnection of the LV DC load resulted in a reduction in LV DC power consumption, which made the HV/LV DC voltage rise. However, the figure shows that no large fluctuation was present for the high and low DC voltages. Since the system power had large fluctuations, the 3-phase incoming current was adjusted for a long time with a duration of 1.2 s. No large change was present for the amplitude of the 3-phase incoming current during the whole transition stage. Obviously, the system remained stable when the LV DC load was disconnected.

6.6 Site Test and Operation of Flexible Substation

To verify the actual operation of the flexible substation, the following tests were conducted: startup for operation, fault ride-through operation and PV station operation together with the flexible substation. The system structure during the tests is shown in Fig. 6.19, which is basically the same as that in the RTDS simulation.

6.6.1 Startup for Operation of Flexible Substation

First, the 380 V LV temporary line was used to charge the module capacitor in the flexible substation via DC/AC. When the voltage of the module capacitor was larger than a certain value (e.g. 700 V), the grid-connecting switch was closed, and the system entered the AC grid-connecting startup. Please note that the PV station did not start up, and the LV AC load power was supplied by the 380 V AC temporary bus.

Figure 6.22 shows the relevant waveforms at the moment when the flexible substation started up. Figure 6.22a shows Phases A/B/C 10 kV AC grid voltage waveforms; Fig. 6.22b shows the upper bridge arm current waveforms of Phases A/B/C; Fig. 6.22c shows the lower bridge arm current waveforms of Phases A/B/C; Fig. 6.22d shows the double-pole-to-ground voltage waveforms of the HV DC ±10 kV bus; and Fig. 6.22e shows the double-pole-to-ground voltage waveforms of the LV DC ±375 V bus.

Note: There were common-mode voltage components in the waveform since the double-pole-to-ground voltage was tested for DC voltage measurement. The voltage between poles, however, was smooth, and the DC voltage stability accuracy was larger than 96%.

It should be clarified that in Fig. 6.22b, c, there was a small impulse current in the upper/lower bridge arms at the moment when the flexible substation started up. The small current was caused by the voltage step-up method. After startup, the soft voltage step-up and stability method was employed with a duration of 5 ms.

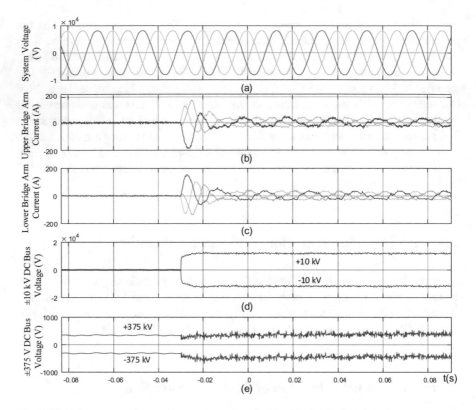

Fig. 6.22 Relevant waveform at the moment when the Flexible Substation started

In Fig. 6.22d, e, there were common-mode voltage components in the waveform since the double-pole-to-ground voltage was tested for DC voltage measurement. The voltage between poles, however, was smooth, and the DC voltage stability error was less than 5%. In addition, it shall be clarified that the ±10 kV DC voltage could be detected only when the flexible substation was unblocked to start because this ±10 kV DC voltage was still zero when the flexible substation was blocked, as shown in Fig. 6.22d.

6.6.2 Fault Ride-Through Operation of Flexible Substation

In the case of recoverable faults (e.g. AC/DC bus over-current, flexible substation module overvoltage, low-voltage ride-through failure and abnormal communications), the flexible substation shall restart for operation in a short time. During these faults, the loads power shall be supplied by the energy storage equipment.

Figure 6.23 shows the relevant waveforms at the moment when the flexible substation restarted after a short-time fault. Figure 6.23a shows AC current waveforms

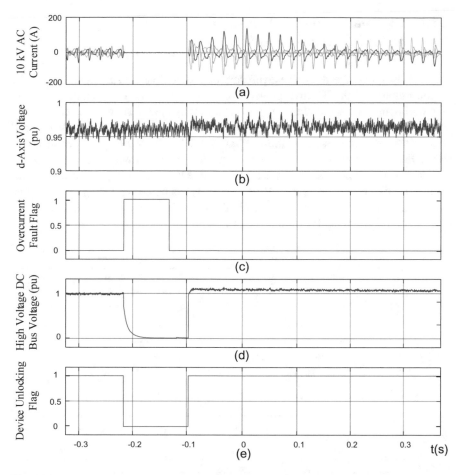

Fig. 6.23 Relevant waveforms at the moment when the Flexible Substation restarted

from the 10 kV bus-bar; Fig. 6.23b shows the nominal waveform of the d-axis voltage after coordinate conversion of the grid voltage; Fig. 6.23c shows the flag at the moment when the fault signal happens; Fig. 6.23d shows the nominal waveform of the HVDC pole-pole voltage; and Fig. 6.23e shows the flag at the moment when the flexible substation restarted and unblocked. Please note that the HV AC bus short-circuit fault was simulated once during operation, which resulted in over-current of the flexible substation bridge arm. At this time, the flexible substation was blocked. After the simulated fault disappeared at the moment of 100 ms, the flexible substation should judge the restarting conditions. The waveforms in Fig. 6.23 show that the flexible substation successfully restarted for operation at the moment of 120 ms, which proves that the designed short-time fault restarting function is feasible.

6.6.3 PV Station Flexible Substation Operating in Grid-Connected Mode

Figure 6.24 shows the relevant waveforms of the flexible substation when the PV station outputs 1.2 MW active power. Please note that the LV system was at no load during the actual test; thus, the power output by the PV station was completely connected to the 10 kV AC grid. Figure 6.24a shows Phases A/B/C 10 kV AC grid voltage waveforms; Fig. 6.24b shows Phases A/B/C 10 kV AC grid current waveforms; Fig. 6.24c shows the double-pole-to-ground voltage waveforms of the HV DC ±10 kV bus; Fig. 6.24d shows the HV DC ±10 kV bus double-pole current waveforms; and Fig. 6.24e shows the LV DC ±375 V bus pole-pole voltage waveforms. Figure 6.24 shows that the HV/LV terminals of the flexible substation also perform well in control when the PV outputs large power.

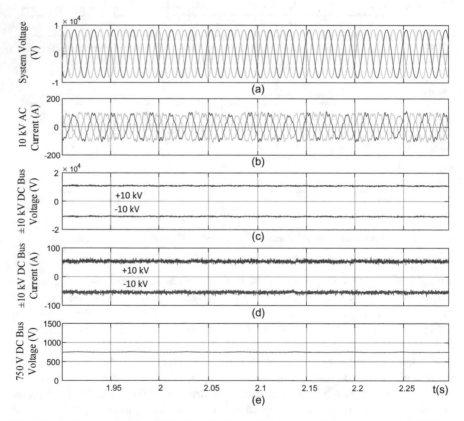

Fig. 6.24 Relevant waveforms of the Flexible Substation when the PV station output 1.2 MW

6.7   Conclusion

This chapter systematically introduces a new generation of substations by taking the demonstration project of the Xiaoertai flexible substation as an example. In the context of the energy internet, the characteristics of a flexible substation, such as flexible power flow control, stability improvement, power quality isolation and mitigation; the increase in the absorption capacity of new energies; and the "multi-function in one" make it highly competitive in many aspects. Field experimental results have verified its fundamental functions. The power system with flexible substations will play an important role in the framework of the energy internet.

References

1. Z. Liu, *Global Energy Interconnection* (China Electric Power Press, 2015)
2. J. Rifkin, The third industrial revolution; how lateral power is transforming energy, the economy, and the world. World Future Rev. **4**(2), 199–202 (2011)
3. G. Cao, Configuration and application of comprehensive automation system for substations. East China Electr. Power **34**(12), 78–80 (2006)
4. H. Zhang et al., Testing technology and practice on 220 kV digital substation. Proc. CSU-EPSA **21**(6), 98–102 (2009)
5. H. Cui et al., Introduction of 220 kV Wushan digital substation. Electr. Power Autom. Equip. **29**(3), 149–152 (2009)
6. Q. Huang et al., Smart substation: state of the art and future development. IEEE Trans. Power Deliv. **32**(2), 1098–1105 (2017). https://doi.org/10.1109/tpwrd.2016.2598572
7. X. Dong et al., Smart power substation development in China. CSEE J. Power Energy Syst. **2**(4), 1–5 (2016). https://doi.org/10.17775/cseejpes.2016.00042
8. H. Li, L. Wang, Research on technologies in smart substation. Energy Procedia **12**, 113–119 (2011). https://doi.org/10.1016/j.egypro.2011.10.016
9. D. Grider, et al., 10 kV/120 A SiC DMOSFET half H-bridge power modules for 1 MVA solid state power substation. in *2011 IEEE Electric Ship Technologies Symposium* (2011), pp. 131–134
10. C. Klumpner, et al., Experimental validation of the solid state substation with embedded energy storage concept. in *8th Annual IEEE Energy Conversion Congress & Exposition* (2016)
11. A. Zhang et al., Power electronic transformer used in AC/DC hybrid distribution network. Electr. Power Constr. **38**(06), 66–72 (2017)
12. S. Xu et al., Review of solid-state transformer technologies and their application in power distribution systems. IEEE J. Emerg. Sel. Top. Power Electron. **1**(3), 186–198 (2013). https://doi.org/10.1109/jestpe.2013.2277917
13. K. Alam, et al., Design and comprehensive modelling of solid-state transformer(SST) based substation. in *2016 IEEE International Conference on Power System Technology (POWERCON)* (2016), pp. 1–6
14. K. Alam, et al., Multi-cell DC-DC converter based solid-state transformer (SST) design featuring medium-voltage grid-tie application. in *2017 IEEE 20th International Conference on Electrical Machines and Systems (ICEMS)* (2017), pp. 1–6
15. T. Guillod et al., Protection of MV converters in the grid: the case of MV/LV solid-state transformers. IEEE J. Emerg. Sel. Top. Power Electron. **5**(1), 393–408 (2017). https://doi.org/10.1109/jestpe.2016.2617620
16. Y. Zhang, Flexible substations: top technologies to realize AC/DC mutual conversion. State Grid News **8** (2017)

17. S. Weng, Future grid will be more flexible—Interview of Teng Letian. State Grid News **09**, 47–49 (2015)
18. Y. Tong, et al., Flexible substation and its control for AC and DC hybrid power distribution. in *13th IEEE Conference on Industrial Electronics and Applications (ICIEA)* (2018)
19. Z. Chai, et al., Modular multilevel converter based bus bar in flexible substation. In *2017 IEEE Transportation Electrification Conference and Expo* (2017), pp. 1485–1489
20. Z. Deng, et al., AC/DC conversion circuit and power electronic transformer (2016). 201610483335.2
21. Z. Li et al., Research on medium and high-voltage smart distribution grid oriented power electronic transformer. Power Syst. Technol. **37**(9), 2592–2601 (2013)
22. L. Xu et al., Study on in-loop testing technologies for RTDS-based SVG controller hardware. Power Electron. Technol. **51**(01), 118–120 (2017)
23. J. Song, et al., A control and protection system in a loop testing technology for a flexible power electronics substation. in *13th IEEE Conference on Industrial Electronics and Applications (ICIEA)* (2018), pp. 2051–2056. https://doi.org/10.1109/iciea.2018.8398047

Chapter 7
Energy "Routers", "Computers" and "Protocols"

Chuantong Hao, Yuchao Qin, and Haochen Hua

Abstract Energy routers are the core units of the Energy Internet. They are an inevitable outcome of the upgrading of energy systems in the advanced development stage of the Energy Internet. Similar to routing devices on the Internet, the energy router is a kind of specialized device for energy transmission, buffering and transactions. Specifically, energy routers should be able to effectively maintain the power quality in the power system. Based on the characteristics of distributed energy resources, energy routers should have plug-and-play functionality and high scalability such that the concepts of the Energy Internet, i.e. open, interconnected, peer-to-peer and sharing, could be achieved. In the advanced development stage of the Energy Internet, energy routers should be able to optimize energy utilization through cyber physical systems integrated in the Energy Internet infrastructure.

7.1 Definition

7.1.1 Definition of Energy Router

The concept of energy router stems from the Energy Internet. The mechanism of the Energy Internet has been explored since 2008. Researchers from the United States, Europe and Japan have put forward their own definitions for the Energy Internet and energy routers.

The concept of energy router was first proposed by American researchers. In 2008, a new power network structure based on renewable energy sources and distributed energy storage devices was investigated in the FREEDM project [1], and this new structure was called the Energy Internet. Inspired by the functionality of core routers

C. Hao
School of Engineering, University of Edinburgh, Edinburgh EH9 3DW, UK

Y. Qin · H. Hua (✉)
Beijing National Research Center for Information Science and Technology, Tsinghua University, Beijing 100084, People's Republic of China
e-mail: hhua@tsinghua.edu.cn

© Springer Nature Switzerland AG 2020 193
A. F. Zobaa and J. Cao (eds.), *Energy Internet*,
https://doi.org/10.1007/978-3-030-45453-1_7

in the information network, researchers proposed the concept of energy router and conducted research on the realization of its prototype. In the same year, the research team at the Swiss Federal Institute of Technology developed a new device named the "energy hub", which is also called the "energy control centre". This device originated from the idea of a hub in computer science. In 2013, the concept of "power router" was proposed by Japanese scientists. Digital power routers developed in Japan are able to coordinate power management in certain areas.

7.1.2 Energy Internet and Energy Router

The development of the Energy Internet has hastened the advancement of theory and technology for energy routers. The Energy Internet is considered a revolution in the field of energy systems. The decentralization, greening and user-driven characteristics of the energy supply are newly brought by the Energy Internet. From the aspects of energy production, transmission, storage and consumption, Energy Internet technology systematically enhances the information and intelligence level of the energy system. A conventional energy management system (EMS) and power trading platform would not be able to adapt to the development of Energy Internet technology. Energy routers were developed in response to the new challenges in the Energy Internet. These routers provide energy management capabilities at both the microgrid level and the large-area interconnection level.

An energy router provides an interface to the Energy Internet for access of renewable energy sources. With the development of renewable energy sources and Energy Internet technologies, the proportion of renewable energy sources in the energy system will increase significantly. An energy router provides a standardized interface and fully featured energy management functionalities for the integration of renewable energy resources. It has strict requirements for energy management in terms of response speed, power accuracy and power quality.

7.2 Research Progress on Energy Routers

7.2.1 Current Research Status

To achieve peer-to-peer interactions between energy routers, power electronics and communication technologies are applied. From the perspective of power electronics technology, FREEDM hopes to realize the concept of the Energy Internet by distributed peer-to-peer systems. Currently, the FREEDM project is led by North Carolina State University, and its partners include Alexandria State University and Florida State University.

Digital power grid routers developed in Japan are able to coordinate power management in their deployed areas and dispatch power flow in different areas. The digital power grid in Japan is based on Internet infrastructures. By gradually restructuring the power system of the whole country, the current synchronization grid has gradually been subdivided into asynchronous autonomous grids.

7.2.2 Routing Decisions

The routing efficiency of the traditional Internet is affected by network structure, network bandwidth and router configurations. In the Energy Internet, apart from the factors mentioned above, the efficiency of energy routing is influenced by factors such as energy storage technology, switching technology and power distribution technology. Existing routing strategies do not fully take all of the above factors into consideration. By ignoring the differences between individual routers, we are able to analyse the routing strategy determined by the nature of the network's topology. In the following analysis, it is stipulated that in the Energy Internet, energy is marked by an information label including the original address, the destination address, the next-hop routing address, etc. An energy flow, along with its information label, is called an energy information flow (EIF). With the information in an EIF, an energy transmission device could be efficiently utilized. The ultimate output of the Energy Internet is in the form of electrical energy. The energy of an EIF is also stored and transferred between energy routers in the form of electrical energy. For congestion control and power balancing in the Energy Internet, a review of the research on routing decisions is provided as follows.

Energy transmission congestion in the Energy Internet means that, in the ideal routing environment, due to the routing strategy, an energy storage unit of an individual energy router has too many EIF transfer tasks, which would result in the overload of the forwarding service in energy routers [3]. Energy transmission congestion leads to a drastic decrease in the overall EIF transmission efficiency of the network and can even cause a network crash. Currently, in many practical communication systems, the shortest path strategy (SPS) is widely used. The SPS can minimize the amount of EIF forwarding and reduce the wait consumption of EIF forwarding. However, in the Energy Internet, the EIF throughput of network transmission is greatly limited by the energy routers in central positions in the network. In a practical application scenario of the Energy Internet, the SPS may easily lead to EIF transmission congestion. In some special conditions, some energy routers in central locations could even be paralyzed, causing serious security risks. Fortunately, this problem can be solved by distributing the EIF forwarding requests on a highly centralized energy router to other low-centralized energy routers via the efficient path strategy (EPS). The research objects of the SPS and EPS are all the energy routers located in the central node of a network. In the Energy Internet, effective marginal transmission lines also affect the transmission efficiency of an EIF. Based on the topological properties of the Energy Internet, an edge-weight strategy (EWS) is proposed to solve the energy transmission congestion issue.

Due to the scale-free nature of the network, some important links become the optimal path for EIF transmission, which brings more EIF transfer tasks to the central nodes in these links. Making full use of these central nodes can effectively improve the efficiency of EIF routing, which requires global or local information of each node in the network.

In addition to the aforementioned traditional routing algorithms, considerable progress has also been made in routing decisions focusing on some typical scenarios in the Energy Internet. In [4, 5], an energy router was used as a flexible interface to integrate a renewable distributed energy source into an active distribution network.

Generally, there are many valuable open problems in the study of routing decisions in the Energy Internet.

7.2.3   Electrical Structure

Energy routers should be able to control the transmission of energy flows. For the integration of distributed energy resources with different parameters and terminals in a backbone network and a local area network, power conversion is an indispensable function. As the basis of an energy router, power conversion greatly enhances the controllability of energy transmission. It also provides corresponding electrical interfaces for different types of AC and DC power devices. A vast number of studies on power conversion architectures have been carried out.

A single-phase structure of a back-to-back multilevel converter is used in the UNIFLEX-PM system in Europe. The core material of the intermediate frequency transformer is amorphous alloy that is insulated by oil immersion. The output inverter is interleaved with the intermediate stage to achieve three-phase equalization. A medium voltage AC bus and a low voltage AC bus are provided in the output. The system has functionalities including multi-directional power flow control, power quality adjustment, etc. Since no DC bus is provided in the system, it is not conducive to the integration of distributed energy sources and energy storage devices.

This structure allows power factor correction and is able to prevent bidirectional flow of harmonics. However, due to the lack of real-time communication modules, it is difficult for these architectures to respond to complex situations.

In view of the rapid development of microgrid technology in recent years, an energy router architecture based on an AC/DC hybrid microgrid was proposed in [7]. The microgrid considered in [7] consisted of an AC/DC bus and a bidirectional AC/DC converter. The impact of the intermittent output of distributed power sources on the power grid can be effectively suppressed via effective cooperation between an inverter and an energy storage unit. Renewable energy sources can be fully utilized to supply power for loads or energy storage devices. By properly utilizing energy storage devices, energy flows can be better controlled. A DC bus can simplify parallel output control of distributed power and energy storage units. Since the AC/DC conversion between a DC bus and AC bus is achieved via a single inverter, loop current and bus voltage instability can be avoided. The structure of the architecture is simple. In this sense, various distributed energy resources and energy storage devices can be integrated conveniently and reliably.

The required functions of energy routers in current power systems are comprehensively summarized in [2]. The following functions are included: basic power conversion, power quality management, bidirectional power flow, auxiliary power supply structure, plug-and-play interface and bus, highly open and real-time information report. To achieve the above functions, an architecture that is capable of high-quality power conversion and has plug-and-play multi-input-multi-output interfaces is proposed. An energy router, along with high-voltage sensors, DC rectifiers, DC converters, auxiliary power supply systems, etc., functions as an adaptive module in the network. This model is capable of AC/DC conversion, and it can supply energy for independent devices as well as provide interfaces for connections with other routers. To monitor and control the dynamics of the system, the architecture design for energy routers relies heavily on the communication system employed in a power network. Such dependency greatly increases the complexity of communication in the system.

A large number of management and protection devices are deployed in the energy system to ensure the reliability and security of the system. All of these devices are coordinated by the communication protocol at the component level. The above two kinds of energy routers rely too heavily on the communication architecture. If they are not well compatible with the communication protocols used in existing devices, the energy network will be faced with enormous security threats.

7.3 Key Technologies

7.3.1 Physical Layer

7.3.1.1 Unified Power Flow Controller (UPFC)

A UPFC is a device that connects a static synchronous compensator (STATCOM) and the DC side of a static synchronous series compensator (SSSC). It is able to control the voltage, impedance, phase angle and other parameters that can affect the power flow of a transmission line either simultaneously or separately. In the absence of external energy storage, UPFC devices can provide active and reactive current compensation for series lines.

Under normal working conditions, the active power of an SSSC is obtained through the STATCOM from the same line. The voltage control function of the STATCOM is achieved by the control of reactive power. In general, a UPFC device is a controller that is able to fully control the active power, reactive power and line voltage in transmission lines [9].

UPFCs have been widely applied in power transmission and distribution grids such as flexible AC transmission systems. By applying UPFCs, the energy loss during transmission can be reduced such that the ultimate transmission capacity of a transmission line can be improved. In addition, the active power, reactive power and line voltage can be completely controlled by UPFCs. In the power transmission grids in Portugal, researchers verified the effect of a double balanced multilevel UPFC DC link coupled with an active reactive power controller, real-time pulse width modulation and DC-side capacitor voltage on voltage stability during line faults. A new pattern recognition method for UPFC transmission line fault analysis was also proposed [9]. The combination of a modular multilevel converter (MMC) and a UPFC in a power grid also shows the superior performance of UPFCs. MMC-UPFCs have good harmonic characteristics. They are compact in size, and it is easy to implement a modular design for MMC-UPFCs, which can be directly connected in series. They are applicable to high-voltage and large-capacity transmission scenarios. Additionally, the transmission capacity, transient stability and grid loss in the power system can be improved by MMC-UPFC devices.

With the development of the Energy Internet, it is difficult for a conventional UPFC to meet the requirements of various complex situations. Researchers have made bold improvements to the structure of UPFCs [11]. For example, for a current-limiting UPFC, the IGCT-based bridge current limiter is combined with the UPFC to reduce the short-circuit cut-off time. The ratings of the components and the impact of the short-circuit current on the power grid are reduced. Conventional UPFCs must be connected to the grid via bulky and tortuous transformers on the series and parallel sides, respectively, which is expensive and space-consuming. Besides, they are prone to failure, and the response speed is quite slow. To solve these problems, a new transformer-less UPFC topology based on a cascaded multilevel inverter shows an ideal application prospect. The new UPFC topology is highly modular with light weight. It has higher reliability and efficiency and a faster dynamic response speed.

7.3.1.2 High-Voltage Direct Current (HVDC) Transmission

HVDC transmission is another key technology for energy routers in the physical layer. HVDC transmission is suitable for high-power long-distance DC transmission scenarios. It has advantages such as no induction and no resistance. An HVDC system consists of two converter stations and a DC transmission line. The converter stations are connected to an AC system. The DC line is used to connect the rectifiers to the inverters. In general, HVDC systems are used for submarine cable transmission and transmission between non-synchronous AC systems.

Compared with conventional AC power transmission, HVDC transmission is more economical. It is mainly used for long-distance and ultra-long-distance power transmission. In HVDC systems, both the power level and directions can be controlled quickly and precisely, which improves the performance and efficiency of the connected AC grids. HVDC does not increase the short-circuit capacity of the system. Thus, it is convenient for achieving networking of power systems with different frequencies. Moreover, the power modulation of the DC system can improve the damping coefficient of the power system, suppress the low-frequency oscillation and improve the transmission capacity of the parallel AC transmission line. However, since an HVDC system is a DC transmission system, it is difficult for an HVDC system to lead out branch lines. Thus, HVDC systems are mainly applied in end-to-end transmission.

As the latest voltage source converter technology, a MMC can be applied to high-voltage and high-power scenarios by using a series connection sub-module technique. It could be widely applied in grid-connected transmission for renewable energy sources [12]. MMCs are also widely employed in applications of HVDC. A novel ordered unlocking starting method, which was designed for two-terminal and multi-terminal flexible DC transmission systems adopting the MMC structure, can effectively reduce the DC voltage drop and the bridge arm current impact at the instant of unlocking the converter valve in the conventional charging mode. The new bidirectional single-stage modular multilevel converter, which consists of interleaved cascade submodules, has a similar blocking capability for a bidirectional DC fault as that of a DC breaker [13].

7.3.1.3 Intelligent Energy Storage

An energy storage device is one of the key components of an energy router. It has three main functions: first, improving power quality and maintaining system stability; second, providing power to the users when the distributed power generation devices cannot work properly; and third, improving the economic benefits for owners of distributed power generation units.

Although the distributed energy storage devices in the system have the required functionalities, the response speed of such devices is not fast enough. A slight disturbance in the system may lead to a severe butterfly effect on the entire energy system. Therefore, energy routers must have a fast-responding energy storage module to achieve real-time regulation of energy flow and maintain the power quality of the system. In addition, energy routers must work continuously, which requires energy storage modules to ensure the reliable operation of energy routers. The development of energy storage technology is of great significance to energy routers. There are two main aspects in the development of energy storage technologies. One is to find an approach to store energy efficiently and massively. The other is to achieve fast and efficient energy conversion.

With regard to the current development of energy storage technologies, energy storage is mainly divided into mechanical energy storage, electromagnetic energy storage and electrochemical energy storage. Mechanical energy storage includes pumped energy storage, compressed air energy storage and flywheel energy storage. Though the rated power for pumped energy storage and compressed air energy storage is large, they are slow and restricted by geographical location. Flywheel energy storage has a smaller response time but has a low energy density. Thus, mechanical energy storage is not suitable for energy routers. Electromagnetic energy storage mainly includes superconducting magnetic energy storage and supercapacitors. They have a small response time and high conversion efficiency. However, superconducting magnetic energy storage requires low temperature, which makes it difficult to meet the requirement of miniaturization for energy routers. Since supercapacitors are easy to install, small in size and can operate in a variety of environments, they can be an option for energy routers. There are many kinds of electrochemical energy storage, including lead-acid batteries, nickel cadmium batteries, sodium sulphur batteries, liquid current batteries, lithium-ion batteries, nickel hydrogen batteries and the latest developed aluminium-ion batteries. Lithium-ion batteries have the advantages of light weight, large energy storage capacity, high power, long lifetime, small self-discharge coefficient and wide temperature adaptation range. Thus, they can be another option for energy storage modules in energy routers. However, lithium-ion battery technology requires further development to improve its reliability and capacity. Compared to lithium-ion batteries, aluminium-ion batteries offer higher reliability and safety. They have faster charging and discharging speeds, lower costs and longer cycle lives and can be ideal choices for energy routers. However, the voltage and energy density of aluminium-ion batteries are relatively low. Improvements are needed before they can be applied in practical scenarios.

Many advances have also emerged in energy management for energy storage. A real-time energy storage management system for microgrids with integrated renewable energy sources was presented in [14]. To reduce the power losses in the system, a distributed cooperative control strategy for energy storage systems in microgrids was proposed in [15]. Additionally, to achieve the optimal allocation of load power in a DC microgrid, a droop control algorithm based on the double-quadrant state of charge for a distributed energy storage system was proposed in [16]. In [17], a frequency-based energy management strategy was proposed. The proposed control method enables an islanded AC grid system to work normally without a communication cable, which makes the system more efficient and reliable.

7.3.2 Information Layer

7.3.2.1 Software-Defined Network (SDN)

In the basic architecture of an SDN, there is only one centralized logical controller, which makes it difficult for the SDN to be adapted to large-scale networks. Thus, it cannot be directly applied to the Energy Internet.

Adding controllers that can utilize global information can achieve precise control of network flow and improve network efficiency. To improve energy dispatching efficiency and to adapt to personalized requirements from end users, a software-defined energy control system designed referring to the concept of an SDN was proposed in [8]. In the Energy Internet, each energy router can manage the energy flow within its jurisdiction with its intelligent control module. Additionally, the personalized requirements can be submitted to the central controller through the communication network. The central controller calculates the optimal policy based on global information and sends corresponding control strategies to different energy routers. The requests from an energy router are generated from the personalization requirements from users and operational constraints. To facilitate the generation of request data files, the energy routers provide a corresponding programming interface. During the operation of the Energy Internet, if there are no proper control strategies stored in an energy router, the system information will be submitted to the central controller. Once the control strategies are generated, they are packed and sent back to the relevant routers.

7.3.2.2 Data Centres

In the future, the Energy Internet will cover many aspects of human life. The large amount of operational data will bring new challenges to the operation of the Energy Internet system. Data collected in the Energy Internet include weather conditions, user requirements, network status, social events, etc. To achieve the target of real-time balancing of energy supply and demand, renewable energy source integration, responses to user requirements and data centres with powerful data processing capabilities will become key units in the Energy Internet [17].

In the Energy Internet, for reliability and security, data redundancy and backup mechanisms are necessary. In addition, the storage model of the data must meet the requirements of fast search and processing. Due to the diversity of applications, the requirements of different applications vary [18]. Thus, data storage should also be divided into two parts: online data and real-time data.

For the predecessor of the Energy Internet, the smart grid, there are currently four main data storage solutions. The first solution employs multiple data concentrators, a single processing node and centralized storage that utilizes a relational database. Each data concentrator is responsible for obtaining data from a certain number of measurement devices. The second solution is similar to the first, except that the centralized storage is replaced by distributed database storage. The third solution abandons the storage model using relational databases. Instead, an XML-based keyword-value model is proposed. A MapReduce-like algorithm is developed to operate the database. The fourth solution uses a distributed file system combined with a database for data storage. In such a solution, only the indexes of the original data are stored in the database. The raw data are saved in a central storage node, which is similar to the mode of search engines. Simulation results show that the scalability of the third scheme is poor. The processing time of the fourth scheme is the longest. The results of the first two solutions are comparable [21].

In large-scale smart grids, a great number of electrical equipment must be monitored, and the collected data are increasing rapidly. Thus, the reliability and real-time requirements are becoming stricter. A conventional data storage method would lead to overload or blockage of the communication network. To deal with the new scenarios, a tiered extended storage mechanism was introduced. The tiered extended storage mechanism is a data-centric storage method that can be applied to data storage and processing in large-scale smart grids. Data transmission, storage and query in large-scale smart grids are taken into consideration in the tiered extended storage mechanism. First, the storage node is dynamically allocated by an extended hash-coding algorithm. The backup of the data is performed in the background at a proper time. Thus, the availability of the system is improved, and the storage performance of the system is enhanced [22]. Second, the data are stored in a distributed manner on multiple storage nodes with the same level via a multi-threshold level storage method. In this sense, the storage and querying of the data are spread across multiple nodes, and load balancing is achieved.

7.3.2.3 Power Trading Platform

As mentioned earlier, energy routers should be able to handle the individual needs of end users. The corresponding control strategy is performed by the intelligent control module of a router and the SDN control system simultaneously. The rationality of the customization requirements of users is a key issue that needs to be addressed in this part. Inappropriate control requirements would directly affect the operation efficiency of the Energy Internet [23]. The essential difference between the Energy Internet and the conventional energy system is that the Energy Internet is able to collect and utilize information more comprehensively. Pushing the current system status to users in real time and tailoring appropriate and scalable customized service are the core goals that the Energy Internet aims to achieve.

Energy routers construct a real-time information database with received information from monitoring devices and data centres in the network. Based on the customized user requirements, energy routers are able to push relevant information to the users in time. The information includes the load, the electricity price, the charging state of energy storage and the efficiency of renewable energy sources.

Customizing templates can help users design a reasonable and personalized energy solution conveniently. Different metrics are involved in these solutions, such as energy costs, energy supply reliability and energy storage status. According to the different requirements of users, a router should generate corresponding control strategies via the SDN network and its own control module.

The advantage of the Energy Internet in information allows full utilization of big data collected from the network. For example, by monitoring energy consumption behaviour, more accurate forecasts for energy consumption in an area can be obtained. An optimal control strategy for multiple metrics can be obtained based on the real-time status of the system and the forecasts.

7.4 Autonomous Microgrid Energy Router Prototype

In the future Energy Internet, energy routers are key equipment for building a bottom-up energy infrastructure. From the basic concepts and requirements of the Energy Internet, a number of relevant studies have been carried out.

7.4.1 Overall Functional Design and Energy Routing Algorithm

The overall architecture of an energy router is shown in Fig. 7.1 [23]. The information support layer of the energy router provides interfaces for various types of communication. An efficient forwarding mechanism for these interfaces is provided simultaneously. Energy routers not only support network protocols such as Ethernet but also support dedicated communication interfaces for remote control, security protection and power quality control. The compatibility of communication protocols and reliability of communication are ensured by energy routers. In a practical implementation, energy routers can be united with an intelligent data centre to provide support for energy exchange between distributed microgrids.

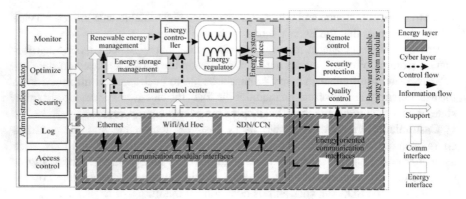

Fig. 7.1 Architecture of energy routers [23]

The energy control layer of an energy router provides energy integration, power quality regulation and system management. Various energy interfaces can be configured through the energy control layer, such that different energy units can be connected to the energy router. Intuitively, configurable interfaces include Energy Internet interfaces, energy storage interfaces, load unit interfaces, renewable energy source interfaces and interfaces for cascaded connections with other energy routers. The control function is realized by the integrated data unit and the special business unit. The control unit supports multiple implementations.

Routing algorithms deployed on energy routers are used to improve routing efficiency for energy flow in the Energy Internet. To ensure the normal operation of the system, the energy routing algorithms must be robust enough. That is, the energy router should be able to operate normally under the most unexpected conditions. As the core equipment of the Energy Internet, energy routers need to ensure long-term stability to better utilize the advantages of open, peer-to-peer and interconnection features.

7.4.2 Research on Open Interface, Interconnection Standards and Source-Load Flexible Connection Technology in the Energy Internet

Energy routers need to implement routing functions for information and energy flow, which requires the cooperation of communication networks and energy networks. The transmission and distribution of energy flow is achieved by adjusting the state of the control devices in the energy network. The information network is responsible for the transmission of system information monitoring and control commands.

Energy routers mainly undertake the interconnection tasks between distributed microgrids. One energy router is connected to the backbone network, the load, the renewable energy unit and the energy storage unit and is interconnected with other energy routers. Users can manage energy routers through communication interfaces and configure personalized energy management schemes. To achieve open and peer-to-peer characteristics like the Internet, an open interface and interoperability model framework for the Energy Internet are essential. For example, the basic features of the TCP/IP protocol for the Internet are open, concise and easy to implement, which leads to the widespread application of the Internet. Therefore, it is necessary to select or design an efficient communication protocol for the exchange of energy and information between energy routers.

Low-voltage energy routers play a major role in low-voltage power grids. In low-voltage power grids, the energy sources and loads are more dynamic. There are more requirements for short-term power balancing. Similar to the cache in routers of the Internet, energy routers need fast energy storage devices to achieve a flexible connection for power sources and loads.

7.4.3 Research on Adaptation Technology for Electric Load Demand and Bidirectional Prediction of Renewable Energy Generation

In line with a bottom-up mechanism for energy switching and routing, when there are faults in the Energy Internet, each distributed microgrid should be able to automatically achieve smooth switching between the islanded operation mode and the grid-connected mode. Each energy router should have the ability to respond independently to most of the events in the Energy Internet based on local information. Conventional energy management systems (EMSs) often encounter bottlenecks when large-scale renewable energy sources are integrated. To solve this problem, a new generation of decentralized collaborative EMSs is urgently needed.

Unlike flexible connection techniques for energy routers, bidirectional prediction and balancing for loads and energy sources are not essential issues for energy management from a macro perspective. However, by achieving a power balance between generation and consumption in a certain period, the difficulty of short-term power balancing for energy routers can be alleviated. Moreover, this helps to reduce the requirements for energy storage devices in energy routers.

7.5 Future Prospects for the Development of Energy Routers

7.5.1 Open and Plug-and-Play Energy Exchange and Routing

Openness is an essential feature of the Energy Internet. The open feature of energy routers is its embodiment. The plug-and-play feature of energy routers is more complicated than that in information networks. Since the power balance issue is a concern for the Energy Internet, the plug-and-play feature leads to rapid power fluctuations in the Energy Internet system. To ensure the security of the entire system, it is necessary to design corresponding control mechanisms to implement this function.

7.5.2 Support for Multi-channel Scalable Renewable Energy Sources and Dynamic Load Integration

Conventional power electronic devices are mostly functional complements to existing power grid architectures. Most of them have a single input and output, e.g. power quality management devices. Energy routers are designed for a new bottom-up energy architecture and allow multi-line integrations and dynamic expansion of the system.

7.5.3 Solving Instantaneous Balanced Energy Internet Energy Management Issues

Energy routers transform the power balance issues in large-scale power grids into decentralized local power balance problems. A large-scale management problem for energy storage devices is decomposed into smaller problems inside individual energy routers.

7.5.4 Realizing the Integration of Information-Energy Infrastructure

Energy routers have powerful communication capabilities that enable the full integration of information and energy systems. A conventional smart grid simply emphasizes the combination of information flow and energy flow. However, there is no special device for the integration of information and energy systems. Energy routers integrate information and energy systems from the hardware level, and they are more suitable for the future Energy Internet.

References

1. A.Q. Huang et al., The future renewable electric energy delivery and management (FREEDM) system: the energy internet. Proc. IEEE **99**(1), 133–148 (2010)
2. J. Cao, H. Hua, G. Ren, *Energy use and the Internet," The SAGE Encyclopedia of the Internet* (Sage, Newbury Park, 2018), pp. 344–350
3. Y. Xu, et al., Energy router: Architectures and functionalities toward Energy Internet, in *Proceedings of 2011 IEEE International Conference on Smart Grid Communications*, October 2011, pp. 31–36
4. H. Hua et al., Stochastic optimal control for energy Internet: a bottom-up energy management approach. IEEE Trans. Indus. Inf. **15**(3), 1788–1797 (2019)
5. X. Han, et al., An open energy routing network for low-voltage distribution power grid, in *Proceedings of 1st IEEE International Conference on Energy Internet, Beijing, China*, April 2017, pp. 320–325
6. H. Hua, et al., Robust control method for DC microgrids and energy routers to improve voltage stability in energy Internet. Energies **12**, Art. no. 1622 (2019)
7. J. Zhang, W. Wang, S. Bhattacharya, Architecture of solid state transformer-based energy router and models of energy traffic, in *Proceedings of 2012 IEEE PES Innovative Smart Grid Technologies*, January 2012, pp. 1–8
8. N. Huin et al., Bringing energy aware routing closer to reality with SDN hybrid networks. IEEE Trans. Green Commun. Netw. **2**(4), 1128–1139 (2018)
9. R.-J. Abbas, F.-F. Mahmud, O. Muhammad, Optimal unified power flow controller application to enhance total transfer capability. IET Gener. Transm. Distrib. **9**(4), 358–368 (2016)
10. Z. Moravej, M. Pazoki, M. Khederzadeh, New pattern-recognition method for fault analysis in transmission line with UPFC. IEEE Trans. Power Deliv. **30**(3), 1231–1242 (2015)
11. S. Yang et al., Modulation and control of transformerless UPFC. IEEE Trans. Power Electron. **31**(2), 1050–1063 (2016)
12. G.J. Kish, M. Ranjram, P.W. Lehn, A modular multilevel DC/DC converter with fault blocking capability for HVDC interconnects. IEEE Trans. Power Electron. **30**(1), 148–162 (2014)
13. R. Li et al., Hybrid cascaded modular multilevel converter with DC fault ride-through capability for the HVDC transmission system. IEEE Trans. Power Deliv. **30**(4), 1853–1862 (2015)
14. K. Rahbar, J. Xu, R. Zhang, Real-time energy storage management for renewable integration in microgrid: an offline optimization approach. IEEE Trans. Smart Grid **6**(1), 124–134 (2014)
15. Y. Xu et al., Cooperative control of distributed energy storage systems in a microgrid. IEEE Trans. Smart Grid **6**(1), 238–248 (2014)
16. X. Lu et al., Double-quadrant state-of-charge-based droop control method for distributed energy storage systems in autonomous DC microgrids. IEEE Trans. Smart Grid **6**(1), 147–157 (2014)
17. A. Urtasun, et al., Frequency-based energy-management strategy for stand-alone systems with distributed battery storage. IEEE Trans. Power Electron. **30**(9), 4794–4808 (2015)
18. X. He et al., QoE-driven big data architecture for smart city. IEEE Commun. Mag. **56**(2), 88–93 (2018)
19. H. Hua et al., Voltage control for uncertain stochastic nonlinear system with application to energy Internet: non-fragile robust H_∞ approach. J. Math. Anal. Appl. **463**(1), 93–110 (2018)
20. R. Wang et al., A graph theory based energy routing algorithm in energy local area network (e-lan). IEEE Trans. Ind. Inf. **13**(6), 3275–3285 (2017)
21. T. Jiang, L. Yu, Y. Cao, *Energy Management of Internet Data Centers in Smart Grid* (Springer, Berlin, Heidelberg, 2015)

22. C. Lin et al., Optimal charging control of energy storage and electric vehicle of an individual in the internet of energy with energy trading. IEEE Trans. Ind. Inf. **14**(6), 2570–2578 (2017)
23. J. Cao, et al, An energy internet and energy routers. **44**(6), 714–727. Science China Press (2014) (in Chinese)

Chapter 8
Two-Stage Optimization Strategies for Integrating Electric Vehicles in the Energy Internet

William Infante, Jin Ma, Xiaoqing Han, Wei Li, and Albert Y. Zomaya

Abstract Electric vehicles (EVs) form an important part of the energy internet, as they connect a transportation network with an electricity network. EV uptake largely depends on the optimization strategies of charging infrastructures such as battery swapping stations (BSSs). These stations can potentially reduce the upfront expenses of EV owners, range anxiety, long charging times and electricity grid strain. Currently, the major challenge in BSSs is the creation of robust business strategies. This chapter proposes BSS stochastic optimization strategies that consider EV uptake uncertainties and power distribution company decisions. Two stochastic optimizations involving two stages are investigated: (a) optimization with recourse and (b) bilevel optimization. The recourse optimization recommends initial battery investment even before the station visits are known in the planning stage and recommends battery allocations in the operation stage. This optimization links a transport network to a distribution line network, providing energy arbitrage and curtailment tractability. The bilevel optimization further links the transport network to a transmission line network using aggregated EV batteries as a form of flexible load to compensate for intermittent renewable source generation. The flexible load is a lower-level decision made by distribution company operators, and the same flexible load is a constraint in the upper-level decisions made by BSS owners. Furthermore, this optimization can link the transportation and electricity networks to a gas network in the presence of

W. Infante · J. Ma (✉)
School of Electrical and Information Engineering, The University of Sydney, Sydney, Australia
e-mail: j.ma@sydney.edu.au

W. Infante
e-mail: william.infante@sydney.edu.au

X. Han
Shanxi Key Laboratory of Power System Operation and Control, Taiyuan University of Technology, Taiyuan, China
e-mail: hanxiaoqing@tyut.edu.cn

W. Li · A. Y. Zomaya
School of Computer Science, The University of Sydney, Sydney, Australia
e-mail: weiwilson.li@sydney.edu.au

A. Y. Zomaya
e-mail: albert.zomaya@sydney.edu.au

© Springer Nature Switzerland AG 2020
A. F. Zobaa and J. Cao (eds.), *Energy Internet*,
https://doi.org/10.1007/978-3-030-45453-1_8

gas as a power source with varying marginal prices. The proposed strategies provide a pathway for integrating EVs in the energy internet.

8.1 Current State of the Electric Power Industry

The electric power industry, moving towards an energy internet, is currently facing challenges (a) on the generation side due to the increased integration of intermittent renewable energy sources and (b) on the distribution side due to the expected growth of electric vehicle (EV) uptake. However, these challenges can lead to sustainable business opportunities such as battery swapping stations (BSSs) given robust strategies.

8.1.1 Rise in Renewable Energies

The renewable energy sector predicts rapid growth following an average energy use increase of 2.6% per year from 2012 to 2040 [1]. Unfortunately, as this sector develops, intermittent renewable energy sources are also causing energy management complexities in the grid [2].

Additional management tasks are prevalent in regions such as the United States and Australia. In California, the electricity net load closely resembles a "duck" curve due to the overgeneration of solar energy and increased penetration of solar photovoltaic (PV) panels [3, 4]. Ideally, a flat line is preferred for the net load, but due to the peaks and valleys formed in the duck curve, the system operator must prepare for upward ramps. In addition, the output generated from renewable sources must be combined with conventional power plants for voltage support and reliability. Similarly, in South Australia, the electricity net load is undergoing changes due to the strong growth in rooftop PV uptake. It was the first region in the National Electricity Market that shifted its minimum demand from the usual overnight to near midday [5]. In a deregulated electricity market, these changes can even cause electricity spot prices to be negative. The distribution grid not only handles large-scale renewable energy resources such as solar PV farms but also manages household prosumers that purchased their rooftop panels and sell excess electricity back to the grid.

Because of renewable energy intermittency, proposals such as energy curtailments, load levelling [6, 7], energy scheduling [2] and energy storage options [8] have been investigated. Currently, research on curtailment, scheduling and storage has not yet been fully investigated. Since additional energy storage infrastructure has associated costs, an alternative is to appropriate existing and underutilized resources such as EV batteries. Such concepts, however, will need innovative approaches.

8.1.2 Rise in Electric Vehicles

The rise of electric vehicles is also building momentum. In terms of energy use by sector, the transportation sector is expected to increase at an average rate of 1.4% [1]. The electricity share of vehicle energy consumption is expected to grow, as electric vehicles are starting to replace conventional vehicles. For example, the number of electric vehicles exceeded 2 million in 2016 after passing the 1 million threshold in 2015 [9]. In Australia, the EV trend in two major energy markets could initially start at a slow pace, but this pace could rise to 20% of vehicle sales by 2020 [10].

Interest in electric vehicles is not only limited to Australia, as it is also experiencing positive trends in other countries such as the United Kingdom, the United States and China. Compared to conventional vehicles, EVs can potentially decrease dependence on fossil fuels and reduce greenhouse gas emissions. In the United Kingdom, the electrification of transport is seen as a vital component of energy self-sufficiency [11]. In the United States, support from the government, such as federal tax credits, is offered. Some states even offer further stimulus for purchasing EVs [12]. In China, research support on electric public taxis is also accessible [13].

Electric vehicles have not only ventured into the transportation network domain but also have become part of the electricity network domain. Both networks need to be planned together in the framework of the energy internet.

On the distribution side of an electricity network, electric vehicles present energy management issues. Uncoordinated EV use can potentially modify the load profile and power source scheduling. A low concentration of EVs already creates a noticeable impact on grid peaks and valleys. One study found that residential peak energy demand can double when a fast-charging EV is present [14]. Currently, EVs only account for a small percentage of the overall electric load, but as EVs reach critical numbers, EV charging will also impact the net load.

Alternatively, EVs also have the potential to become part of distributed energy storage. Using their inductive power can improve grid security without committing further grid capacity expansion just to accommodate peaks [15]. Integrating electric vehicles in the future energy internet can transform the increasing load demand to an increasing flexible load through managed and aggregated EV batteries. Prime examples of EV aggregators are battery charging stations and battery swapping stations (BSSs).

8.1.3 Charging and Swapping Stations

With the rise of electric vehicles, public and private charging infrastructures are continually growing [9], but robust business models are still needed to ensure long-term benefits.

Both charging and swapping stations play key roles in managing electric vehicles. In this chapter, battery swapping stations are given focus where the business model may be preferred in several situations. For example, if the price of batteries significantly increases vehicle ownership, a BSS model can own the batteries and, in turn, reduce the upfront cost for EV owners. Moreover, given conventional charger types, swapping stations offer a faster means of providing fully charged batteries to EV owners. Finally, in places where electricity supply is intermittent, swapping stations have time flexibility in scheduling battery charging and can even include complimentary grid services.

8.1.4 Swapping Station Business Strategies

Although BSSs may aid in addressing some of the energy internet challenges, specifically in the integration of renewable energies and the growth of electric vehicles, planning and operational strategies are necessary for a BSS business to be sustainable. However, these strategies have not been fully investigated so far.

To date, not all BSS ventures have been successful commercially. Battery swapping stations still need robust and detailed operational strategies to mitigate failure due to high infrastructure investments, possible detrimental strain on the grid [16] and uncertain EV station visits. For example, Better Place's battery swapping stations have become bankrupt [17]. This potentially could have been mitigated with complementary businesses and if uncertainties in both planning and operations were considered.

Different BSS operational strategies have been proposed, such as battery allocations [18, 19], dynamic operation in the electricity market [20] and initial discounts for customer battery delivery delay [18, 21]. Although these investigations have contributed to BSS designs, they have not yet thoroughly explored the interactions between BSS owners and power distribution company operators in terms of power limits. Research investigations have also started on the service availability of stations in the planning stage [22], but these still need robustness to EV station visit uncertainties. To deal with power limits and uncertainties, two-stage stochastic optimization techniques, a subset of multilevel programming, are presented in this work.

8.2 Two-Stage Stochastic Optimization

8.2.1 Overview

Optimization is a key tool for efficiently allocating limited resources for a given problem scenario. A standard mathematical program provides a set of decisions, usually for a single objective, made by a single decision maker. In practice, many optimization problems are modelled by considering multiple decision makers and

Fig. 8.1 Examples of
two-stage stochastic program
cases

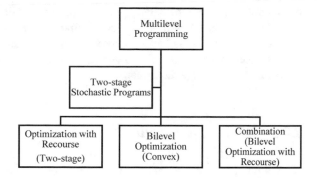

multiple stages. Multilevel programming involves multiple levels or stages in which each stage can have multiple objectives that can be made by multiple decision makers.

Two-stage stochastic optimization programs are a subset of multilevel programming. Stochastic programs usually model uncertain parameters based on probability distributions [23, 24]. For battery swapping stations where the EV station visit demand is uncertain, a stochastic approach that considers robustness might be appropriate.

Uncertainties occur in two stages based on decisions when the uncertainties are made known, such as in optimization with recourse, or based on decisions when upper-level decisions are influenced by lower-level decisions, such as in bilevel optimization. Figure 8.1. provides an overview of two-stage stochastic program cases showing optimization with recourse, convex bilevel optimization and a combination of these.

For two-stage optimization, the objective function $\Psi(x)$ can be divided into the initial solution $r^{T}x$ and the risk measure $\Xi(x)$ due to uncertainty or the second-stage risk function, as presented in (8.1). The risk function can then arise due to the stochasticity exemplified in two-stage optimization with recourse or due to the response to a lower-level optimization, as in the case of bilevel optimization.

$$\Psi(x) = r^{T}x + \Xi(x) \tag{8.1}$$

8.2.2 Optimization with Recourse

Two-stage optimization with recourse is a special case of multilevel programming where a single decision maker may need to optimize the objective in two stages. Recourse programs refer to stochastic programs with recourse actions taken after the uncertainty is disclosed. In cases where decisions must be made before the uncertainty is made known and where the distribution of scenarios can be approximated, stochastic optimization with recourse is suited to solve these [25, 26]. This context occurs in BSSs where the battery purchase investment in the planning stage must be decided before the EV station visit uncertainties take place.

Decisions can then be divided into two stages. Decisions made before the realization of a scenario are called first-stage decisions, and decisions after this realization are called second-stage decisions.

8.2.3 Bilevel Optimization

Bilevel optimization is a technique where an outer optimization contains an inner optimization in its constraints. The optimization follows a hierarchical nature, where the decisions made by an upper-level decision maker are affected by lower-level decision makers seeking to perform their own optimizations [27]. Decisions in the lower-level problem would restrict the feasible set in the upper-level problem.

Bilevel optimization can have two decision makers with potentially conflicting objectives. It is also different from multi-objective optimization with two decision makers in that bilevel optimization happens at different levels, effectively embedding one problem in another problem [28].

In the expanded form [29], a bilevel optimization can be represented by the equation set in (8.2). The objective function of the upper-level problem is represented by $F(x^u, x^l)$ with inequality constraints $G(x^u, x^l) \leq 0$ and equality constraints $H(x^u, x^l) = 0$. Similarly, the objective function of the lower-level problem is represented by $f(x^u, x^l)$ with inequality constraints $g(x^u, x^l) \leq 0$ and equality constraints $h(x^u, x^l) = 0$. x^u are the upper-level decisions, and x^l are the lower-level decisions.

If the lower-level problem formulation is convex and continuously differentiable, such as in linear optimization, a standard bilevel program can be reformulated to a single-level optimization using the Karush–Kuhn–Tucker (KKT) optimality conditions [30].

$$
\begin{aligned}
& \min F(x^u, x^l) \\
& s.t. \, G(x^u, x^l) \leq 0; \, H(x^u, x^l) = 0 \\
& x^l = \operatorname*{argmin}_{x^l \in X^L} f(x^u, x^l) \\
& s.t. \, g(x^u, x^l) \leq 0; \, h(x^u, x^l) \leq 0
\end{aligned}
\tag{8.2}
$$

8.2.4 Karush–Kuhn–Tucker Optimality Conditions

For cases when the lower-level problem is a linear problem with a bounded optimization, the Karush–Kuhn–Tucker optimality conditions are necessary and sufficient for an optimal solution. These conditions can be grouped into stationarity,

Table 8.1 KKT optimality conditions

Condition	Main equations	
Stationarity	$\nabla f(x^*) + \sum\limits_{i \in I} \lambda_i^* \nabla g_i(x^*) +$ $\sum\limits_{j \in J} \mu_j^* \nabla h_j(x^*) = 0$	
Primal feasibility	$g_i(x^*) \leq 0$	$\forall i \in I$
	$h_j(x^*) = 0$	$\forall j \in J$
Complementary slackness	$\lambda_i^* g_i(x^*) = 0$	$\forall i \in I$
Dual feasibility	$\lambda_i^* \geq 0$	$\forall i \in I$

primal feasibility, complementary slackness and dual feasibility conditions [31–33]. Given that the KKT conditions are satisfied with strong duality, the optimization can be transformed using the optimality conditions into a single-level optimization. Table 8.1 presents the optimality conditions with I inequality constraints and J equality constraints with x^* as the optimal solution and λ_i^* as the optimal Lagrangian multiplier.

8.3 Battery Swapping Station with Power Limit Strategies

An overview of the two stochastic optimization techniques and how these strategies can respond to power limits are presented in this section. These will be elaborated in further sections.

8.3.1 Power Limits in Optimization with Recourse

An optimization with recourse including power constraints has been proposed for stochastic planning and operations.

The formulation is presented in Sect. 8.4 with use case scenarios in Sect. 8.5. The scenarios demonstrate how a BSS can adapt to static and dynamic power curtailments in terms of the recommended battery investment, level of grid services and service availability to EV owners. The optimization can also be used by power distribution company operators for the power curtailment decisions that they provide to the battery swapping station.

8.3.2 Power Limits in Bilevel Optimization

To further explore the interactions between the battery swapping station owners and the power distribution company operators, a bilevel optimization is formulated in Sect. 8.6. The upper-level optimization decisions of the battery swapping station are constrained by the lower-level optimization decisions of the power distribution company operators.

The optimization links the transport network through EV station visits and the electricity network through power transmission constraints.

Aggregated batteries constitute the flexible load needed by the distribution and transmission networks, as exemplified in a modified IEEE 14-bus test system scenario in Sect. 8.7. Furthermore, gas power is also seen as a potential power source in the bilevel optimization linking the gas network to the electricity and transport network in the energy internet.

8.4 Optimization with Recourse Formulation

Using optimization with recourse, battery swapping station owners can incorporate stochastic planning and operational strategies that consider EV station visit uncertainties and power limit constraints.

The formulation can provide first-stage planning decisions even before the EV station visit uncertainties are made known and second-stage operation decisions when the station visit uncertainties are made known.

The problem formulation has also been framed such that the decisions are limited by the power curtailment from the power distribution company; the maximum allowed power for the BSS limits the number of chargers that can simultaneously perform charging and discharging. The decisions in both the first and second stages are inhibited by this power limit. To help present the optimization with recourse formulation, a list of parameters is included in Table 8.2.

In the first stage, BSS owners must prepare for the initial battery investment x°. In the second stage, BSS owners must decide on the battery allocations, namely, swapping $x_{t,\omega}^s$, charging $x_{t,\omega}^c$, and discharging $x_{t,\omega}^d$ at every time interval t of the planning horizon T in every scenario ω of all scenarios in Ω . Given that the probability φ_ω of a scenario can be approximated, the two-stage stochastic optimization can be represented by (8.3).

$$\min\left\{\pi^\circ x^\circ + \sum_{t\in T}\sum_{\omega\in\Omega}\varphi_\omega\left(-\pi^s x_{t,\omega}^s + \pi^m\left(D_{t,\omega} - x_{t,\omega}^s\right) - \pi_t^e\left(\eta_t^d x_{t,\omega}^d - \frac{x_{t,\omega}^c}{\eta_t^c}\right)\right)\right\}$$
(8.3)

The objective can be further analysed in three parts related to battery investment, swapping and grid decisions. For battery investment, the strategy prefers lower costs

Table 8.2 Nomenclature for the recourse formulation

Symbol	Meaning
Sets	
T	Set of time slots with index t
Ω	Set of scenarios with index ω
Parameters	
π°	Price of a battery purchase
π_t^e	Price of time-of-use electricity at time t
π^s	Price of a battery swap
π^m	Price of missing a swap
$D_{t,\omega}$	Electric vehicle station visit demand at time t for scenario ω
δ, ζ	Charging and discharging durations for one time slot
C^δ, C^ζ	Charging and discharging power for one time slot
η_t^c, η_t^d	Efficiency coefficients related to charging and discharging at time t
φ_ω	Probability of scenario ω
ρ_t	Power curtailment imposed by the distribution company at time t
Decision Variables	
x°	Initial battery investment
$x_{t,\omega}^s, x_{t,\omega}^c, x_{t,\omega}^d$	Battery allocations for swapping, charging and discharging at time t for scenario ω
$R_{t,\omega}$	Resource number of fully charged batteries in the station at time t for scenario ω

for the investment battery purchase π°. The second-stage decisions involve both swapping and grid decisions. For swapping, the aim is to maximize the gains from the price of swaps π^s, and at the same time, minimize the price of missing a swap π^m. This number of missed batteries is dependent on how many swaps $x_{t,\omega}^s$ are made compared to the current EV station visit demand $D_{t,\omega}$. For grid decisions, the BSS manages charging and discharging given the time-of-use electricity pricing π_t^e. Efficiencies related to charging η_t^c and discharging η_t^d at time t are also considered in the formulation.

The associated constraints are presented in (8.4)–(8.11). As a start, the decisions are real integers, as shown in (8.4), and the swaps should not exceed the number of EV station visits, as shown in (8.5). To aid in operation decisions, the number of fully charged batteries available in the station $R_{t,\omega}$ needs to be defined. This value cannot be negative and should not exceed the number of invested batteries, as shown in (8.6). $R_{t,\omega}$ is also dependent on its previous state, as shown in (8.7). For a battery to be added to $R_{t,\omega}$, a depleted battery needs to wait for the charging duration interval δ. The remaining batteries in the station $(x^\circ - R_{t,\omega})$ and the recently swapped batteries become the limit for the number of potential batteries that can be charged, as shown in (8.8). Furthermore, $R_{t,\omega}$ is also the maximum value of the number of batteries that can be swapped and discharged at any time interval, as shown in (8.9).

$$x^\circ, x_{t,\omega}^s, x_{t,\omega}^c, x_{t,\omega}^d \geq 0 \quad \begin{array}{l} \forall t \in T, \omega \in \Omega \\ x^\circ, x_{t,\omega}^s, x_{t,\omega}^c, x_{t,\omega}^d \in \mathbb{Z} \end{array} \tag{8.4}$$

$$x_{t,\omega}^s \leq D_{t,\omega} \quad \forall t \in T, \omega \in \Omega \tag{8.5}$$

$$0 \leq R_{t,\omega} \leq x^\circ \quad \forall t \in T, \omega \in \Omega \tag{8.6}$$

$$R_{t,\omega} + x_{t-\delta,\omega}^c - x_{t,\omega}^d - x_{t,\omega}^s = R_{t+1,\omega} \quad \forall t \in T, \omega \in \Omega \tag{8.7}$$

$$x_{t,\omega}^c \leq \left(x^\circ - R_{t,\omega}\right) + x_{t,\omega}^s \quad \forall t \in T, \omega \in \Omega \tag{8.8}$$

$$x_{t,\omega}^s + x_{t,\omega}^d \leq R_{t,\omega} \quad \forall t \in T, \omega \in \Omega \tag{8.9}$$

The last two constraints involve the power limits. Considering the charging duration δ and discharging duration ζ, the number of batteries that are simultaneously charging and discharging should not exceed the initially invested batteries x°, as shown in (8.10). As shown in (8.11), the total charging and discharging power should not exceed the power curtailment ρ_t imposed by the power distribution company. Each charging and discharging decision will have an associated consumed power for charging C^δ and discharging C^ζ.

$$\forall t \in T, \omega \in \Omega \sum_{f=0}^{\delta-1} x_{t-f,\omega}^c + \sum_{f=0}^{\zeta-1} x_{t-f,\omega}^d \leq x^\circ \quad \forall t \in T, \omega \in \Omega \tag{8.10}$$

$$C^\delta \sum_{f=0}^{\delta-1} x_{t-f,\omega}^c - C^\zeta \sum_{f=0}^{\zeta-1} x_{t-f,\omega}^d \leq \rho_t \quad \forall t \in T, \omega \in \Omega \tag{8.11}$$

8.5   Optimization with Recourse Scenario Analysis

8.5.1   Power Limit Environment

The optimization with recourse formulation is explored specifically for how the strategies adapt based on the power limits. The analysis is performed in a week horizon with hour resolution.

Representative scenarios were mainly based on Australian data sources. For example, the electricity pricing was based on an Australian operator [34], and the electric vehicle station visits $D_{t,\omega}$ were based on a K-means clustering technique [35] performed on data sourced from the Australian Energy Market Commission [10] and Australian traffic services [36]. For the renewable energy profile, the California

Independent System Operator California ISO [37] was used since it presented a more diverse profile.

Managing the power consumed in the BSS also depends on the charger-level specifications, battery capacities and EV acceptance rates. The acceptance rate is the maximum power allowed by the EV energy management system.

Different levels of EV charging are in use and can be grouped into three categories. Level 1 chargers, also known as slow chargers, are mainly used at home or at a convenience outlet, but this process could take overnight. Level 2 chargers are used by private and public facilities. Level 3 chargers, known as DC fast charging, are mainly used for commercial applications where the on-board charger system is being bypassed [38, 39]. Although Level 3 chargers can charge quickly, the infrastructure needed to support them is still at an early stage. Table 8.3 presents some of the charger-level specifications [40].

Even if the BSS environment has Level 2 and 3 chargers, the charging time will be limited by the acceptance rates of EV technology. Table 8.4 shows examples of EVs with their acceptance rates and charger levels based on collated data [41]. A similar method is used for the discharging duration ζ.

Table 8.3 Example charger power level specifications

Classification level	Charger location	Common usage	Predicted power level
Level 1	On-board	Charging at offices and homes	1.4 kW, 12 A 1.9 kW, 20 A
Level 2	On-board	Charging at private and public outlets	7.7 kW, 32 A 19.2 kW, 80 A
Level 3	Off-board	Commercial applications	50 kW 100 kW

Table 8.4 Sample electric vehicle battery charging times

Electric vehicle type	Acceptance rate (kW)	Battery size (kWh)	Expected power level	
			Level 1 charger (1.4 kW)	Level 2 charger (7.7 kW)
BMW i3 (2017, 60 Ah)	7.4	23	16.5	3
Chevy Spark	3.3	23	16	7
Ford Focus EV	6.6	31	22	4.5
Mitsubishi i-MiEV	3.3	16	11	5
Nissan Leaf (2017)	6.6	30	21.5	4.5
Tesla Model S 60	9.6	60	8	43

If battery swapping stations consider the constraints placed by the power infrastructure in the energy internet, the initial battery investment, grid services, and service availability may change in the stochastic optimization strategies, as is observed in the next subsections.

8.5.2 Threshold Curtailment

The threshold curtailment ρ_t is modified to investigate its impact on BSS planning and operational strategies. In this subsection, ρ_t is time-invariant. In Sect. 5.3, ρ_t is dynamic based on the renewable energy source availability.

To compare the effects of threshold values, a relaxed threshold curtailment and a restricted threshold curtailment are used for the Level 3 charging (fast-charging) scenario and the Level 2 charging scenario.

In a relaxed threshold curtailment, the number of batteries that can simultaneously charge is more than the number of batteries needed by the optimization with recourse formulation without power constraints. Even if the relaxed threshold curtailment is increased, the recommended number of batteries does not change. On the other hand, a restricted threshold curtailment limits the number of simultaneously charging batteries below the recommended number. The restricted threshold curtailment is specified in the use case scenarios.

For the fast-charging example with a low EV station visit uptake, Fig. 8.2 presents how curtailing the power in the BSS will impact the grid services and the number of initial batteries needed by the station. If the BSS has a relaxed threshold curtailment, as in Fig. 8.2a, the optimal number of batteries is 6. If we still have a higher power limit than the value set in Fig. 8.2a, the optimal battery allocation will not change.

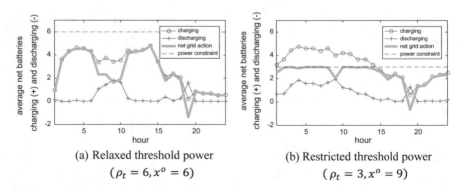

(a) Relaxed threshold power
$(\rho_t = 6, x^o = 6)$

(b) Restricted threshold power
$(\rho_t = 3, x^o = 9)$

Fig. 8.2 Average net charging and discharging allocations, with more grid services and initial batteries needed in the restricted curtailment than in the relaxed curtailment

As shown in Fig. 8.2b, the restricted threshold curtailment only allows 3 batteries to charge simultaneously in every period. Therefore, the optimal initial battery investment increases to 9 batteries. Because missed swaps π^m cost a substantial amount, it is better for the system to buy more batteries so that it can stock enough batteries to fulfill an appropriate demand that will maximize the business operator gains.

More grid services and battery-to-battery services are recommended by the optimization with recourse shown in Fig. 8.2b. Charging above the power constraint limit is an indication that the station is performing battery-to-battery services. Fully charged batteries are partially discharged to provide energy to depleted batteries. Although charging and discharging inefficiencies η_t^c and η_t^d are present whenever grid services are performed, grid services are still preferred compared to missing EV swaps.

From the fast-charging scenarios, the analysis will now investigate Level 2 scenarios. In the succeeding scenarios, 7.7 kW chargers are used. In addition, the electric vehicle type has a battery capacity of 32 kWh and an acceptance rate of 7.0 kW. To further investigate the setup, the Quality of Service (QoS) is defined as in (8.12). The QoS is a measure of the satisfaction of EV customers based on the percentage of successful swaps $x_{t,\omega}^s$ in the EV station visit demand $D_{t,\omega}$.

$$QoS = \sum_{t \in T} \sum_{\omega \in \Omega} \frac{x_{t,\omega}^s}{D_{t,\omega}} \times (100\%) \tag{8.12}$$

Visualizations of a relaxed threshold curtailment gradually moving to a restricted threshold curtailment are presented in Fig. 8.3. From the relaxed power threshold shown in Fig. 8.3a, a lower threshold, as presented in Fig. 8.3b, can have the QoS remain close to the QoS value in Fig. 8.3a, but the recommended number of initial batteries x° has to increase. This allocation is similar to that of the fast charger

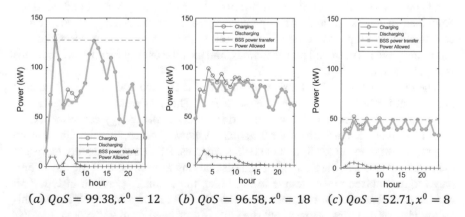

(a) $QoS = 99.38, x^0 = 12$ (b) $QoS = 96.58, x^0 = 18$ (c) $QoS = 52.71, x^0 = 8$

Fig. 8.3 Constant threshold curtailment response showing (**a**) the relaxed mode where no modification is made, (**b**) increased grid services and higher batteries, and (**c**) in extreme cases, lower battery investment but with many missed swaps reflected in the QoS

analysed at the start of this subsection. Interestingly, if the threshold is restricted further, as shown in Fig. 8.3c, the number of initial batteries decreases to $x° = 8$ but with a sharp decrease in the QoS.

Based on the optimization with recourse formulation, a lower power threshold may implement a BSS strategy by increasing the battery investment, as shown in Fig. 8.3b, or by lowering the battery investment with a sharp decrease in the QoS, as shown in Fig. 8.3c.

If we further restrict the threshold curtailment shown in Fig. 8.3c, such as allowing only 2 batteries to be charged simultaneously, the system provides a no-action response. In that case, the business is better off not operating.

Given these charging scenarios, the interactions between BSS owners and distribution company operators play an important part in the energy internet design. The BSS investment, profit and QoS depend on the contract agreements between the BSS owners and the power distribution company operators regarding the allowable threshold curtailment.

8.5.3 Renewable Energy Curtailment

Curtailment can also depend on the power source that varies at each time interval, especially when the system is connected through a microgrid. In a microgrid, the aggregated charging and discharging of EVs can provide a significant load to the grid [42]. Thus, observing the coordination of the renewable energy source with the demand side is imperative.

For this analysis, the scenarios were based on the normalized distributions of renewable power sources from an independent system operator California ISO [37], as presented in Fig. 8.4. The individual power sources include solar PV, solar thermal, wind and biomass. As part of the illustration, solar PV, solar thermal and wind are combined in the "solar and wind" power source. A mixed renewable source is a combination of all power sources. In generating the dynamic power curtailment, a minimum power is set for all power sources, and the total energy for all power sources is kept constant.

The first scenario in this subsection involves the relaxed and restricted dynamic curtailment for one power source type (solar PV). The next scenario investigates the optimization with recourse response of different renewable energy curtailments.

For the dynamic curtailment for the solar PV power source, Fig. 8.5 presents the varying allowed power levels and the BSS strategies. BSS power transfers represent the mean net grid action of charging and discharging at each hour of the day. Compared to the relaxed power source in Fig. 8.5a, the optimal operation in Fig. 8.5b needed more batteries to maintain a similar QoS. As shown in Fig. 8.5c, the initial number of batteries is lower, but the QoS is sacrificed from 98.56% to 59.20%, leading to a low profit. The two strategies adopted by the BSS are like the threshold curtailment strategies described in Sect. 5.2.

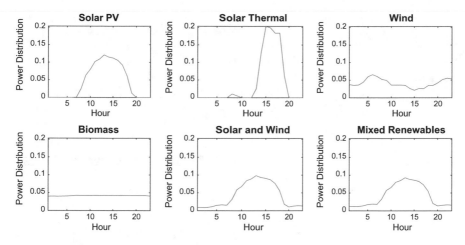

Fig. 8.4 Normalized renewable energy sources for power curtailments

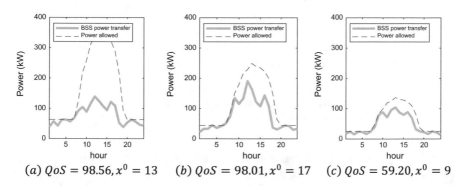

$(a)\ QoS = 98.56, x^0 = 13$ $(b)\ QoS = 98.01, x^0 = 17$ $(c)\ QoS = 59.20, x^0 = 9$

Fig. 8.5 BSS power transfer for different levels of the solar PV power source

As shown in Fig. 8.5b, c, the net power transfer closely adapted and resembled the shape of the power curtailment for solar PV. The strategy also showed improvement areas in the early morning and late evening. At noon, due to the optimal scheduling, the power demand is at its peak, but there is still excess unused power.

The improvement areas can also be used by the power distribution company operators to choose the threshold level. For example, the battery swapping station does not need all the power resources at midday given the Level 2 charger specifications, but it would need more power other than solar PV for activities early in the morning and late in the evening.

For the next scenario, different renewable energy sources were used to model the energy curtailments. Figure 8.6 presents the different renewable energy curtailments and their responses to the recommended battery investment and QoS.

Although wind and biomass are promising power sources, the system needs to assume a high penetration of wind or biomass energy. Wind and biomass, however,

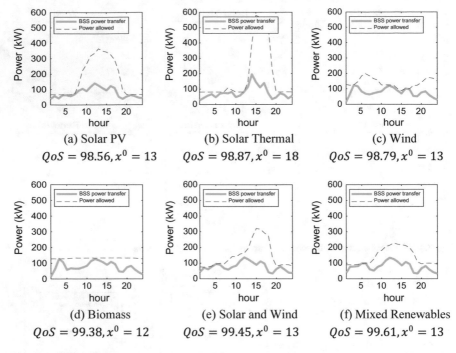

(a) Solar PV
$QoS = 98.56, x^0 = 13$

(b) Solar Thermal
$QoS = 98.87, x^0 = 18$

(c) Wind
$QoS = 98.79, x^0 = 13$

(d) Biomass
$QoS = 99.38, x^0 = 12$

(e) Solar and Wind
$QoS = 99.45, x^0 = 13$

(f) Mixed Renewables
$QoS = 99.61, x^0 = 13$

Fig. 8.6 BSS power transfer given optimal battery allocations for different renewable energy sources with (**d**) biomass, (**e**) combination of solar and wind energy and (**f**) mixed renewable energy having the highest QoS with minimal batteries needed

are not always available in all regions. If a mix of renewable energy is used, as in the case of the CAISO distribution, the initial battery investment will be less compared to a purely solar source. A diversified renewable energy source may be a key point to reduce the initial battery investment needed without sacrificing the QoS to EV owners.

8.6 Bilevel Optimization Formulation

To further examine the interactions between the battery swapping station owners and the power distribution company operators, specifically on the impact of power limits and renewable energy sources, a bilevel optimization has been proposed. In this formulation, the BSS upper-level decisions are constrained by the power distribution company lower-level decisions. Aside from using the parameters listed in Table 8.2, the bilevel optimization has additional parameters presented in Table 8.5.

For the upper-level optimization, the objective is the same as the optimization with recourse formulation denoted in (8.6), and the constraints remain almost the same.

Table 8.5 Additional nomenclature for the bilevel formulation

Symbol	Meaning
Sets	
G	Set of predecessor bus nodes with index g
I	Set of power sources with index i
J	Set of successor bus nodes with index j
K	Set of constrained transmission lines with index k
Parameters	
$A_{i,t}$	Marginal cost for the power source i at time t
$\tilde{A}_t$	Marginal cost for the flexible load at time t
$B_{g,j}$	Element in the susceptance coefficient matrix from node g to node j
C^α, C^β	Linear flexible load coefficients
$H_{j,k}$	Element in the branch flow matrix from the diagonal and node-arc incidence matrix associated with node j and line k
$M_i^{P-}, M_i^{P+}, M^{y-}, M^{y+},$ $M_j^{\theta-}, M_j^{\theta+}, M_k^{H-}, M_k^{H+}$	Large positive values used in the Big-M reformulation associated with $\mu_{i,t,\omega}^{P-}, \mu_{i,t,\omega}^{P+}, \mu_{t,\omega}^{y-}, \mu_{t,\omega}^{y+},$ $\mu_{j,t,\omega}^{\theta-}, \mu_{j,t,\omega}^{\theta+}, \mu_{k,t,\omega}^{H-}, \mu_{k,t,\omega}^{H+}$
$\hat{P}_{i,t}$	Maximum amount of power allowed from source i at time t
$\hat{P}_k^H$	Maximum transfer capacity for line k
$P_{g,t}^L$	Power needed by the load at node g at time t
$U_{g,i}$	Element in the binary matrix for the connected power source i at node g
$\tilde{U}_g$	Element in the binary vector for the connected flexible load at node g
ϕ_t	Guaranteed power from the distributing company without elastic load
$\hat{\theta}_j$	Maximum bus angle difference for node j
ξ	Maximum flexible load percentage based on the total load
Decision Variables	
$b_{i,t,\omega}^{P-}, b_{i,t,\omega}^{P+}, b_{t,\omega}^{y-}, b_{t,\omega}^{y+},$ $b_{j,t,\omega}^{\theta-}, b_{j,t,\omega}^{\theta+}, b_{k,t,\omega}^{H-}, b_{k,t,\omega}^{H+}$	Binary variables used in the Big-M reformulation associated with $\mu_{i,t,\omega}^{P-}, \mu_{i,t,\omega}^{P+}, \mu_{t,\omega}^{y-}, \mu_{t,\omega}^{y+}, \mu_{j,t,\omega}^{\theta-},$ $\mu_{j,t,\omega}^{\theta+}, \mu_{k,t,\omega}^{H-}, \mu_{k,t,\omega}^{H+}$
$P_{i,t,\omega}$	Amount of power needed from each power source i at time t for scenario ω

(continued)

Table 8.5 (continued)

Symbol	Meaning
$y_{t,\omega}$	Amount of flexible loaded needed at time t for a scenario ω
$\theta_{j,t,\omega}$	Change in bus angle in node j at time t for scenario ω
$\lambda_{t,\omega}^{P}$	Lagrange multiplier for the energy balance at time t for scenario ω
$\lambda_{t,\omega}^{B}$	Lagrange multiplier for the economic dispatch at node g for time t for scenario ω
$\mu_{i,t,\omega}^{P-}, \mu_{i,t,\omega}^{P+}$	KKT multipliers for generator boundaries of source i at time t for scenario ω
$\mu_{t,\omega}^{y-}, \mu_{t,\omega}^{y+}$	KKT multipliers for flexible load boundaries at time t for scenario ω
$\mu_{j,t,\omega}^{\theta-}, \mu_{j,t,\omega}^{\theta+}$	KKT multipliers for bus angle boundaries at time t for scenario ω
$\mu_{k,t,\omega}^{H-}, \mu_{k,t,\omega}^{H+}$	KKT multipliers for transmission capacity boundaries at time t for scenario ω

Constraints (8.4)–(8.10) still apply, but (8.11) is modified to reflect the energy curtailment decisions by the power distribution company operators presented in (8.13). ϕ_t signifies the minimum guaranteed power that the distribution company operators will always provide the BSS in every time interval, and $y_{t,\omega}$ represents the BSS flexible load. $y_{t,\omega}$ presented in (8.13) is part of the upper-level constraints, but later in this formulation, $y_{t,\omega}$ is also a lower-level optimization decision made by the power distribution company operators.

$$C^\delta \sum_{f=0}^{\delta-1} x_{t-f,\omega}^c - C^\zeta \sum_{f=0}^{\zeta-1} x_{t-f,\omega}^d \leq \phi_t + y_{t,\omega} \quad \forall t \in T, \omega \in \Omega \quad (8.13)$$

For the lower-level optimization, the objective of the power distribution company operators is to distribute the energy and meet the load demand at the lowest cost, as shown in (8.14). $P_{i,t,\omega}$ represents the amount of power from each power source i of the set I with associated marginal costs $A_{i,t}$ for every power source i at every time t. The flexible load $y_{t,\omega}$ is also included in the optimization with an associated cost $\tilde{A}_t$
.

$$\min_{y_{t,\omega}, P_{i,t,\omega}, \theta_{j,t,\omega}} \sum_{\omega \in \Omega} \sum_{t \in T} \left(\sum_{i \in I} A_{i,t} P_{i,t,\omega} + \tilde{A}_t y_{t,\omega} \right) \quad (8.14)$$

The flexible load marginal cost $\tilde{A}_t$ is computed in (8.15), where C^α and C^β are used to provide linear dependence to the current total load $\sum_{g \in G} P_{g,t}^L$, where $P_{g,t}^L$ is the load at each node g of the set G at time t. Power distribution company operators would prefer that the flexible load $y_{t,\omega}$ be reduced to a low value during the peak load demand so that the company can direct its power use to its main load. During high-load demand, $\tilde{A}_t$ will have a high positive value. Similarly, during low-load demand, $\tilde{A}_t$ will have a high negative value.

$$\tilde{A}_t = C^\alpha \sum_{g \in G} P^L_{g,t} + C^\beta \quad \forall t \in T, \omega \in \Omega \tag{8.15}$$

If coordinated properly, instead of having to build a new infrastructure to support some of the peak loads, the power distribution company can use a managed flexible load from the battery swapping station.

The constraints for the lower-level optimization are presented from (8.16) to (8.21). The generated power should match the overall load representing the main energy balance, as shown in (8.16). Similarly, each node should also satisfy the energy balance, as shown in (8.17), where $U_{g,i}$ and $\tilde{U}_g$ are elements in the binary matrices expressing the presence of power sources and flexible load connected to the node g of the set G. The bus angle difference $\theta_{j,t,\omega}$ and the coefficient matrix from load susceptance $B_{g,j}$ from node g of the set G to node j of the set J are included in the nodal energy balance.

The generator boundaries in (8.18) are considered with $\hat{P}_{i,t}$ representing the maximum allowed generation. The flexible load boundary is also included in (8.19), where the maximum flexible load is based on the percentage ξ of the maximum load $\sum_{g \in G} P^L_{g,t}$. Similarly, the bus angle differences also have limits $\hat{\theta}_j$, as shown in (8.20). The transmission capacity $H_{j,k}$ is also limited by the maximum transmission power capacity limit $\hat{P}^H_k$ for each transmission line k of the set K, as shown in (8.21).

$$\sum_{i \in I} P_{i,t,\omega} = \sum_{g \in G} P^L_{g,t} - y_{t,\omega} \quad \forall t \in T, \omega \in \Omega \tag{8.16}$$

$$\sum_{i \in I} U_{g,i} P_{i,t,\omega} + \tilde{U}_g y_{t,\omega} - P^L_{g,t} = \sum_{j \in J} B_{g,j} \theta_{j,t,\omega} \quad \forall g \in G, t \in T, \omega \in \Omega \tag{8.17}$$

$$0 \le P_{i,t,\omega} \le \hat{P}_{i,t} \quad \forall i \in I, t \in T, \omega \in \Omega \tag{8.18}$$

$$0 \le y_{t,\omega} \le \xi \sum_{g \in G} P^L_{g,t} \quad \forall t \in T, \omega \in \Omega \tag{8.19}$$

$$\left| \theta_{j,t,\omega} \right| \le \hat{\theta}_j \quad \forall j \in J, t \in T, \omega \in \Omega \tag{8.20}$$

$$\sum_{j \in J} \left| H_{j,k} \theta_{j,t,\omega} \right| \le \hat{P}^H_k \quad \forall k \in K, t \in T, \omega \in \Omega \tag{8.21}$$

To solve the convex bilevel optimization, the lower-level optimization used the KKT optimality conditions of stationarity, primal feasibility, dual feasibility and complementary slackness to form a single-level optimization.

In defining the stationarity, each lower-level constraint will have dual variables. The variables presented in Table 8.6 are the multipliers in the KKT conditions.

Forming the Lagrange function using the multipliers, the first KKT optimality condition of stationarity can be represented from (8.22) to (8.24). These formulations are based on the partial derivatives of the Lagrange function in terms of the power source (8.22), flexible load (8.23) and bus angle differences (8.24). The optimal condition is achieved in a convex problem formulation when the formulation has no more feasible descent that will give the optimization an improved value.

$$\frac{\partial \mathcal{L}}{\partial P_{i,t,\omega}} : A_{i,t} - \lambda_{t,\omega}^{P} - \mu_{i,t,\omega}^{P-} + \mu_{i,t,\omega}^{P+} - \sum_{g \in G} U_{g,i} \lambda_{g,t,\omega}^{B} = 0 \quad \forall i \in I, t \in T, \omega \in \Omega$$

$$(8.22)$$

$$\frac{\partial \mathcal{L}}{\partial y_{t,\omega}} : \tilde{A}_{t} - \lambda_{t,\omega}^{P} - \mu_{t,\omega}^{y-} + \mu_{t,\omega}^{y+} - \sum_{g \in G} \tilde{U}_{g} \lambda_{g,t,\omega}^{B} = 0 \quad \forall t \in T, \omega \in \Omega \quad (8.23)$$

$$\frac{\partial \mathcal{L}}{\partial \theta_{j,t,\omega}} : \sum_{g \in G} \lambda_{g,t,\omega}^{B} B_{g,j} - \sum_{k \in K} \mu_{k,t,\omega}^{H-} H_{j,k} + \sum_{k \in K} \mu_{k,t,\omega}^{H+} H_{j,k} = 0 \quad \forall j \in J, t \in T, \omega \in \Omega \quad (8.24)$$

The primal feasibility is represented by the lower-level constraints from (8.16) to (8.21). In this KKT condition, the lower-level constraints should still hold true in the single-level equivalent.

For the dual feasibility to be satisfied, the Lagrange and KKT multipliers are non-negative integers as presented from (8.25) to (8.29).

$$\lambda_{t,\omega}^{P}, \mu_{t,\omega}^{y-}, \mu_{t,\omega}^{y+} \geq 0 \lambda_{t,\omega}^{P}, \mu_{t,\omega}^{y-}, \mu_{t,\omega}^{y+} \geq 0 \quad \begin{matrix} \forall t \in T, \omega \in \Omega \\ \lambda_{t,\omega}^{P}, \mu_{t,\omega}^{y-}, \mu_{t,\omega}^{y+} \in \mathbb{Z} \end{matrix} \quad (8.25)$$

$$\lambda_{g,t,\omega}^{B} \geq 0 \quad \begin{matrix} \forall g \in G, t \in T, \omega \in \Omega \\ \lambda_{g,t,\omega}^{B} \in \mathbb{Z} \end{matrix} \quad (8.26)$$

$$\mu_{i,t,\omega}^{P-}, \mu_{i,t,\omega}^{P+} \geq 0 \quad \begin{matrix} \forall i \in I, t \in T, \omega \in \Omega \\ \mu_{i,t,\omega}^{P-}, \mu_{i,t,\omega}^{P+} \in \mathbb{Z} \end{matrix} \quad (8.27)$$

Table 8.6 Lagrange (λ) and KKT multipliers (μ) for lower-level constraints

Equation description	Reference equation	Multipliers
Main energy balance	(8.16)	$\lambda_{t,\omega}^{P}$
Economic dispatch	(8.17)	$\lambda_{g,t,\omega}^{B}$
Generator boundaries	(8.18)	$\mu_{i,t,\omega}^{P-}, \mu_{i,t,\omega}^{P+}$
Flexible load boundaries	(8.19)	$\mu_{t,\omega}^{y-}, \mu_{t,\omega}^{y+}$
Bus angle boundaries	(8.20)	$\mu_{j,t,\omega}^{\theta-}, \mu_{j,t,\omega}^{\theta+}$
Transmission capacity boundaries	(8.21)	$\mu_{k,t,\omega}^{H-}, \mu_{k,t,\omega}^{H+}$

$$\mu_{k,t,\omega}^{H-}, \mu_{k,t,\omega}^{H+} \geq 0 \quad \begin{array}{l} \forall k \in K, t \in T, \omega \in \Omega \\ \mu_{k,t,\omega}^{H-}, \mu_{k,t,\omega}^{H+} \in \mathbb{Z} \end{array} \tag{8.28}$$

$$\mu_{j,t,\omega}^{\theta-}, \mu_{j,t,\omega}^{\theta+} \geq 0 \quad \begin{array}{l} \forall j \in J, t \in T, \omega \in \Omega \\ \mu_{j,t,\omega}^{\theta-}, \mu_{j,t,\omega}^{\theta+} \in \mathbb{Z} \end{array} \tag{8.29}$$

The complementary slackness equations are presented from (8.30) to (8.37). All four KKT conditions have now been presented. However, the conditions for the complementary slackness are non-linear.

$$\mu_{i,t,\omega}^{P-}(P_{i,t,\omega}) = 0 \quad \forall i \in I, t \in T, \omega \in \Omega \tag{8.30}$$

$$\mu_{i,t,\omega}^{P+}(\hat{P}_{i,t} - P_{i,t,\omega}) = 0 \quad \forall i \in I, t \in T, \omega \in \Omega \tag{8.31}$$

$$\mu_{t,\omega}^{y-}(y_{t,\omega}) = 0 \quad \forall t \in T, \omega \in \Omega \tag{8.32}$$

$$\mu_{t,\omega}^{y+}\left(\xi \sum_{g \in G} P_{g,t}^{L} - y_{t,\omega}\right) = 0 \quad \forall t \in T, \omega \in \Omega \tag{8.33}$$

$$\mu_{j,t,\omega}^{\theta-}(-\hat{\theta}_j + \theta_{j,t,\omega}) = 0 \quad \forall j \in J, t \in T, \omega \in \Omega \tag{8.34}$$

$$\mu_{j,t,\omega}^{\theta+}(\hat{\theta}_j - \theta_{j,t,\omega}) = 0 \quad \forall j \in J, t \in T, \omega \in \Omega \tag{8.35}$$

$$\mu_{k,t,\omega}^{H-}\left(-\hat{P}_k^H + \sum_{j \in J} H_{j,k}\theta_{j,t,\omega}\right) = 0 \quad \forall k \in K, t \in T, \omega \in \Omega \tag{8.36}$$

$$\mu_{k,t,\omega}^{H+}\left(\hat{P}_k^H - \sum_{j \in J} H_{j,k}\theta_{j,t,\omega}\right) = 0 \quad \forall k \in K, t \in T, \omega \in \Omega \tag{8.37}$$

To convert the complementary slackness conditions into equivalent linear forms, the two-factor property can be used in the Big-M reformulation [43]. For the complementary slackness condition to be satisfied, at least one of the two factors should be set to zero. In the reformulation, the non-linear constraints are converted into a set of constraints with the same feasible set but with additional variables and constants. These additional binary variables and large boundary constants are introduced for each KKT multiplier, as presented in Table 8.7.

The equivalent reformulations are shown in (8.38)–(8.53), where each equation in the complementary slackness conditions is replaced by two inequality equations.

$$\mu_{i,t,\omega}^{P-} \leq M_i^{P-} b_{i,t,\omega}^{P-} \quad \forall i \in I, t \in T, \omega \in \Omega \tag{8.38}$$

Table 8.7 Big-M additional
variables and constants for
each KKT multiplier

KKT multiplier	Binary variable	Large boundary constant
$\mu_{i,t,\omega}^{P-}$	$b_{i,t,\omega}^{P-}$	M_i^{P-}
$\mu_{i,t,\omega}^{P+}$	$b_{i,t,\omega}^{P+}$	M_i^{P+}
$\mu_{t,\omega}^{y-}$	$b_{t,\omega}^{y-}$	M^{y-}
$\mu_{t,\omega}^{y+}$	$b_{t,\omega}^{y+}$	M^{y+}
$\mu_{j,t,\omega}^{\theta-}$	$b_{j,t,\omega}^{\theta-}$	$M_j^{\theta-}$
$\mu_{j,t,\omega}^{\theta+}$	$b_{j,t,\omega}^{\theta+}$	$M_j^{\theta+}$
$\mu_{k,t,\omega}^{H-}$	$b_{k,t,\omega}^{H-}$	M_k^{H-}
$\mu_{k,t,\omega}^{H+}$	$b_{k,t,\omega}^{H+}$	M_k^{H+}

$$-P_{i,t,\omega} \leq M_i^{P-}\left(1 - b_{i,t,\omega}^{P-}\right) \quad \forall i \in I, t \in T, \omega \in \Omega \tag{8.39}$$

$$\mu_{i,t,\omega}^{P+} \leq M_i^{P+}b_{i,t,\omega}^{P+} \quad \forall i \in I, t \in T, \omega \in \Omega \tag{8.40}$$

$$-\hat{P}_{i,t} + P_{i,t,\omega} \leq M_i^{P+}\left(1 - b_{i,t,\omega}^{P+}\right) \quad \forall i \in I, t \in T, \omega \in \Omega \tag{8.41}$$

$$\mu_{t,\omega}^{y-} \leq M^{y-}b_{t,\omega}^{y-} \quad \forall t \in T, \omega \in \Omega \tag{8.42}$$

$$-y_{t,\omega} \leq M^{y-}\left(1 - b_{t,\omega}^{y-}\right) \quad \forall t \in T, \omega \in \Omega \tag{8.43}$$

$$\mu_{t,\omega}^{y+} \leq M^{y+}b_{t,\omega}^{y+} \quad \forall t \in T, \omega \in \Omega \tag{8.44}$$

$$-\xi \sum_{g \in G} P_{g,t}^L + y_{t,\omega} \leq M^{y+}\left(1 - b_{t,\omega}^{y+}\right) \quad \forall t \in T, \omega \in \Omega \tag{8.45}$$

$$\mu_{j,t,\omega}^{\theta-} \leq M_j^{\theta-}b_{j,t,\omega}^{\theta-} \quad \forall j \in J, t \in T, \omega \in \Omega \tag{8.46}$$

$$-\hat{\theta}_j + \theta_{j,t,\omega} \leq M_j^{\theta-}\left(1 - b_{j,t,\omega}^{\theta-}\right) \quad \forall j \in J, t \in T, \omega \in \Omega \tag{8.47}$$

$$\mu_{j,t,\omega}^{\theta+} \leq M_j^{\theta+}b_{j,t,\omega}^{\theta+} \quad \forall j \in J, t \in T, \omega \in \Omega \tag{8.48}$$

$$\theta_{j,t,\omega} - \hat{\theta}_j \leq M_j^{\theta+}\left(1 - b_{j,t,\omega}^{\theta+}\right) \quad \forall j \in J, t \in T, \omega \in \Omega \tag{8.49}$$

$$\mu_{k,t,\omega}^{H-} \leq M_k^{H-}b_{k,t,\omega}^{H-} \quad \forall k \in K, t \in T, \omega \in \Omega \tag{8.50}$$

$$-\hat{P}_k^H + \sum_{j \in J} H_{j,k}\theta_{j,t,\omega} \leq M_k^{H-}(1 - b_{k,t,\omega}^{H-}) \quad \forall k \in K, t \in T, \omega \in \Omega \tag{8.51}$$

$$\mu_{k,t,\omega}^{H+} \leq M_k^{H+}b_{k,t,\omega}^{H+} \quad \forall k \in K, t \in T, \omega \in \Omega \tag{8.52}$$

$$\hat{P}_k^H - \sum_{j \in J} H_{j,k}\theta_{j,t,\omega} \leq M_k^{H+}\left(1 - b_{k,t,\omega}^{H+}\right) \quad \forall k \in K, t \in \mathrm{T}, \omega \in \Omega \qquad (8.53)$$

The complete reformulated bilevel optimization problem then has an objective given by (8.3) and has the following constraints: (8.4)–(8.10), (8.13), (8.16)–(8.29), and (8.38)–(8.53).

8.7 Bilevel Optimization Scenario Analysis

8.7.1 Test System

To showcase a bilevel optimization application, the modified IEEE 14-bus test system presented in Fig. 8.7 is used in the scenarios. This bus test system has four different power sources $P_{i,t,\omega} = \left\{ P_{t,\omega}^{\mathrm{solar}}, P_{t,\omega}^{\mathrm{wind}}, P_{t,\omega}^{\mathrm{gas}}, P_{t,\omega}^{\mathrm{coal}} \right\}$ and a flexible load $y_{t,\omega}$ contributed by the battery swapping station. The coal source is connected to bus 1, the solar PV source to bus 2, the wind source to bus 3, and the gas source to bus 6. Furthermore, the flexible load is connected to bus 8. Bus 1 is used as the reference bus, and the proportional load values and the susceptance matrix are based on the Manitoba IEEE 14-bus model [44].

Gas prices were proportionally based on the changing gas prices from AEMO [34], as illustrated in Fig. 8.8.

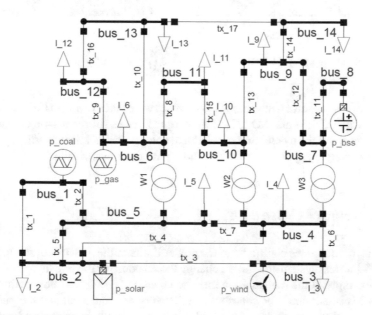

Fig. 8.7 Modified IEEE 14-bus test system

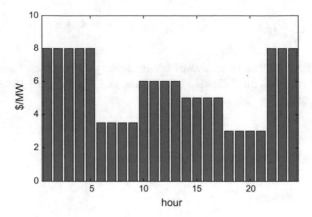

Fig. 8.8 Gas marginal price

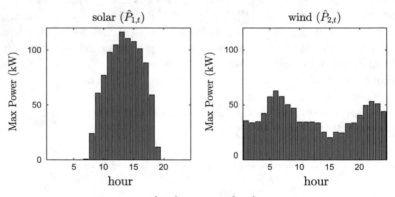

Fig. 8.9 Sample max power for solar $\left(\hat{P}_{1,t}\right)$ and wind $\left(\hat{P}_{2,t}\right)$

For the power sources, the proportional values were taken from the ISO Renewable Energy Watch California ISO [37]. The marginal prices for renewable energies are close to zero, but the availability of wind and solar power is intermittent. Figure 8.9 presents a sample day for the solar PV and wind power availability.

8.7.2 Network Constraints

The transmission lines contribute to the power limits in the network of the modified IEEE 14-bus test system and further change the response of the bilevel optimization.

For example, if the transmission line shown as tx_1 in Fig. 8.7 connecting bus 1 and bus 2 is constrained, the contribution of each power source changes. Figure 8.10 presents the responses of the system with a relaxed, slightly restricted and restricted

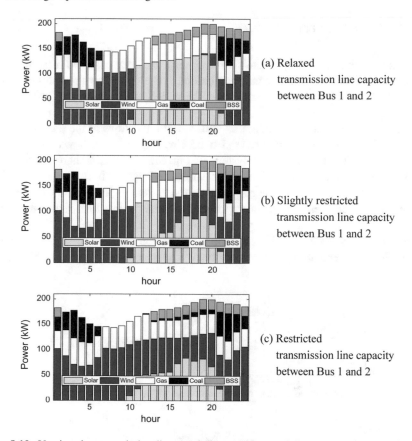

Fig. 8.10 Varying the transmission line restriction could potentially generate the same flexible allotted power but different power source type responses

capacity for transmission line tx_1. The supply of solar power is restrained as the value of the transmission line capacity is restricted.

The bilevel optimization still has a feasible solution and similar flexible load response $y_{t,\omega}$ because the wind power source can produce more power to compensate for the reduced solar power source.

Variations also happen for the coal and gas power sources, as observed in Fig. 8.10b, c. More of the coal power source is used as the restriction increases. The next section discusses what-if scenarios when the power source locations are interchanged.

8.7.3 Power Source Changes

After the power limit in the transmission capacity has been explored, this time, the positioning of the power source is investigated to determine how this influences the upper-level decisions of the battery swapping station by changing the flexible load $y_{t,\omega}$.

Figure 8.11 presents three diagrams with varying locations of the power sources where the transmission capacity of tx_1 is 65 kW. Figure 8.11a shows the response using the modified IEEE bus test system in Fig. 8.7. After interchanging the coal and solar power, the response changes, as shown in Fig. 8.11b. If the gas and wind power are also interchanged, the response of the system is presented in Fig. 8.11c. Aside from the different power source allocations, the flexible load response for the three systems also changes.

In the bus system in Fig. 8.11a, if the transmission capacity of tx_1 is reduced to 40 kW, the bilevel optimization may have a varying power source response, as

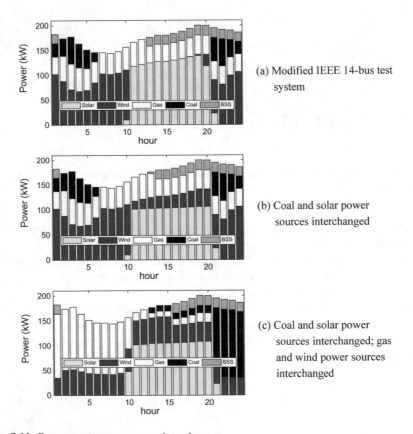

(a) Modified IEEE 14-bus test system

(b) Coal and solar power sources interchanged

(c) Coal and solar power sources interchanged; gas and wind power sources interchanged

Fig. 8.11 Responses to power source interchanges

demonstrated in Sect. 7.2, but still produce a feasible solution. If the same transmission capacity of 40 kW is applied when the solar power and coal power have been interchanged, the bilevel optimization does not produce any feasible solution. The placement of the power source types coupled with the transmission line constraints should be included in the planning and operations between the BSS and the power distribution company.

8.8 Conclusion

This work has shown how two-stage optimization techniques can be used as a step towards the integration of electric vehicles in the energy internet. Specifically, this integration is presented through alternative business models such as battery swapping stations that aggregate and allocate electric vehicle batteries through optimization with recourse and bilevel optimization.

The power limit strategies performed with the optimization techniques involve stakeholders such as electric vehicle owners through station visit uncertainties and the power distribution company operators through threshold curtailments and allowed flexible load decisions. Two forms of two-stage optimization have been presented in the planning and operations of battery swapping stations.

In optimization with recourse, BSS owners can decide on the battery purchase investment in the planning stage and the optimal allocation of charging, discharging and swapping in the operation stage.

The interactions between battery swapping station owners and power distribution company operators through energy curtailments play a role in planning and operational decisions. A restricted threshold curtailment either increases the amount of battery investment and performs more battery-to-battery services or potentially lowers the optimal quality of service and misses more battery swaps to EV customers. A dynamic curtailment depending on the power source was also investigated. A mixture of renewable energies such as the combination of wind and solar power sources can better respond to BSS demands in terms of the quality of service and initial battery investment compared to a single source power such as a solar PV power source.

The initial assessment through optimization with recourse is needed, especially in the long run, when the integration of electric vehicles into the energy internet is desired. To further examine the interaction of the battery swapping station owners with the power distribution company operators in terms of the power limits, the next optimization technique is also necessary.

In the bilevel optimization, the BSS owners perform upper-level optimization decisions that are constrained by the allowed flexible load. This flexible load is a lower-level optimization decision made by the power distribution company operators. Battery swapping stations use the aggregated batteries as a form of flexible load to manage electric vehicle station visits and at the same time manage the peak loads from the overall bus system load.

In the scenario analysis, the transmission line and power source locations must be considered in optimizing the appropriate allowed flexible load. The BSS optimization decisions then meet the demand for electric vehicle batteries while also helping the power distribution company in managing the peak loads in the bus system.

This work provides an alternative way of integrating electric vehicles in the energy internet by providing a business model such as a battery swapping station with two-stage optimization strategies that consider the electric vehicle station visit uncertainties and the decisions of power distribution company operators.

Acknowledgments This work has been supported in part by the University of Sydney FEIT Mid-Career Research Development Scheme.

References

1. J. Conti, P. Holtberg, J. Diefenderfer, A. LaRose, J.T. Turnure, L. Westfall, *International Energy Outlook 2016 With Projections to 2040. USDOE Energy Information Administration (EIA)* (Washington, DC, 2016)
2. W. Su, J. Wang, J. Roh, Stochastic energy scheduling in microgrids with intermittent renewable energy resources. IEEE Trans. Smart Grid **5**, 1876–1883 (2014). https://doi.org/10.1109/TSG.2013.2280645
3. P. Denholm, M. O'Connell, G. Brinkman, J. Jorgenson, *Overgeneration from Solar Energy in California: A Field Guide to the Duck Chart* (National Renewable Energy Laboratory, Golden CO, 2015)
4. F.P. Sioshansi, California's 'Duck Curve' arrives well ahead of schedule. Electr. J. **29**, 71–72 (2016). https://doi.org/10.1016/j.tej.2016.07.010
5. Australian Energy Market Operator, South Australian Electricity Report Australian Government (2015)
6. A. Henriot, Economic curtailment of intermittent renewable energy sources. Energy Econ. **49**, 370–379 (2015). https://doi.org/10.1016/j.eneco.2015.03.002
7. A.Y. Saber, G.K. Venayagamoorthy, Plug-in vehicles and renewable energy sources for cost and emission reductions. IEEE Trans. Ind. Electron. **58**, 1229–1238 (2011). https://doi.org/10.1109/TIE.2010.2047828
8. S.S. Raza, I. Janajreh, C. Ghenai, Sustainability index approach as a selection criteria for energy storage system of an intermittent renewable energy source. Appl. Energy **136**, 909–920 (2014). https://doi.org/10.1016/j.apenergy.2014.04.080
9. E.V. Global, *Outlook 2016: Beyond One Million Electric Cars* (International Energy Agency, Paris, France, 2016)
10. K. Feeney, D. Brass, D. Kua, A. Yamamoto, E. Tourneboeuf, D. Adams, *Impact of Electric Vehicles and Natural Gas Vehicles on the Energy Markets* (Australian Energy Market Commission, AECOM, Sydney, Australia, 2012)
11. D.J. MacKay, *Sustainable Energy—without the Hot Air* (2009)
12. W.R. Fuqua, *Cost Benefit Analysis of the Federal Tax Credit for Purchasing an Electric Vehicle* (ProQuest Dissertations Publishing, 2012)
13. Q. Dai, T. Cai, S. Duan, F. Zhao, Stochastic modeling and forecasting of load demand for electric bus battery-swap station. IEEE Trans. Power Deliv. **29**, 1909–1917 (2014). https://doi.org/10.1109/TPWRD.2014.2308990
14. S. Networks, *White Paper: How the Smart Grid Enables Utilities to Integrate Electric Vehicles* (Redwood City, CA, 2013)

15. S. Mohagheghi, B. Parkhideh, S. Bhattacharya, Inductive power transfer for electric vehicles: potential benefits for the distribution grid, in *2012 IEEE International Electric Vehicle Conference* (2012), pp. 1–8. https://doi.org/10.1109/ievc.2012.6183266

16. I.S. Bayram, G. Michailidis, M. Devetsikiotis, F. Granelli, Electric power allocation in a network of fast charging stations. IEEE J. Sel. Areas Commun. **31**, 1235–1246 (2013). https://doi.org/10.1109/JSAC.2013.130707

17. L.P. Zulkarnain, T. Kinnunen, P. Kess, The electric vehicles ecosystem model: construct analysis and identification of key challenges. Manag. Glob. Transit. **12**, 253–277 (2014)

18. W. Infante, J. Ma, A. Liebman, Operational strategy analysis of electric vehicle battery swapping stations. IET Electr. Syst. Transp. **8**, 130–135 (2018). https://doi.org/10.1049/iet-est.2017.0075

19. T.H. Wu, G.K.H. Pang, K.L. Choy, H.Y. Lam, An optimization model for a battery swapping station in Hong Kong, in *2015* (IEEE, 2015), pp. 1–6. https://doi.org/10.1109/itec.2015.7165769

20. S. Yang, J. Yao, T. Kang, X. Zhu, Dynamic operation model of the battery swapping station for EV in electricity market. Energy **65**, 544–549 (2014). https://doi.org/10.1016/j.energy.2013.11.010

21. M.R. Sarker, H. Pandzic, M.A. Ortega-Vazquez, Optimal operation and services scheduling for an electric vehicle battery swapping station. IEEE Trans. Power Syst. **30**, 901–910 (2015). https://doi.org/10.1109/TPWRS.2014.2331560

22. L. Xinyi, L. Nian, H. Yangqi, Z. Jianhua, Z. Nan, Optimal configuration of EV battery swapping station considering service availability, in *2014 International Conference on Intelligent Green Building and Smart Grid (IGBSG)* (2014), pp. 1–5. https://doi.org/10.1109/igbsg.2014.6835217

23. J.R. Birge, F. Louveaux, *Introduction to Stochastic Programming*, vol. Book, Whole (Springer, Berlin, 2011)

24. A. Georghiou, W. Wiesemann, D. Kuhn, *The Decision Rule Approach to Optimisation under Uncertainty: Methodology and Applications in Operations Management* (2011)

25. J.L. Higle, Stochastic programming: optimization when uncertainty matters, in *Emerging Theory, Methods, and Applications. Informs* (2005), pp. 30–53

26. J. Linderoth, A. Shapiro, S. Wright, The empirical behavior of sampling methods for stochastic programming. Ann. Oper. Res. **142**, 215–241 (2006). https://doi.org/10.1007/s10479-006-6169-8

27. A. Sinha, P. Malo, K. Deb, A review on bilevel optimization: from classical to evolutionary approaches and applications. IEEE Trans. Evol. Comput. (2017)

28. G. Eichfelder, Multiobjective Bilevel Optim. Math. Program. **123**, 419–449 (2010). https://doi.org/10.1007/s10107-008-0259-0

29. T. Stoilov, K. Stoilova, V. Stoilova, Bi-level formalization of urban area traffic lights control, in *Innovative Approaches and Solutions in Advanced Intelligent Systems*, ed by S. Margenov, G. Angelova, G. Agre G (Springer International Publishing, Cham, 2016), pp. 303–318. https://doi.org/10.1007/978-3-319-32207-0_20

30. S. Alizadeh, P. Marcotte, G. Savard, Two-stage stochastic bilevel programming over a transportation network. Transp. Res. Part B: Methodol. **58**, 92–105 (2013)

31. G. Bouza Allende, G.J. Still, Solving bilevel programs with the KKT approach. Math. Program. **138**, 309–332 (2013). https://doi.org/10.1007/s10107-012-0535-x

32. S. Dempe, F.M. Kue, Solving discrete linear bilevel optimization problems using the optimal value reformulation. J. Global Optim. **68**, 255 (2016). https://doi.org/10.1007/s10898-016-0478-5

33. S. Dempe, A.B. Zemkoho, On the Karush–Kuhn–Tucker reformulation of the bilevel optimization problem nonlinear analysis: theory. Methods Appl. **75**, 1202–1218 (2012). https://doi.org/10.1016/j.na.2011.05.097

34. Australian Energy Market Operator, NSW Electricity Price and Demand Australian Government: Australian Energy Market Operator NSW Electricity Price and Demand (2017)

35. W. Infante, J. Ma, A. Liebman, Optimal recourse strategy for battery swapping stations considering electric vehicle uncertainty. IEEE Trans. Intell. Transp. Syst. **21**, 1369–1379 (2020). https://doi.org/10.1109/TITS.2019.2905898
36. New South Wales Roads and Maritime Service, Traffic Volume Viewer in Area 19065 Road Traffic Volume 2016. (2017). http://www.rms.nsw.gov.au/about/corporate-publications/statistics/traffic-volumes/aadt-map/index.html#/?z=6
37. California ISO, California ISO Renewable Watch. California ISO. http://www.caiso.com/informed/Pages/CleanGrid/TodaysRenewables.aspx. Accessed 11 Sept. 2017
38. IEEE, *IEEE Std* 2030.1.1–2015: *IEEE Standard Technical Specifications of a DC Quick Charger for Use with Electric Vehicles* (2016)
39. M. Shin, H. Kim, H. Jang, Building an interoperability test system for electric vehicle chargers based on ISO/IEC 15118 and IEC 61850 standards. Appl. Sci. **6**, 165 (2016). https://doi.org/10.3390/app6060165
40. M. Yilmaz, P.T. Krein, Review of battery charger topologies, charging power levels, and infrastructure for plug-in electric and hybrid vehicles. IEEE Trans. Power Electron. **28**, 2151–2169 (2013). https://doi.org/10.1109/TPEL.2012.2212917
41. ClipperCreek Inc. Determining Charging Times (2015). https://www.clippercreek.com/charging-times-chart/
42. X. Jiang, J. Wang, Y. Han, Q. Zhao, Coordination dispatch of electric vehicles charging/discharging and renewable energy resources power in microgrid. Procedia Comput. Sci. **107**, 157–163 (2017). https://doi.org/10.1016/j.procs.2017.03.072
43. M.H. Zar, J.S. Borrero, B. Zeng, O.A. Prokopyev, A note on linearized reformulations for a class of bilevel linear integer problems. Ann. Oper. Res. 1–19 (2017). https://doi.org/10.1007/s10479-017-2694-x
44. H.V.D.C. Manitoba, *Research Centre. IEEE 14 Bus System* (2014)

Part III
Information and Communication for Energy Internet

Chapter 9
Key Data-Driven Technologies in the Energy Internet

Ting Yang, Yuqin Niu, and Haibo Pen

Abstract Monitoring and measurement technology is very important for the energy internet. As a complex network system, there are a large number of state variables that need to be monitored continuously in the Energy Internet. To control the whole network reasonably and efficiently, it is indispensable to acquire these state variables in a timely and accurate manner. Data monitoring and measurement technologies are a set of information technologies, including data acquisition technology, data transmission technology, data analysis technology and data security technology. In this chapter, the above technologies and their applications in the Energy Internet are introduced in detail, which can help readers fully understand the basic role of monitoring and measurement technology in the Energy Internet architecture.

List of Abbreviations

Advanced Metering Infrastructure	AMI
Measurement Data Management System	MDMS
Automated Data Collection System	ADCS
Network Management System	NMS
Security Management System	SMS
Home Area Network	HAN
Remote Terminal Unit	RTU
Radio-Frequency Identification	RFID
Quick Response	QR
Local Area Network	LAN
Phasor Measurement Unit	PMU
Global Positioning System	GPS
Alternating Current	AC
Wide-Area Measurement System	WAMS
Phasor Data Concentrator	PDC

T. Yang (✉) · Y. Niu · H. Pen
School of Electrical and Information Engineering, Tianjin University, Tianjin, China
e-mail: yangting@tju.edu.cn

© Springer Nature Switzerland AG 2020
A. F. Zobaa and J. Cao (eds.), *Energy Internet*,
https://doi.org/10.1007/978-3-030-45453-1_9

241

Power Line Carrier	PLC
Wireless Sensor Networks	WSNs
Low-Rate Wireless Personal Area Network	LR - WPAN
Physical Layer	PHY
Media Access Control	MAC
Internet Engineering Task Force	IETF
Advanced Meter Reading	AMR
Supervisory Control and Data Acquisition	SCADA
Condition Information Acquisition Controller	CAC
Condition Information Acquisition Gateway	CAG
Fuzzy C-Means	FCM
Infrastructure as a Service	IaaS
Platform as a Service	PaaS
Software as a Service	SaaS
Direct Hashing and Pruning	DHP
Artificial Neural Network	ANN
Hopfield Neural Network	HNN
Back-Propagation	BP
Radial Basis Function	RBF
Self-Organizing Map	SOM
Deep Neural Network	DNN
Feed-Forward Deep Network	FFDN
Multi-Layer Perceptron	MLP
Convolutional Neural Network	CNN
Feedback Deep Network	FBDN
Deconvolutional Network	DN
Hierarchical Sparse Coding	HSC
Bidirectional Deep Network	BDDN
Deep Boltzmann Machine	DBM
Deep Belief Network	DBN
Stacked Auto-Encoder	SAE
Restricted Boltzmann Machine	RBM
Research and Development	R&D
Virtual Private Network	VPN
Combined Cooling, Heating and Power	CCHP
Uninterrupted Power Supply	UPS
Battery State of Charge	SoC

9.1 Introduction of Energy Internet Monitoring and Measurement System Architecture

The Energy Internet is usually composed of a power system, natural gas networks, heating networks and information networks. The power system, as the hub of energy conversion, is the core of the Energy Internet. The Energy Internet and a smart grid are highly similar, and the Energy Internet is a further development and deepening of the concept of a smart grid. However, compared with a smart grid, the allocation of energy in the Energy Internet is much more extensive. First, the physical entities of the smart grid are mainly electrical systems. The physical entities of the Energy Internet comprise a power system, natural gas networks, distributed energy systems and energy storage facilities. Second, in the smart grid, energy can only be transferred and used in the form of electrical energy. However, in the Energy Internet, energy can be converted into electricity, chemical energy, thermal energy and other forms. Finally, smart grid research mainly adopts local consumption and control for distributed devices such as distributed generation, energy storage and controllable loads. However, we need to plan energy distribution and flow globally in the Energy Internet.

In the concept of the Energy Internet, advanced metering infrastructure is an important supporting infrastructure. As an interface between energy users and the Energy Internet, advanced metering infrastructure is the basis for the coupling of information and energy flow. On the basis of various smart meters, advanced metering infrastructure realizes wide interconnection and open interactions between users and energy interconnection through two-way communication networks.

An Advanced Metering Infrastructure (AMI) system includes smart meters for measuring various types of energy (such as smart electric meters and smart gas meters), and the communication network and information management system are dedicated to improving the management efficiency of energy enterprises and guiding users to utilize all kinds of energy efficiently. AMI builds a two-way connection between the energy system and the users, enabling the users to support the running of the energy network. Through measurement of users' energy usage data and real-time price feedback, alarms and other relevant information, AMI realizes rational allocation and efficient utilization of resources. However, with the increasing demand for more measurement information by the Energy Internet, AMI systems have gradually shifted to multi-service platforms. These systems will transmit the collected users' energy information to a measurement data management system in real time and use data analysis tools to process and store energy consumption data of energy users to analyse the consumption characteristics and load distributions of users. This section briefly reviews the architecture and technology of the Energy Internet detection and measurement system.

9.1.1 Introduction to Monitoring and Measurement Systems

AMI is a complete network and system for measuring, collecting, storing, analysing and using information about users' energy usage. It mainly includes intelligent energy meters, a communication network, a measurement data management system, etc. With the support of two-way measurement and two-way real-time communication technology, AMI can realize remote meter reading for various energy users, energy remote transmission, energy parameter monitoring and other functions and can support two-way interactions between the energy Internet and users [1].

In recent years, in response to increasing demand measurement management needs, AMI has also incorporated the content of home area networks (HANs), which consist of mass smart meters and network devices that connect them. These smart meters can achieve multiple metering and fixed metering intervals; they also have two-way communication functions and support remote setting, switching and disconnecting; bidirectional measurement; and timing or random measurement reading. Additionally, some smart meters can also serve as gateways to a user's indoor network and customer portal, providing real-time energy prices and information to users. They will control the user's indoor energy consumption device load to achieve the purpose of demand-side management. Because of the multi-metering with a timestamp, the smart meters are actually the system sensors and measuring points distributed in the network. Therefore, AMI can not only provide all kinds of energy enterprises with a communication network and facilities throughout the system but also provide an objective measurement of the system scope, which is the first step towards the Energy Internet. AMI can directly make users participate in a real-time energy market and can bring great benefits to system operation and asset management.

Comprehensive and accurate situational awareness is the basis of efficient management and dispatching of the Energy Internet. Compared with a smart grid, the Energy Internet needs to manage more forms of energy and a wider range of physical devices with finer granularity, higher frequency and higher requirements for plug and play. Therefore, more advanced intelligent sensing technology, advanced measurement sensors, communication technology, sensor network systems, related identification technology and measurement sensor technology standards need to be developed for the Energy Internet.

9.1.2 AMI System Architecture

As the basic function module of the Energy Internet, the implementation of AMI requires the efficient integration of an energy network, a communication network, related information systems and other infrastructure. There are three main functions in AMI:

(1) Energy market application: The wide use of intelligent terminals can reduce or eliminate the costs of personnel, transportation and infrastructure generated by meter reading personnel and equipment maintenance.

(2) Client application: AMI is beneficial for improving the convenience and satisfaction of users' energy consumption; it also supports demand response and load management and improves system stability and operational efficiency.

(3) Energy network operation: Based on the data collected by AMI, by optimizing the energy network structure and implementing distributed energy management, it supports fault location and recovery to reduce the accident handling time and power outage time.

In a smart grid, the data collection objects of an AMI system include dedicated line users, all kinds of large- and medium-sized special transformers and various types of 380/220 V power supply for industrial, commercial and residential users. In the Energy Internet, we need to measure the energy data of distributed energy access and energy storage devices to build a perfect information data platform. Therefore, an AMI system is the source of system status monitoring data and various user data in the Energy Internet.

On the logical level, AMI is composed of an application layer, a network layer and a perception layer. The system architecture is shown in Fig. 9.1. The application layer mainly includes data collection, data storage, data management and business

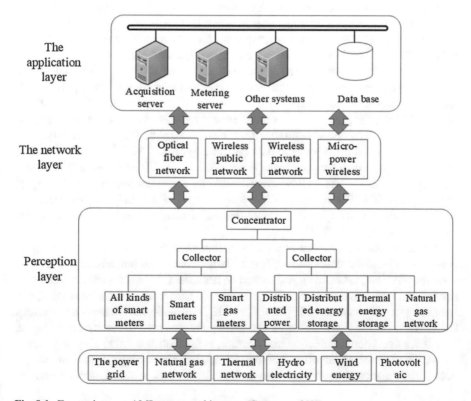

Fig. 9.1 Energy internet AMI system architecture (Courtesy of [2])

applications. The database system is responsible for information storage and processing. The network layer is responsible for data transmission and routing between the perception layer and application layer, providing available communication channels including optical fibre networks, 3G/4G networks, 230 MHz broadband networks, etc. The perception layer is the lowest level of the AMI architecture and is responsible for collecting and providing energy usage information for the entire system [3].

The sensing layer can be further divided into measuring equipment and terminal equipment. The measuring equipment includes various energy meters, which are intelligent terminal equipment used for measuring energy parameters such as electricity, gas and heat. Intelligent terminals, such as concentrators and collectors, are responsible for collecting user metering device information and processing and freezing the related data. The terminal equipment realizes the aggregation of data and interacts with the upper main station. The measurement data management system (MDMS) is the core of the whole application layer and is a database with analytical capability that handles and stores the metering values of smart meters by cooperating with other systems. The MDMS is responsible for validating, editing and estimating AMI data. Even if an underlying interrupt occurs, the accuracy and completeness of the information from the user to the management module can be ensured. For the current AMI system, the data collection interval is 15 min; the data volume can be considered "big data", and the MDMS is also the key application for the analysis and utilization of these "big data" [4].

The physical architecture of the AMI system can be divided into three components: a measurement data management system (MDMS), communication channel and user terminal (smart meters). Among them, the MDMS of the system is a separate network. It is isolated from other application systems and public network channels to ensure the information security of the system. The MDMS consists of a database server, application server, pre-acquisition server and related network equipment. It mainly completes the functions of AMI data acquisition, storage, analysis and network security management. At present, the deployment mode of the main station system may be classified as centralized and distributed. The MDMS is mainly composed of four parts: an automated data collection system (ADCS), a metering data management system, a network management system (NMS) and a security management system (SMS). The ADCS is responsible for the collection and processing of measurement data. The NMS and SMS cooperate to build a multi-service measurement network security system to ensure that the reading table and other businesses can operate safely and reliably.

A communication channel is an information channel connecting an intelligent energy metre and the main station. It is a hierarchical communication system composed of WAN and LAN. In an AMI based on IPv6, the LAN adopts the IPv6 network instead of a traditional private collection network. Currently, international standardization organizations such as IEEE, IEC and IETF provide complete protocol stacks for each layer of the LAN. There are two main points we need to focus on in this area. One is the open communication architecture, which forms a "plug and play" environment and enables networked communication between components in an energy network. The second is the uniform technical standard, which enables seamless communication and interoperability between all smart energy meter, concentrators, sensors, and application systems.

As the basic unit of AMI, the field terminal is the most important link in the system. The main advantage of a smart energy meter is programmable and bidirectional measurement. Energy users can implement energy management based on energy usage and energy conservation principles. In the implementation of AMI, an intelligent energy meter supports instant reading, remote connection and disconnection. By integrating advanced sensing and measurement technology, an intelligent energy meter can also realize fault location and energy quality monitoring.

Conclusion

This section introduces advanced metering infrastructure based on the Energy Internet and each sub-module of the AMI system. The concept of AMI first appeared in the field of smart grids. At the forefront of information collection for energy suppliers, AMI can efficiently measure the load distributions of users and realize the rational allocation of power resources and high efficiency. Therefore, the application of AMI in the Energy Internet can not only achieve more comprehensive and accurate energy situational awareness but also promote users to participate directly in a real-time power market; thus, it can bring very large benefits to the operation of energy systems and greatly facilitate asset management.

9.2 Data Acquisition Technology in the Energy Internet

Sensors and metering and measurement systems provide the necessary data for an energy system and its market operation. In the Energy Internet environment, these systems need to be comprehensively upgraded at all levels. For instance, in a smart grid, deployed sensors may provide the following applications: the detection and processing of power outages, the assessment of equipment health status and power integrity, eliminating errors caused by estimated meter reading and providing stolen electricity protection and support to power users in demand-side management.

In metering and measurement, new digital technology will adopt two-way communication, multiple inputs (e.g. all kinds of energy price information), various types of outputs (real-time energy consumption data, energy quality and parameters), grid interconnection and the ability of controlled islanding, which greatly enhances the application of digital technology in measurement and advances the development of intelligent energy meter technology on the user side. These data collection techniques are described in detail below.

9.2.1   Multi-energy Acquisition Terminal Technology

Energy data collection and monitoring is the basis of Energy Internet operation, the accuracy and completeness of the acquired monitoring results determine the overall performance of the Energy Internet, and all kinds of collection terminal technologies are the basis for the acquisition systems.

The acquisition and monitoring techniques applied to the Energy Internet include sensing, concentration and field control techniques.

(1)   Sensing technology

The sensor function monitors and controls the main equipment, lines and environment in the Energy Internet and collects the state quantity of the equipment generally by an embedded sensor (or sensor network). Sensor networks that can be used in the Energy Internet have been widely studied, including the information model and its application, such as the routing protocol that can help a packet to avoid a congested area and the security problem of the network. Based on the sensor network, a web service for energy management of the Energy Internet on the demand side is also proposed, which can save energy for smart homes. Sensing is the basis of intelligent sensing and intelligent measurement and will be widely used in the Energy Internet [5]. Sensor devices will be developed in the direction of self-organized network communication, high bandwidth utilization and high robustness against environmental fluctuations.

In terms of user power measurement, centralized meter reading terminals include concentrators and collectors, which are used by low-voltage non-residential users and residential users to collect electricity information and manage and monitor abnormal information. In addition to the above equipment, other monitoring devices and systems can be installed, for example, transmission line dynamic capacity expansion, various wire sensors, insulator pollution discharge monitoring, radio frequency identification and electronic transformers. There are also online monitoring systems for circuit breakers, cables, batteries, temperatures, current frequencies, etc. Studying and applying these technologies to smart grids is important and enduring work.

In terms of user gas metering, an intelligent gas network is an integration between the Internet of Things and a gas pipe network by using various sensing devices to collect information about gas pipe network facilities (including location, status, operation status, etc.); it can realize the intelligent management of a gas transportation system by communicating with a database system through wireless networks, wired networks and information management centres [6]. Advanced sensor technology is one of the keys to building intelligent gas networks. It combines intelligent meters with two-way communication, gas stations and transmission pipelines to build an intelligent gas network monitoring and management system, which realizes real-time information extraction, timely judgement, rapid response, intelligent monitoring and prevention. In the construction of an intelligent gas network, sensors for measuring pressure, temperature, humidity, concentration of CO and other parameters should be deployed.

(2)   Data concentration technology

To rationally utilize network spectrum resources and time resources to reduce transmission overhead, the local information concentration is of great significance. The data centralization of different grades also affects the performance of remote monitoring and control systems in the Energy Internet, such as end-to-end communication delay. As the amount of data collected by the Energy Internet is very large, data acquisition is usually combined with noise and loss, so data concentration will be a valuable job. The future focus of data concentration will be improving local information processing performance and efficiency, noise reduction and extraction of essential characteristic data.

(3)   Field Control Technology

Using the communication networks in various energy systems, the related equipment can be controlled automatically. For example, in a smart grid, the main control equipment includes transmitters, protection devices, relay protection and automation equipment. The controlled equipment can receive the network fault location result in time, automatically realize fault isolation and protect related equipment. A smart power management system can achieve fully automatic control of a non-human station, and there is a mutual promotion between the control of the smart system and the Energy Internet. Automatic field control is more important in the Energy Internet, as it needs to guarantee the smooth operation of a network with access to distributed energy. Field control systems will develop in the directions of fully covered information networks, full automation and low communication delay.

9.2.2   Smart Meter Technology

Smart meters are composed of metering and data processing, storage units, communication units and interface units and have operation, display and exchange interfaces. Smart meters are not only smart network sensors and controllers but also two-way energy and information transfer gateways. In addition to the power measurement functions of traditional electricity meters, smart meters also have the following functions according to specific needs.

(1)   Measurement of electrical parameters with a time stamp. The measurement includes voltage, current, reactive and active power, frequency, harmonic and other electrical parameters, which are marked with time. The measurement is the basis of evaluating the power quality and the operating state of the power grid. Power quality can be monitored in real time by smart meters. To respond to customer complaints in a timely and accurate manner, it is necessary to take precautionary measures to prevent power quality problems.

(2)   Form of electricity price. To guide users to avoid peak hours, ladder electricity price and time-of-use electricity price will be executed in the future and will promote the safe operation of smart grids and elevate their utilization rates.

Electricity meters must then support complex pricing systems. An intelligent electric meter can realize accurate and real-time cost settlement information processing, simplifying the complexity of past account processing. In a power market environment, the dispatching personnel will be able to switch suppliers in a timely manner on the basis of electricity price.

(3) Storage. It is possible to record measurement data within a certain time as well as events such as loss of pressure, loss of flow, and power failure.

(4) Communication. The communication function is the most significant feature of smart meters compared with traditional electric meters. It includes sending electrical parameters to the power grid, obtaining the form of price from the grid, and communicating with and controlling household electrical equipment. The communication function can be realized by power line carriers, ethernet, wireless LAN and other technologies.

(5) Illegal electricity detection. A smart meter can detect opening of the box, a change in wiring, and the updating of the meter software, which can be used to detect the phenomenon of stealing in time. In the case of a high-voltage area, by comparing the meter data of the whole area with the data from a single meter, potential theft can be detected.

GE's smart meters, for example, add two technical features to smart meters compared with conventional meters. The first is two-way communication, which means that the power grid can not only collect electricity information but also send power grid information (the true time price) and control orders to the electric energy meter, and the electricity meter receives the information and makes an intelligent response. Two-way communication also includes information transmission and control commands for smart home appliances and other electricity meters. The second feature is standards-based, built-in smart programs. As long as the meter receives information in accordance with the default logic, it will be able to make independent judgement and response without waiting for the main centre to issue instructions again, and it can implement remote modification strategies and upgrade programs for software and maintenance.

9.2.3 Synchronous Phasor Measurement Technology

Based on the timing of a GPS system, a phasor measurement unit (PMU) can monitor, analyse and protect the power system. At present, PMUs have become the basic means of power system monitoring and power state estimation. Since GPS timing can achieve a high time synchronization accuracy ($0.1\ \mu s$), a PMU can measure the current and voltage parameters of substation nodes with phase error less than $0.018°$ at 50 Hz power frequency. The PMU also supports the remote transmission of data to ensure that the control centre can obtain highly real-time monitoring data.

Phasor analysis is an important tool for analysing AC power grids. The change in a phasor can accurately describe the change in the running state of the power system.

By observing the synchronized phasor of a wide-area system, the dynamic process changes at various nodes in the system can be clearly monitored. Synchronous phasor measurement can measure every busbar voltage and line current phasor (amplitude and phase) of the power grid synchronously based on the time synchronization signal provided by a GPS satellite to achieve the goal of measuring power system state variables in real time. PMUs are the basic devices for completing synchronized phasor measurement and can be installed in different locations selected in the power system to accurately measure the voltage and current values of the system synchronously in real time; then, they can transfer the phasor to a remote monitoring centre. Based on the analysis and calculation of the collected synchronous phasor data, various protection and control devices can be used to achieve the goal of ensuring the safe operation of the power grid.

9.2.4 Wide-Area Measurement System

A wide-area measurement system (WAMS) consists of a PMU, a control centre and a communication network. First, the PMU measures the state information of the power grid with high precision; then, this information is transmitted to the data centre through the private network for analysis. The purpose of a WAMS is to ensure that the control centre can obtain a large amount of real-time information related to the steady state of the power system and then make timely and accurate regulation.

Time synchronization is crucial for the measurement of a wide area in space. In addition, the results of synchronized phasor measurement can be obtained according to the GPS timestamp, which improves the observation accuracy of the power system. There are three characteristics of WAMS data: first, synchronization in the time domain; the wide-area interconnection of a power grid brings new problems for the grid, such as transient problems, which cannot be solved by the present state. However, the characteristics of the WAMS for controlling synchronization in the time domain can effectively improve these problems. Second, the wide area in space; the WAMS obtains wide-area power grid data on the basis of time synchronization to enable synchronous monitoring and processing. Third, the phase angle can be directly measured and provides a database for system control. The wide-area measurement system consists of three major parts: the synchronous phasor measurement unit, the communication network covering the whole network and the phasor data concentrator (PDC) installed at the dispatcher. The overall system composition is shown in Fig. 9.2.

In recent years, WAMSs have made great progress and have been widely used in power systems [8].

(1) Online analysis of low-frequency oscillation

With the increasing scale and transmission power of the grid, the low-frequency oscillations between regions have become a problem that cannot be ignored. To identify low-frequency oscillations accurately, a WAMS must be able to send alarm information to the dispatcher when there is a strong weak damping oscillation component

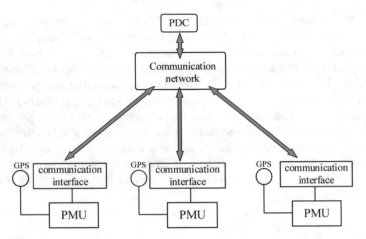

Fig. 9.2 WAMS architecture (Courtesy of [7])

in the 0.2–2.5 Hz range. In addition, the abnormal area is shown on the area map of the power grid, and the data platform is triggered to record the real-time data. Online analysis of low-frequency oscillations requires the WAMS as a real-time data platform to provide unified data, quickly identify which units oscillate in the centre of the data, and finally analyse the phenomenon.

(2)   State Estimation

State estimation combines the synchronous phase measured by the WAMS with the measurement values of SCADA, which greatly improves the state estimation speed and accuracy of the system. When the system is not fully observable, it can be used as the true value of the measured result directly. Alternatively, we can use the WAMS data to increase the redundancy of measurement to improve accuracy. Therefore, the key is determining how to combine the PMU and SCADA data for state estimation.

(3)   Continuous Control of the Power System

Continuous control refers to similar control actions as a protection switch and the control methods of reactance and capacitor switching. Because the output of the controller is not in one-to-one correspondence with the input, the persistence of the input is not strictly required. Therefore, it is easy to obtain measurement information through the WAMS. For example, the wide-range stability and voltage control system developed by BPA in the United States has realized feedback control based on a WAMS. The switching action between the cutter and the reactive compensation device is carried out according to the real-time response of the power system. In the WAMS platform, the input signal of the wide-area controller is no longer limited to the local scope, so it has better observability for the interval low-frequency oscillation mode.

(4) Wide-Area Protection

Wide-area protection relies on multiple points of power system information to remove faults quickly, reliably and accurately. It can also analyse the impact of fault removal on system security and stability and then take corresponding control measures to improve the availability or system reliability of transmission lines. By collecting and analysing the wide-area information and evaluating the state of the system, the power system can be protected against disturbances, and this approach is more effective than that of traditional protection systems. This is because the goal of wide-area protection is to maintain system stability immediately after system failure and the removal of the equipment.

Conclusion

Real-time and efficient data acquisition is extremely important for the Energy Internet. With the rapid development of sensor technology and communication technology, real-time and accurate physical information measurement is becoming possible and is increasingly widely used in power systems. This section introduces the mainstream data terminal acquisition technologies used in the Energy Internet. Including sensor technology, data centralization technology and field control technology, this section mainly focuses on smart meter technology and wide-area measurement systems based on PMU technology, as well as some application scenarios of these technologies in the Energy Internet.

9.3 Data Transfer Technology in the Energy Internet

Information exchange technology is the basis for realizing the intelligentization and interaction of the Energy Internet, mainly including communication and information. Communication technology focuses on processing the transmission of information, including transmission access, network switching, wireless communications, optical communication, private network communications and other technologies. The demand for the communication capability of AMI is closely related to network scale and network complexity. For the Energy Internet, a communication system should meet the following basic needs:

(1) The network communication rate needs to be able to meet the data bandwidth requirements of various basic services in the future, such as the application of remote meter reading and billing systems.
(2) A large number of services in the Energy Internet need the support of bidirectional communication capability. Therefore, for communication systems, both the abilities of data acquisition in the uplink and control information transmission in the downlink are needed.
(3) High coverage. There are many scenarios in Energy Internet applications. Communication networks should fully consider the different needs for each scenario and provide communication coverage as wide as possible.

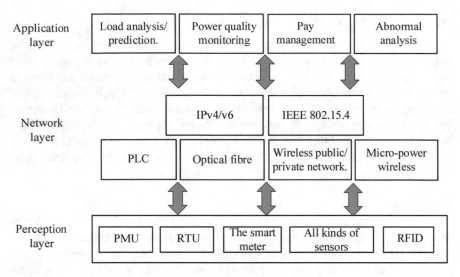

Fig. 9.3 Communication network Structure in the monitoring and measurement system (Courtesy of [9])

The communication network in a monitoring and measurement system adopts a layered system similar to a computer network and is designed to be composed of perception, network and application layers; its structure is shown in Fig. 9.3. Among them, the perception layer includes all kinds of sensor terminals; it also includes the physical layer and data link layer. It uses the transmission medium to establish, manage and release physical connections between hosts on the communication network to realize transparent transmission and ensure the correct sequence of the bitstream through the transmission medium. The function of the network layer is to realize transparent transmission between two communication terminals, including addressing and routing, establishment, maintenance and termination of connection. The application layer mainly includes all kinds of application processes. In the monitoring and measurement system of the Energy Internet, this layer mainly includes data management and business application functions, such as data storage, data validation, load analysis, energy quality monitoring and pay management.

9.3.1 Perception Layer Communication Technology

By using a fixed two-way communication network, AMI can read intelligent energy meter data many times every day and transmit the meter information approximately in real time to the data centre, including fault alarms and device interference alarms. The structure of common communication systems includes a layered system and stellate and mesh networks, and different media can be used to implement wide-area communication to data centres, such as optical fibres, power line carriers (PLCs) and wireless networks. In a hierarchical communication network, LAN connects intelligent terminals, smart meters, various sensors and data concentrators, and the data concentrators are connected to data centres by WAN. In LAN, the main consideration is to connect users at a lower cost. Common communication modes are PLCs and latticed radio-frequency networks. The following section briefly introduces the communication technology in the physical layer of the monitoring and measuring system communication network.

(1) Power line carrier. PLC communication serves as information transmission media by using cables in the existing power distribution lines. Compared with other communication methods requiring re-establishment of communication networks, PLC technology has lower equipment and installation costs and has been widely used in the internal communication facilities of power systems. With the development of electronic and communication technologies, broadband power line carrier technology has developed rapidly, which effectively promotes the communication rate of traditional PLC technology. However, PLC technology is still susceptible to interference, and the transmission distance is relatively short without relay.

(2) Optical fibre communication. Optical fibre communication features high bandwidth, strong anti-jamming capability and good confidentiality, so it has become the preferred communication mode of backbone communication networks in monitoring and measurement systems. However, the optical fibre communication mode has certain limitations. For example, its construction wiring cost is high, so it needs more expensive construction equipment. It is impossible to supply the equipment with power while transmitting the signal. However, with the improvement of optical fibre technology and the decreasing production cost, optical fibre communication will be adopted more widely.

(3) Wireless public network communication. The adoption of wireless public networks avoids the investment of network infrastructure construction. In addition, network maintenance costs will also be significantly reduced. At present, the real-time requirement for data transmission is relatively low; thus, wireless public networks can meet the existing requirements. However, with the expansion of the Energy Internet, the user demand for data transmission rates is becoming much more urgent. Therefore, it may be difficult for present wireless public networks to meet the future data transmission needs of the Energy Internet.

(4) Micro-power wireless technology. Wireless sensor networks (WSNs) have made much progress in recent years. WSNs have low cost and wide detection coverage compared with traditional wireless networks and can be quickly deployed in some areas where wiring is difficult in energy network scenarios; thus, WSNs can greatly expand the coverage of wireless networks. Management nodes are essential for managing WSNs, monitoring tasks, collecting the monitoring data, fusion and collaborative control of tasks. In monitoring and measurement systems, WSNs have been applied in the fields of remote meter reading, substation equipment condition monitoring and power grid asset automation management,

effectively improving the operation and maintenance efficiency and reducing the human cost.

9.3.2   Network Layer Communication Technology

The service of the network layer is to provide connections and path choices between communication nodes in different geographical networks. Choosing the best transmission path for data transmission and ensuring the reliability of the transmission process are the main challenges of the network layer. Compared with a usual computer network, the data collection of the measurement system in the Energy Internet has the following characteristics.

(1) Massive terminals. In the perception layer of the Energy Internet, there are a large number of intelligent terminal devices, including various intelligent meters and control terminals installed for ordinary energy users. In addition, the geographical distribution of these intelligent terminal devices is scattered, which increases the difficulty of establishing a communication network for these terminals.

(2) Short communication distance. In an independent monitoring and measurement system, the distances between intelligent terminals such as concentrators and other data-gathering centres are short. As a result, the monitoring and measurement system generally combines the main channel with a cell branch communication network. The intelligent terminals in the network are usually at a distance of 2–3 km from remote aggregation sites.

With the development of the Energy Internet, it is difficult for traditional communication protocols to properly process various intelligent terminal data and complicated communication data in future monitoring and measurement systems. IPv6 and micro-power wireless communication technology have provided new means to solve these problems [10].

IPv6 Technology

With the development of the Energy Internet, the demand for reliable transmission and efficient data processing has become more urgent. In the foreseeable future, traditional IPv4 communication networks will not provide enough address space for the connection requirements. Due to the limited data processing capability of an intelligent terminal, it is generally required to have a lightweight network protocol, and the self-organizing capability to support node mobility management is also needed. The next-generation IPv6 protocol can solve the above problems, making it possible to build the Internet of Things. IPv6 technology has been put forward as a next-generation IP protocol to solve the problems of IPv4 and is also known as a next-generation Internet protocol; this technology not only greatly extends the address capacity but also enhances the network management and security abilities.

Compared to IPv4, IPv6 has significant improvements in security, service quality assurance and mobile IP. IPv6 technology has the following advantages:

(1) The address volume is greatly expanded, and each intelligent terminal can be assigned an IP address.
(2) Transmission to the end has higher efficiency and faster data transmission speed.
(3) IPv6 has stricter management standards and cooperates with the unique IP address agreement to ensure a smooth network.
(4) IPv6 has higher network security to prevent hackers and virus attacks.

These characteristics of IPv6 enable broad application in the Energy Internet. Some basic functions of the Energy Internet, such as demand-side management, energy price implementation and supply quality assurance, all require data information transmission and IP technology. IPv6 can provide enough addresses. In an energy network intelligent meter reading system, each energy equipment (such as terminals, concentrators, smart meters and collectors) can be assigned an IP address, which simplifies the meter reading network software structure and facilitates maintenance work. In addition, this technology can reduce the parameter settings and traffic because of the realization of point-to-point communication.

The IEEE 802.15.4 Standard

The IEEE 802.15.4 protocol is designed for low power consumption and low data transfer rate applications and has a maximum data rate of 2.5×105 bit/s and a maximum output power of 1 MW. IEEE 802.15.4 mainly defines the physical layer and the media access control (MAC) layer, and other parts of the protocol mainly adopt existing standards that are designed for precise, low-energy and low-cost embedded devices. IEEE 802.15.4 is aimed at short-distance communication networks with low complexity, low power and low data rates. The specification focuses on transceivers with low complexity and low cost, and the goal is to extend the service life of ordinary batteries to a few years. In 2003, IEEE launched the IEEE 802.15.4-2003 standard, which formed the ZigBee protocol and was the foundation of Wireless HART, ISA100a and 6LoWPAN. The IEEE 802.15.4-2006 standard was also officially released in September 2006.

6LoWPAN Agreement

In November 2004, the Internet Engineering Task Force (IETF) established a task group for the 6LoWPAN protocol to study the implementation of IPv6 in IEEE 802.15.4, and it put forward several drafts marking the integration of IPv6 and wireless sensor networks into the standardization process. The physical and MAC layers of IEEE 802.15.4 constitute the basis of 6LoWPAN technology. Unlike ZigBee technology, 6LoWPAN acts as the interlayer between the IPv6 network layer and the MAC layer, which can effectively support the IPv6 protocol stack. A model of the 6LoWPAN protocol stack is shown in Fig. 9.4.

At present, proprietary protocols are generally adopted in energy communication networks. Due to the high memory and bandwidth required by IP protocols, designing devices that adapt to microcontrollers and low-power wireless connections based on

Fig. 9.4 6LoWPAN stack
reference model (Courtesy of
[11])

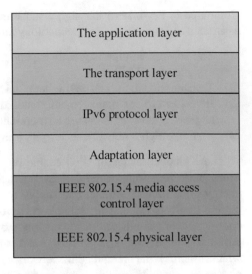

The application layer
The transport layer
IPv6 protocol layer
Adaptation layer
IEEE 802.15.4 media access control layer
IEEE 802.15.4 physical layer

operating environment requirements has become a widespread desire. The low-power operation of 6LoWPAN makes it possible to introduce IP to wireless communication. The 6LoWPAN technology enables seamless connectivity between IPv6 and wireless sensor networks and access to the next generation of Internet via the IPv6 protocol; thus, the objective of "IP Anywhere" can be achieved. Currently, the IETF 6LoWPAN workgroup is planning to add compatibility for IP communication in IEEE 802.15.4, making it a truly open standard. Thus, complex gateway support can be eliminated (only requiring the local IEEE 802.15.4 protocol gateway), and the management process can be simplified.

9.3.3 Application Layer Protocol and Standard

In the Energy Internet, AMI is made up of measurement, communication and data management subsystems, which can provide two-way data and control signal transmission functions between customers and the utility control centre. The application of the AMI concept in a power grid was first embodied in an AMR (advanced meter reading) system, which was used to automatically record power users' electricity bills. In the AMI, it not only improved the data collection ability but also increased the control ability of remote intelligent energy meters from the control centre. The realization of the AMI functions cannot be separated from the support of communication standards and protocols, which define the transmission order, format and

Table 9.1 AMI information models and communication protocols (Courtesy of [12])

Name	Category	Content	Area
ANSI C12.19	Information	Model industrial terminal equipment datasheet	Gas, water, electricity
ANSI C12.22	Communication protocol	Protocol specification for data communication network interface	Gas, water, electricity
IEC 62065-61	Information protocol	Instrument object recognition system	Gas, water, electricity
IEC 62065-61	Information protocol	Data exchange interface	Gas, water, electricity

content of information. In recent years, the industry has put forward a variety of communication protocols applied in AMI. This section briefly introduces these protocols (Table 9.1).

ANSI C12.19

ANSI C12.19 was originally the cooperation achievement of a partnership between power companies and automation instrument manufacturers. At present, ANSI C12.19 comprises two versions: ANSI c12.19-1997 and ANSI c12.19-2008. ANSI c12.19-2008 is designed to be compatible with the recently proposed smart meter concept.

ANSI C12.19 defined a group of standard tables and functions: The tables standardize the method of data storage and instrument data and control, while the functions are the operation methods for calling the data and parameters defined in ANSI C12.19. The tables in C12.19 can be divided into multiple sub-tables, each of which involves a specific functionality set and related functions. In the C12.19 standards, data transmission between intelligent terminals requires read-write operations for specific tables. C12.19 is a general standard. Through a series of standard operations, this information module can be used to serve all areas of the Energy Internet, including the measurement of energy such as electricity, water and natural gas.

ANSI C12.22

ANSI C12.22 defines and standardizes the transmission of smart devices through telephone line modems. This protocol is an extension of the concept of ANSI C12.18 (end-to-end communication protocol), which enables data transmission reliably in communication networks.

The ANSI C12.22 standard has been proposed to customize network infrastructures based on AMI. The purposes of this standard are listed as follows:

(1) Protocol datagram definition: defining ANSI C12.19 data forms, which can communicate through any network, including specific AMI networks and universal networks;
(2) Providing a basic seven-layer infrastructure when C12.22 equipment is connected; and

(3)   Providing interfaces (fibre interfaces and modems) for point-to-point communication using local ports.

A C12.22 network defines a specific AMI network architecture that includes several C12.22 network segments. Similar to the network model of the Internet, C12.22 also includes seven layers: the application layer, presentation layer, session layer, transport layer, network layer, data link layer and physical layer. However, C12.22 is specifically designed for data transmission of smart meters; thus, its application layer only supports measured data encapsulated in ANSI C12.19.

IEC 62056

IEC 62056 defines the instrument interface class that is compatible with the energy metering model and defines relevant standards for smart meter reading, charging and data exchange. Similar to ANSI C12.22, the definition of IEC 62056-53 is based on other series of IEC 62056 protocols, which include IEC 62056-21, 42, 46 and 47. In addition to IEC 62056-21, which is used for data exchange between handheld devices and local instrumentation, the other protocols define different layers of communication networks to support the communication of the application protocols, which include the physical layer (IEC62056-42); the data link layer (IEC 62056-46); and the transport layer (IEC 62056-47).

There are three types of services in IEC 62056-53: access service (request, confirm), setup service (request, confirm) and action service (request, confirm). Although both IEC 62056-53 and ANSI C12.22 can be applied to AMI, they have different targeted markets: IEC 62056 is mainly designed for the European market, while ANSI C12.22 is usually adopted by North American manufacturers. To maximize their markets, most commercial products provide support for these two protocols. Compared with traditional protocols, C12.22 and IEC 62056-53 will have prominent advantages in the future.

IEC 62056-62

The ANSI C12.19 protocol uses data tables to save instrument measurements; however, IEC 62056-62 takes another approach and models data from intelligent meters by defining several interface classes. Although the encapsulation method is different, the exact data dealt with by the two protocols are exactly the same. As a universal instrument measurement data model, IEC 62056 is designed to support ammeters, gas meters and water meters. For most AMI manufacturers, there are strong geographical differences in the choice of protocols. For example, most smart meter suppliers in North America tend to use the ANSI standard series, while vendors from Europe are most likely to choose IEC standards in their products. With the continuous advancement of smart grid standardization, most AMI manufacturers have begun to adopt the above protocols and data models.

9.3.4 Energy Router Technology

The information system is the core of the Energy Internet and is responsible for the transmission and analysis of massive amounts of data in the network. The timely acquisition, processing and accurate transmission of information are necessary abilities that the communication system must possess. However, existing computer networks use the basic processing modes of link sharing, receiving and forwarding, which cause data congestion and other serious problems affecting the efficiency of information processing and affect the timeliness and compatibility of data processing. Therefore, in the Energy Internet, the direct utilization of computer networks for scheduling and control will face huge challenges. On the other hand, the communication networks of existing energy systems mainly use reserved special channels to meet the information demands of key businesses, which is a huge waste of resources. Therefore, a basic requirement of the Energy Internet is to make comprehensive use of computer networks and current communication system facilities and to construct information architectures with guarantees of bandwidth, mass computing and sensitive response ability. In view of the challenges faced by the Energy Internet in energy access, energy control and energy transmission, by referring to the idea of switching equipment in a computer network, an intuitive and feasible scheme for the construction of the Energy Internet is designing an energy router that can setup interconnections, perform scheduling and take control of an energy network [13]. The Energy Internet architecture of the energy router is shown in Fig. 9.5.

Analogous with network routers, energy routers can be stratified into energy hubs, switches, routers, routing stations, etc. Analogous with a substation that realizes energy conversion with the power grid, the key point of power conversion is a substation or transformer, but it cannot control the decoupling to realize the "source" and "use" flexibly. In contrast, in the Energy Internet architecture, the energy router can be regarded as a carrier of energy exchange, and it can play the role of distributed energy management and operation scheduling. To accomplish the above functions, the energy router needs not only all kinds of energy storage technologies and power electronic technologies to support the energy exchange but also a data centre to support the collection and processing of dynamic information. As the core equipment in the centralized optimization management layer in a local Energy Internet, the energy router is mainly responsible for centralized energy management, power electronic variable pressure, optimal power flow control, risk assessment and early warning and information aggregation services.

The system structure of the energy router is designed with modular components and consists of a power electronic transformer and server clustering. The energy router can realize flexible allocation and optimal power flow and has the function of analysing and managing data. The energy router will run in multiple and high-power modes with advanced functions such as information fusion situational awareness ability, balancing system power, power quality optimization control, efficient demand-side energy utilization and multi-energy cooperative interactive control. The energy router also has distributed power generation access, voltage transformation, power

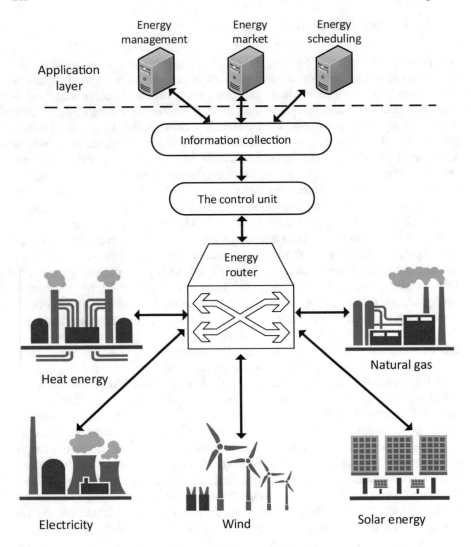

Fig. 9.5 Energy router and energy internet (Courtesy of [14])

exchange, power quality control and other customized advanced power functions. The main functions of the energy router include monitoring and control of distribution networks, automatic addressing functions, power energy exchange functions, power flow optimization control functions and optimal power flow control.

Conclusion

Communication technology is one of the most important basic technologies for the Energy Internet. Because of the huge scale and complex network topology, the real-time performance and communication bandwidth of network communication are required to be higher for the Energy Internet. In addition, there are several levels of architecture in the Energy Internet, and different communication modes are needed at different levels. This section describes in detail the communication technologies applied in the perception layer, network layer and application layer. In addition to hardware technology, the communication protocol is also an important part; thus, this section also provides a brief introduction to the communication protocols applied in AMI systems proposed by the industrial community. Finally, this section also briefly introduces an energy router based on communication technology, which can couple electricity, heat and gas in the Energy Internet.

9.4 Data Processing Technology in the Energy Internet

This section first introduces the monitoring and measurement big data sources and characteristics in the energy internet, then introduces the data analysis process of monitoring and measurement of big data, and finally briefly summarizes a typical analysis mining algorithm and data fusion algorithm applied to big data in the above systems.

Monitoring and measurement big data cover all aspects of the energy internet, including energy production, energy transmission, trading and consumption, involving hundreds of millions of devices and systems. Due to the large number of data sources and various measurement sensors, the data types are different, and the data contents are complicated. Among them, the energy supply side comprises conventional energy equipment, wind turbines, photovoltaics and other distributed renewable energy equipment. The energy consumption side comprises integrated energy utilization devices such as triple heat supply, heat pumps, industrial waste heat and residual pressure utilization; energy storage (including various scattered and redundant energy storage batteries, uninterruptible power supplies, electric vehicles, etc.); storage energy access for many types of energy storage, such as cold storage and heat storage; smart homes, smart buildings, and smart factories; and various energy-consuming equipment, such as charging stations, charging piles and gas stations. These data are monitored and have the following characteristics:

(1) The data volume in the energy internet is very large. The energy internet has deployed a large number of smart meters and other monitoring devices, generating massive and growing amounts of data. The data coverage includes the production and application of energy sources such as power systems, heating systems, cooling systems and gas systems. It also includes data on various auxiliary systems, such as transportation systems and meteorological systems. Therefore, the data volume of the energy internet has been further expanded.

(2) There are various data structures in the energy internet. In addition to traditional structured data, it also contains a large amount of semi-structured and unstructured data. With the continuous development of monitoring and metering technologies and the increasing types of intelligent devices, unstructured

data such as images and videos are increasing, such as voice data from service centre information systems and video data and image data from equipment online monitoring systems.

(3) The real-time requirements (speed) are high. Energy production, conversion and consumption are required to be instantaneously completed. Many real-time data must be analysed in real time. With the improvement of the level of intelligence and the gradual completion and application of smart meters, energy equipment online monitoring systems, smart substations, online mobile maintenance systems and information management systems serving various energy companies, the size and variety of data generated by the energy system has grown rapidly.

(4) The value of energy internet big data is very high. The application of big data runs through every part of the energy internet. Through big data processing technology, scientific predictions can be made for various stages of energy production, distribution, conversion and consumption. Potential risks can be discovered in time to ensure safety and economy. Big data applications support the energy internet. The emergence of new business forms provides new services for all participants.

Therefore, with the continuous development of the energy internet, an increasing amount of monitoring and measurement data have been collected and stored. It is important to extract useful knowledge about the operation of the energy internet from these energy big data, such as assisting multi-source system collaborative decisions and supporting the safe and stable economic operation of the energy internet. Next, the content further describes in detail big data intelligent efficient analysis and mining technology for the energy internet and data fusion technology for the energy internet. The summary of big data processing technology provides important support for the reliable and efficient operation of the energy internet.

9.4.1 Big Data Analysis Technology

The steps of big data processing are shown in Fig. 9.6 and contain major steps, such as big data acquisition, big data import/preprocessing, big data analysis statistics and big data mining.

Big Data Acquisition: The collection of big data requires the support of Internet and Internet of Things technologies, including identification, sensing and data concentration. The identification technologies include RFID, barcodes, QR codes, biometrics (iris, fingerprint and voice), etc. Among them, RFID can identify devices within a certain distance without human participation. Therefore, it can be widely applied to energy systems. The sensing function generally uses embedded sensors to form a sensor network and collect various indicators and data that affect or reflect the operating status of the power grid. The acquisition data type includes a state quantity, electrical quantity or quantity measurement. The acquisition result can be used

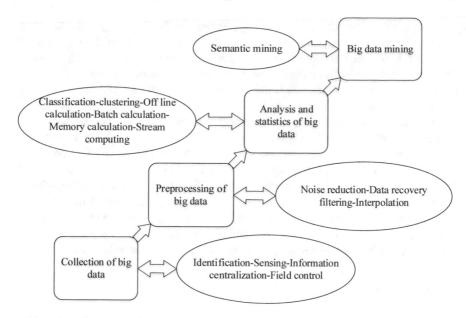

Fig. 9.6   Big data processing (Courtesy of [15])

for monitoring systems such as SCADA (supervisory control and data acquisition), WAMSs and CAC/CAG (Condition Information Acquisition Controller/Condition Information Acquisition Gateway).

Big data import/preprocessing: Data import also requires preprocessing; due to the effects of the physical environment, weather and the ageing or failure of the monitoring equipment, noise or erroneous data is inevitably present in the collected data. In addition, a bad communication environment will also lead to data errors and losses. Therefore, it is necessary to perform noise reduction on the collected data and recover lost data, which is also referred to as data cleaning. Noise reduction is mainly performed by a smoothing filter. For a stationary system, the high-frequency part is likely to correspond to a noise component, and processing of the high-frequency part can effectively reduce noise. Additionally, a smoothing filter can also be used as a means to recover lost data. In addition, lost data can be efficiently recovered through interpolation techniques.

Big data analysis/processing: The specific techniques include classification, clustering, correlation and prediction. According to the time characteristics of processing, it can be divided into offline computing, batch computing, memory computing and stream computing. In data analysis, it is often necessary to classify data. The algorithms used in big data classification include nearest neighbor algorithms, support vector machines, boosted tree classification, Bayes classification, neural networks and random forest classification. Classification algorithms can fuse with fuzzy theory to improve the classification performance. Clustering can be expressed as an unsupervised classification, including k-means, FCM clustering and other algorithms. Correlation analysis is also one of the main methods of data analysis. It is mainly based on the support and confidence to mine the relationship between objects. The basic algorithms include A priori and FP-Growth. Similarly, in the data analysis process, data prediction is indispensable. The main algorithms are extreme learning machines, deep learning and decision trees. Furthermore, these data mining algorithms can be parallelized, which greatly reduces the computational delay.

Big data mining: Big data analysis results are used for data mining. Because the previous analysis is only data-centric, the results obtained are not easily understood by humans and do not necessarily match the research purpose. They may result in useless or even seemingly opposite results. Therefore, people's participation is needed for filtering and purifying the results, translating the results into a semantic form that people can understand, and finally achieving the purpose of data mining.

Energy internet big data analysis technology mainly includes two parts: a big data processing platform and a big data analysis algorithm. We introduce these two key links separately below.

(1) Big Data Processing Platform

Cloud computing platform: Big data systems require very large data processing, transmission and storage capabilities. In particular, in recent years, cloud computing is a new computing model that has developed rapidly. It is a collective term for several new computing technologies. First, cloud computing integrates various wide-area heterogeneous computing resources by leveraging the Internet to form an abstract, dynamically expandable computing resource pool. Then, it provides storage space, platforms and application software through the Internet to users. In conclusion, cloud computing has the following main features: (1) integration of large-scale heterogeneous computing resources; (2) dynamical expansion; (3) virtualized resources; (4) three different service levels: IaaS, PaaS and SaaS; (5) Internet-based communication platform; and (6) economic advantages.

A cloud computing platform in an energy internet monitoring system can realize the virtualization of computing resources and physical resources. Through the resource pool, the processing capacity can be dynamically allocated and scheduled. This can maximize the use of existing computing capacity and storage capacity, reduce operating costs and save expenses. Additionally, the cloud platform also has high security, which can ensure that user data are not stolen. The combination of big data and cloud platforms will become the basic performance support for the energy internet. A cloud platform for energy monitoring and measurement systems is shown in Fig. 9.7.

In the figure, each cloud server must have a cloud computing infrastructure, operating environment, service application and service access. The infrastructure and operating environment layers provide basic hardware and software environments for the operation of the cloud computing platform, respectively. The service access layer is used to provide users of the cloud computing platform with the means of interaction with the energy internet, such as a terminal, web browser and dedicated

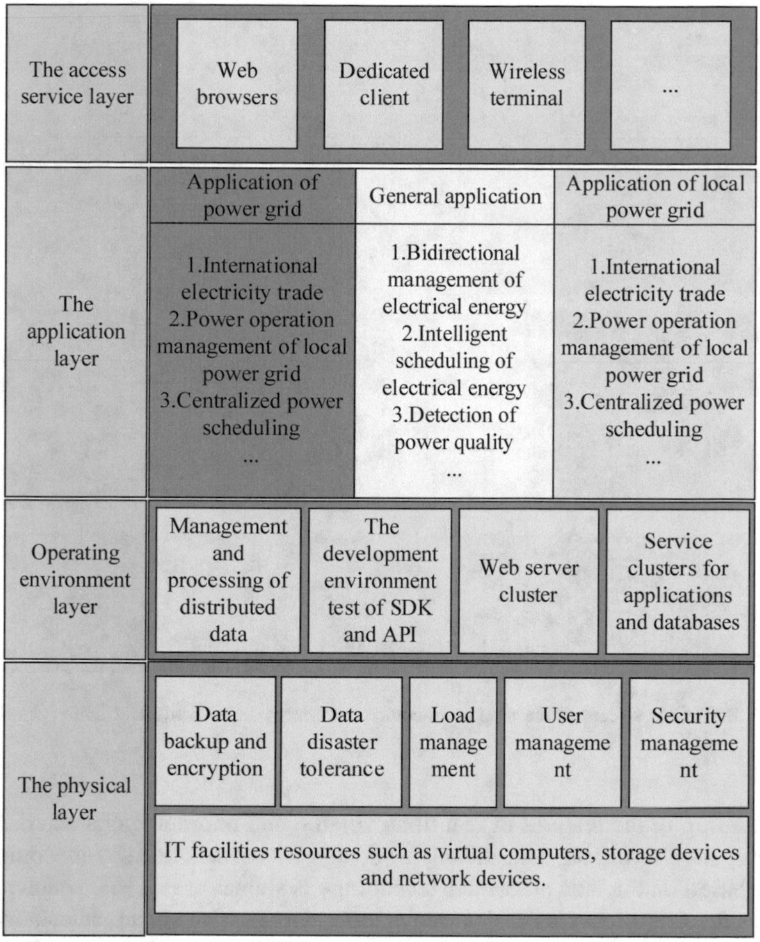

The access service layer	Web browsers	Dedicated client	Wireless terminal	...

	Application of power grid	General application	Application of local power grid
The application layer	1.International electricity trade 2.Power operation management of local power grid 3.Centralized power scheduling ...	1.Bidirectional management of electrical energy 2.Intelligent scheduling of electrical energy 3.Detection of power quality ...	1.International electricity trade 2.Power operation management of local power grid 3.Centralized power scheduling ...

Operating environment layer	Management and processing of distributed data	The development environment test of SDK and API	Web server cluster	Service clusters for applications and databases

The physical layer	Data backup and encryption	Data disaster tolerance	Load manage ment	User manageme nt	Security manageme nt

IT facilities resources such as virtual computers, storage devices and network devices.

Fig. 9.7 Cloud computing platform hierarchical architecture (Courtesy of [16])

client. The business application layer must deploy custom application software and general-purpose application software for the wide-area cloud and local area cloud according to the similarities and differences between the functions of the wide-area energy monitoring system and the local area energy monitoring system.

Stream data processing: In the energy internet, to continuously monitor, manage and optimize the operation of the system, the energy monitoring centre will receive a large number of data processing requests. Different monitoring data have different urgency and timeliness requirements. If they cannot be processed within the specified time limit, they will lose value. These require continuous real-time processing, and traditional data processing frameworks are incompetent. As a result, the concept of streaming computing was raised.

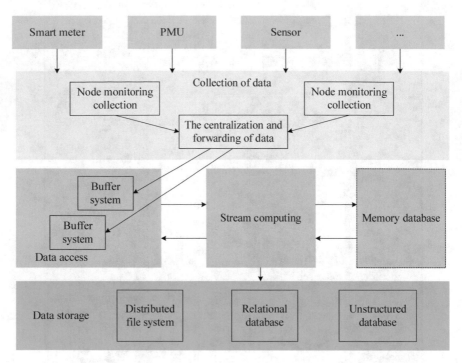

Fig. 9.8 Real-time stream processing architecture for energy monitoring Big Data (Courtesy of [17])

According to the features of real-time, volatile and unordered characteristics of an energy monitoring big data stream, the design framework of a stream computing system based on big data processing technology is shown in Fig. 9.8, which mainly includes the four processes of data acquisition, data access, stream calculation and data storage.

Based on this architecture, it is possible to carry out anomaly detection of equipment status in the energy internet. It works through monitoring all state information data of multiple devices at different time points, such as partial discharge of transformers, dissolved gases in oil, temperature measurements of winding fibres, partial discharges of switchgears, characteristics of operating mechanisms and the working state of energy storage motors, and the dielectric loss factor, capacitance and leakage current of the device, such as the total current, harmonic, capacitive current and resistive current of the arrester. In the Storm stream computing system, a threshold value judging method is used to detect the anomaly of the timing flow data of the smart grid condition monitoring. The sliding window model for design condition monitoring of anomaly detection is shown in Fig. 9.9. The sliding distance of the sliding window model is 1 s, and the sliding window size is 1 min. Then, the sliding window is divided into 60 basic time windows. The attribute set represents the collection of state monitoring amounts for each substation. Each basic window judges

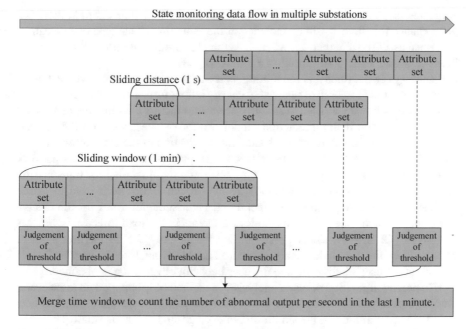

Fig. 9.9 Sliding window model for abnormal detection of energy equipment condition monitoring (Courtesy of [17])

the outliers of data in parallel and only caches the 60 result tuples after the threshold judgement of each basic window rather than cache each tuple in the original window one by one. After the calculation in the basic window, the number of occurrences of outliers in the most recent 60 basic windows is counted; that is, the number of occurrences of data outliers in the most recent one minute is output every second. Then, the device status recognition results are obtained.

(2) Big Data Mining Algorithm

The goals of data mining are descriptive and predictive. The descriptive goal is to analyse correlations between data and find relevant trends or rules. The predictive goal is to establish a model based on the historical data's laws and relationships and predict future results. Big data mining algorithms often use parallelization ideas to improve efficiency. Typical data mining algorithms include association analysis, clustering algorithms, classification algorithms, neural networks, deep learning and other intelligent algorithms.

Association analysis: Association analysis extracts the relationship between a series of variables or factors from a large data set to tap the existing relationships between two or more variables in the data set. The association rule can be interpreted as $I = \{i_1, i_2, \ldots, i_m\}$, representing a collection of all distinct items, called a transaction database, which is the main analysis aim. T represents a set of items in a transaction record with a unique identification, and $T \subseteq I$. If, in set D, item X and item Y

generate association rules, when the transaction record T contains item X, there is a great chance that item Y will be included at the same time. This type of rule can be described as $X \cap Y$ (if X then Y), where X is the precondition item set, Y is the result item set, X and Y are both subsets of I and $X \cap Y = \emptyset$.

In energy big data analysis and mining, correlation analysis is everywhere. On the source side, in the research of renewable energy output forecasting, the grey correlation analysis method was used to analyse the various factors affecting the distributed energy output, and then a prediction model was established based on the correlation analysis results. On the network side, the application of the correlation analysis method can analyse the proximity between each plan of the transmission network planning and the ideal planning plan, thus obtaining the final plan for transmission network planning. On the charge side, the FP-Growth algorithm is used to search the historical data frequently and generate strong association rules for the change law of power load. The influence of relevant factors in the load forecasting process is analysed. On the user side, by enumerating the factors affecting the user's power consumption, the correlation analysis method is used to analyse the various factors that affect the user's behaviour and the user behaviour evaluation is obtained.

Clustering algorithm: Cluster analysis is a method of grouping data into several clusters based on data similarity or dissimilarity, making the data in a group of data or individuals similar, and the similarity between groups of data is small. The same group of samples sometimes form different grouping results because of different purposes, data input methods, selected group characteristics or data attributes. Clustering is based on the similarity between samples. Different distance calculation methods achieve different classification effects. Some typical similarity distance calculation methods have been developed in recent years, such as Manhattan distance, Euclidean distance, Minkowski distance, weighted distance, normalized distance, Mahalanobis distance and Hamming Distance. Cluster analysis uses grouping to find possible hidden features, patterns or associations behind each sub-cluster. The cluster analysis implementation does not know the number of clusters, but the characteristics of the clustering results and the significance they represent can only be explained later. Therefore, cluster analysis can be viewed as unsupervised learning.

Cluster analysis algorithms are widely used in energy monitoring and measurement big data. For example, the parallel k-means algorithm and massive energy use data are used in the energy user behaviour analysis process to complete the categorization task for resident users. In short-term load forecasting, the fuzzy clustering analysis method is adopted to classify the historical load data into several categories and to determine the prediction categories that match the prediction data. In the power system network loss evaluation process, the characteristics of power data are fully considered. The clustering algorithm is selected by dividing the clustering algorithm and the hierarchical clustering algorithm, and the clustering results of the sub-problems are integrated. Finally, a typical grid is generated based on the hybrid clustering results. It is used for network loss evaluation, and the implementation process of the network loss evaluation method is shown in Fig. 9.10. In addition, the cluster analysis method plays a key role in the research process of equipment fault identification, system steady-state fault identification and bad data identification.

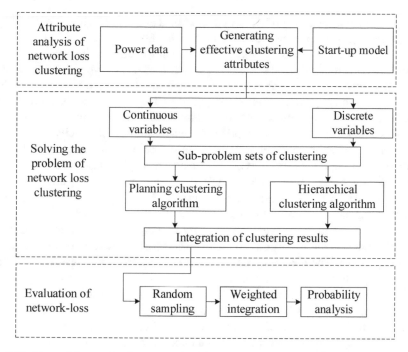

Fig. 9.10 Network loss evaluation process based on hybrid cluster analysis (Courtesy of [18])

Classification algorithm: After forming specific criteria and rules, classification algorithms can be used for data analysis. In the application process of the classification algorithm, through the analysis of the known class of training sets, the classification rules are discovered and the categories of new data are predicted. The classification includes two processes, training and classification, and can continue to implement incremental learning. Compared with clustering, classification has more specific goals. In the face of massive data, classification algorithms also need to be implemented in parallel, such as Mahout-related classification algorithms. When the training sample number is relatively small, compared with the Mahout classification algorithm, the performance of traditional data mining methods will be better. However, with an increasing sample number, the processing time required for the traditional non-scalable classification algorithm increases rapidly. At this time, the advantages of Mahout's scalable and parallel algorithms become apparent.

The classification algorithm is indispensable in the energy monitoring big data analysis process and is involved in power consumption prediction, equipment fault identification, and reliability evaluation. Specifically, in the process of customer electricity analysis and research in the energy internet, the researched customer is divided into five categories, and the characteristics of the five types of users' electricity consumption data of A, B, C, D and E are studied. Class A customer is a special transformer user whose capacity is 100 kVA and above. The main features of these customers include the use of relatively concentrated power loads; high-density, high-voltage and high-power equipment; and high safety factors for electrical equipment. Class B customers are users of small and medium-sized transformers, especially those with capacities below 100 kVA. The power load of this type of dedicated user has characteristics such as stable time, large load and diversified load composition, and its power consumption has strong regularity. Class C customer is a three-phase general industrial and commercial user and refers to a three-phase power user who performs a non-resident electricity price. Such a customer's electricity load has the characteristics of stable load, relatively regular load composition and the use of electricity has a strong regularity. Class D customers are single-phase general industrial and commercial users. It refers to a single-phase power user who performs a non-resident electricity price. D customers' electricity load has characteristics such as irregular time, large number of loads, relatively simple load components and most secondary loads and has a weaker regularity in electricity consumption. Class E customers are resident users and refer to electricity users who implement residents' electricity prices. The electricity consumption characteristics of resident users are significantly different from those of other types of users. The main electricity load of resident users is the heat load; the peak time of electricity consumption for general resident users is 12:00–14:00 and 18:00–24:00, which is contrary to the electricity consumption characteristics of general industry and commerce; the electricity consumption curve for 24 h There will be large fluctuations, there are obvious peak and valley, generally not a flat steady state; the electricity consumption during holidays will be greater than the working day; the different users have different degrees of response to the demand side and have a strong dispersion Some users have distributed power generation equipment. The power and electricity bills flow in both directions. The strong randomness will have an effect on the power quality.

Artificial Neural Network: Artificial neural networks (ANNs) are information processing systems that mimic biological neural networks. It addresses complex problems and obtains information from other artificial neurons or external environments. The artificial neural network is divided into different phases: 1. The learning phase is mainly to establish the connection pattern between neurons, to modify the weights among the neurons and to adjust the threshold in the activation function of neurons; 2. When the neural network receives an input stimulus, the recall stage generates a corresponding output value based on the established neural network architecture; 3. The induction stage provides efficient memory and storage mode for the process to derive the overall characteristics from the local observation.

The artificial neural network structure is composed of many neurons or nodes in various connection modes. The typical three-layer network topology is shown in Fig. 9.11, including three layers. Among them, the input layer receives external environment information. Based on the problem characteristics, the input data may be converted into a signal adapted to the network by using a linear conversion function. The hidden layer acts as an intrinsic structure of interacts between processing units to solve nonlinear problems. The output layer processing unit processes information output to the external environment, and its functions include normalized output, competitive output and competitive learning.

Fig. 9.11 Basic architecture
of an artificial neural
network (Courtesy of [19])

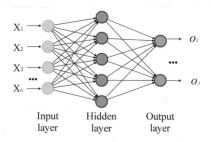

Input layer   Hidden layer   Output layer

An ANN is also an important tool for energy monitoring and measurement big data analysis. It promotes the progress of the field in energy equipment fault diagnosis, real-time intelligent control, system stability analysis and load forecasting. For example, in the prediction process of the energy internet load model, in the power system analysis and calculation, the load model is a physical or mathematical description of the load characteristics, and the load characteristics reflect the load power with the system operating parameters (mainly the voltage and frequency). The law of change, so there is an essential link between the load model and the load. The specific load model parameter prediction process is shown in Fig. 9.12.

Deep Learning Algorithm: Deep Learning is essentially an artificial neural network with multiple hidden layers. It generally refers to the network structure of high-level learning. This structure usually combines linear and nonlinear relationships. The deep network structure obtained by deep learning contains a large number of neurons, and neurons are interconnected. In the learning process, the connection strength between neurons, namely, weight, is modified and determines the function of the network. The deep network structure obtained by deep learning conforms to the feature of the neural network. Therefore, the deep network is the deep neural network (DNN).

A deep neural network model is complicated, and there are many training data and a large amount of calculation. On the one hand, a DNN needs to simulate the computational power of the human brain. The human brain contains more than 10 billion neurons, which requires more neurons in the DNN, and the number of connections between neurons is quite alarming. Many parameters that need to be calculated. In speech recognition and image recognition applications, there are tens of thousands of neurons, tens of millions of parameters and the complexity of the model leads to a large amount of computation. On the other hand, to avoid over-fitting, the DNN requires large amounts of data to train high-accuracy models. Take speech recognition as an example. Currently, the industry usually uses a sample size of several billion. It usually takes several days to perform a training. Therefore, deep learning requires the support of big data. To reduce training time, the current acceleration of deep learning mainly includes GPU acceleration, data parallelism and computational parallelism. In the field of energy, the use of deep learning methods for transient stability assessment, load forecasting and equipment image recognition after power system failures is used as a deep learning model that can better simulate human

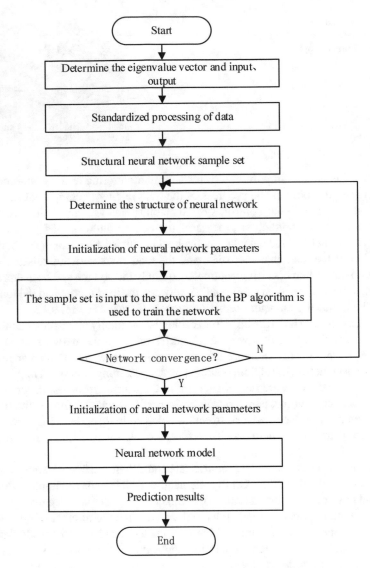

Fig. 9.12 Load model prediction process based on an artificial neural network (Courtesy of [20])

intelligence. With the further improvement of technology, it will be applied widely in the energy internet.

In the system stability, there are two parts of offline learning and online evaluation. The offline learning phase has recently been completed. Under each typical operation mode, "sample generation" → "expression learning" → "classification evaluation", etc. It is used to learn the transient stability rules corresponding to each typical operating mode. The online evaluation phase is event-driven. Similar to the interrupt

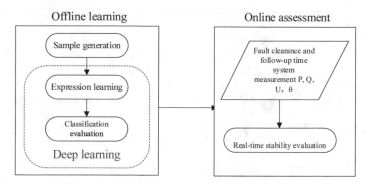

Fig. 9.13 Transient stability process based on deep learning (Courtesy of [21])

mechanism, the system begins to operate in the event of a failure and receives measurement data from the PMU system for stable evaluation and control. The specific process of the transient stability method based on deep learning is shown in Fig. 9.13.

As shown in the figure, at the stage of expression learning, a deep belief network is used to implement nonlinear transformation on the original features and to learn the abstract expression hierarchically. Through expression learning, the deep learning model maps transient stability data from the original input space to a binary linearly separable representation space. At the stage of training, the deep belief network (DBN) is a deep expression learning model that uses restricted Boltzmann machine (RBM) as the basic unit and uses a training method combined with RBM pre-training and DBN inverse fine-tuning to train layer by layer from the bottom (input layer) to upper layer. The typical DBN training process is shown in Fig. 9.14. Since the actual state variables are difficult to measure, the process uses the active power, reactive power and bus voltage as well as the phase angle of the system circuit at the time of fault clearance as the inputs of the DBN, and the DBN is monitored uses the transient stability or not as the measurement of DBN supervision to train. The introduction of constraints on the RBM structure allows it to learn some local structures and features and adapt to the characteristics of the power grid. The constraint on the connection matrix is embodied by adding a penalty function to the RBM loss evaluation formula (9.1) and implicitly increasing the constraint on the connection matrix through the penalty function.

$$\max P(x|h, \theta) = \sum_h e^{-E(x,h|\theta)} \tag{9.1}$$

(1) Sparsity constraint for networks. In power systems, the range of systems affected by transient faults is usually limited, so properly neglecting the less affected areas helps highlight local features. The sparse constraint for networks refers to that the weight amplitude of the connection matrix and the weight value of the constraint connection area are close to zero, thereby highlighting some important connections to help the RBM capture the local features of transient

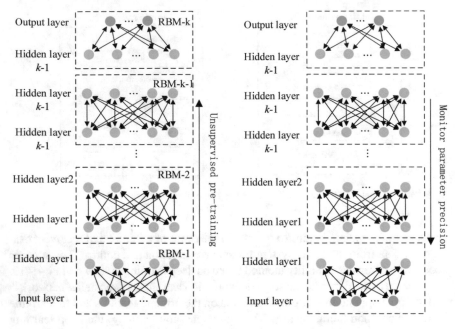

Fig. 9.14 Deep belief network training method (Courtesy of [21])

faults. The penalty function of the sparsity constraint for networks is shown in formula (9.2).

$$\Omega_1(\theta) = \sum_{i,j} \left| W_{ij} \right| = \| W \|_1 \tag{9.2}$$

(2) Network smoothness constrained. In a power system, the elements whose electrical distances are close to each other exhibit similar behaviour in transient accidents; therefore, the corresponding RBM network weights should be similar as well. Network-constrained smoothing refers to that by increasing the constraints between the connection matrices, the difference between input matrix weight values of adjacent nodes is close to zero, so that the characteristics learned by adjacent nodes are similar. The penalty function of network-constrained smoothing is shown in formula (9.3). $\rho \in (0,1)$ represents the electrical distance between the input features. The closer the electrical distance, the stronger the correlation of the features. Among them, the admittance matrix between the nodes is used as a measurement of the electrical distance as $\rho_{ij} = \left| Y_{ij} \right|$.

$$\Omega_2(\theta) = \sum_{ij} \rho_{ij} \sum_k (W_{ik} - W_{jk})^2 \tag{9.3}$$

The sparsity constraint for the network equals the L^1 regularization, and the network smoothness constrained also has commonality with the common L^2 regularization. The sparsity constraint can limit the number of non-zero weights of the connection matrices. At the same time, both the sparsity constraint and the network smoothness constrained have the effects of constraining the network weight and reducing the weight. Therefore, the deep belief network model considering power system constraints can effectively reduce the risk of over-fitting and make the model have more generalization ability. It can be further used for online transient stability.

In addition, due to the intelligent, real-time operation of the operation decision of the energy internet, the amount of required operation data is huge, so it is impossible to be completed by man and it is incompetent by common machine learning algorithms. Based on the combination of deep learning, ultra-large-scale neural networks and big data, the performance of machine learning will be greatly improved. Under these circumstances, it is feasible to use computers to simulate the status of ultra-large-scale neural network technology in the energy internet. The research results of ultra-large-scale neural networks are expected to be effectively applied to energy information systems, which will have a significant and positive impact on the promotion and improvement of online real-time energy operation status and energy dispatch.

9.4.2 Data Fusion Technology

During the data measurement and acquisition process of the energy internet, multi-sensor systems are used to continuously or periodically collect and detect the electrical, physical and chemical characteristics of the equipment, and at the same time, the real-time perception and dynamic control of the smart energy system are realized through the combination of information technologies. However, energy supervision and the data flow of all major elements of measuring big data also need to overcome difficulties in the data fusion process. Specifically, the information system of each business department in the same company is designed and developed by different R&D teams and around different application requirements; the cross-type, data redundancy, and data inconsistency in various types of collected data; large differences in collection frequency and storage frequency; inconsistent data formats; and other issues. Additionally, the entire energy industry lacks industry-level data model definitions and master data management. More severely, data sharing has not yet been fully achieved between service chains, which has brought great technical challenges to data fusion. Therefore, research on multi-source data fusion is needed to significantly improve the time effectiveness, accuracy and reliability of data collection. Horizontally collaborative and vertically connected data platforms should be built to support geographic information systems, energy system analysis, maintenance, energy markets, mobile applications, cloud environment business applications and integration service modes to improve the integrity, timeliness, effectiveness and consistency of data. Then, we further use big data analysis technology to analyse and support decision-making, which helps to ensure smart and safe production and distribution of energy.

Data fusion is also called information fusion or multi-sensor data fusion, and a more accurate definition of multi-sensor data fusion can be expressed as follows: taking full advantage of multi-sensor data resources and using computer technology to obtain sensor observation data according to the time series, under certain criteria, we can analyse, integrate, dominate and use this approach to obtain a consistent interpretation of the measured object so that corresponding decisions and estimates can be achieved, which makes the system able to obtain more full-scale information than its components.

At present, data fusion is divided into three basic structural systems: data-level fusion, feature-level fusion and decision-level fusion. The data fusion process structure is shown in Fig. 9.15.

Data-level fusion is shown in Fig. 9.15a. First, the observation data of all sensors are fused, and then the feature vector is extracted from the fused data while performing judgement and recognition. This requires the sensor to be homogeneous (the sensor observes the same physical phenomenon). If multiple sensors are heterogeneous (not observing the same physical characteristic), the data can only be used at the feature level or decision level. There is no problem of data loss in the data-level fusion, and the result is also the most accurate, but the system communication bandwidth requirements are very high.

Feature-level fusion is shown in Fig. 9.15b. Sensors provide representative features that are extracted from observation data. These features are mixed into a single feature vector and then processed by using pattern recognition. This method has a lower demand for communication bandwidth, but its accuracy is reduced due to the loss of data.

Decision-level fusion refers to merging the recognition results of multiple sensors after each sensor identifies its target, as shown in Fig. 9.15c. Since the sensor's data have been condensed, the result of this method is relatively the least accurate, but it has the lowest communication bandwidth demand.

From the data fusion process, it can be seen that the use of multiple sensors to obtain comprehensive information from the measured object and the environment mainly depends on the fusion algorithm. Therefore, choosing a suitable fusion algorithm is a core problem. For multi-sensor systems, information is diverse and

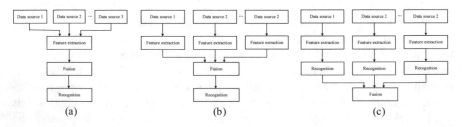

Fig. 9.15 Three hierarchical structures of multi-source data fusion (Courtesy of [22]). **a** Data-Level Fusion **b** Feature-Level Fusion **c** Decision-Level Fusion

complex. Therefore, the robustness and parallel processing capabilities are basic requirements for information fusion methods. Furthermore, it also includes methods' calculation speed and accuracy, the interface performance, etc. At present, typical data fusion algorithms include random methods and artificial intelligence methods.

(1) Random methods

Weighted Average: As the most simple and intuitive method, the weighted average method weights and averages the redundant information provided by a group of sensors. This method is a direct operation on the data source.

D-S Evidential Reasoning: D-S evidential reasoning is an extension of Bayesian reasoning. The three basic points are the basic probability assignment function, trust function and likelihood function. The reasoning structure of the D-S method is divided into three levels. Level 1 is the target synthesis. Its role is to synthesize the observations from the independent sensors into a total output. Level 2 is the inference. The purpose is to obtain the sensor observations and infer the results to expand the sensor observations into target reports. The basis of this reasoning is that certain sensor reports will generate certain trusted target reports logically with a certain degree of certainty. Level 3 is the update. Random errors usually exist in various sensors. Therefore, in terms of time, a set of independently consecutive reports from the same sensor is more reliable than any single report. Therefore, the sensor observation data need to be combined (updated) before inference and multi-sensor synthesis.

(2) Artificial Intelligence Methods

Fuzzy logic reasoning: A fuzzy logic method is a multi-valued logic. Specifically, a real number between 0 and 1 represents the authenticity, which is equal to the premise of implicit operators. The uncertainty of a multiple sensor information fusion process is allowed to be directly expressed in the reasoning process. If a systematic approach is used to infer the uncertainty in the fusion process, consistent fuzzy reasoning can be produced. Compared with probabilistic statistical methods, logical reasoning has many advantages: (1) information representation and processing is close to human thinking; (2) it is generally more suitable for high-level applications (such as decision making). However, logical reasoning is not systematic. In addition, since logical reasoning has a large subjective factor in the information description, representation and processing still lack objectivity.

The practical value of the fuzzy set theory for data fusion is its extension to fuzzy logic. Membership degree can be regarded as an inaccurate representation of the true data value. In the MSF (Microsoft Solutions Framework), the existence of uncertainty can be directly represented by fuzzy logic. Then, using multi-valued logic inference, various propositions are combined according to various calculus of fuzzy set theory to realize data fusion.

Conclusion

This section outlines the sources and characteristics of monitoring and measurement big data analysis and mining in the energy internet and then introduces the corresponding data analysis processes for monitoring and measurement big data. Finally, this section summarizes the existing typical data analysis and mining algorithms and data fusion algorithms that are applied to big data in monitoring and measurement systems.

9.5 Information Security in Monitoring and Measurement Systems

This section first introduces the basic concepts and requirements of information security in the monitoring and measurement systems of the energy internet and then introduces the security threats and attack methods faced by energy measurement information security. Finally, it analyses the defence strategies of energy measurement information security against the aforementioned security threats.

9.5.1 Information Security Requirements

The volume of data in energy internet monitoring and measurement systems is constantly increasing, while big data involves the privacy of many users, thus putting higher requirements on information security. The energy internet has wide geographical coverage. The protection systems of various units are not balanced, the level of information security is inconsistent and the security needs to be improved. It is necessary to constantly perfect protection measures and improve the level of protection for safe transmission and safe storage. Security protection measures and key protection measures need to be further strengthened.

One definition of information security is the security protection established for a data processing system technically and administratively: the protection of computer hardware, software and data so that they are not destroyed, altered or leaked for accidental or malicious reasons. Energy measurement information security refers to the safety of energy measurement information during transmission and application. It can guarantee the reliability, confidentiality, integrity, availability and controllability of the information in the energy measurement equipment, such as the main station, terminals and smart meters, in energy monitoring and measurement systems.

(1) Reliability refers to the characteristic that energy monitoring and measurement systems can perform specified functions within the specified conditions and time.
(2) Confidentiality refers to the characteristics of energy monitoring and measurement systems that prevent illegal disclosure of measurement information.

(3) Integrity refers to the characteristic that energy measurement information cannot be changed without authorization in the process of storage and transmission.
(4) Controllability refers to the characteristic that energy monitoring and measurement systems have the ability to control the content and transmission of energy measurement information.
(5) Validity refers to the characteristics that energy measurement information can be accessed by the authorized entity and used as required and that the energy monitoring and measurement system can continue to operate at an acceptable quality level, providing people with a valid information service.
(6) Non-repudiation refers to the non-repudiation or denial of completed operations and commitments by both parties to a communication, usually through a digital certificate signature mechanism to achieve non-repudiation of both sides.

With the further improvement of information integration in energy monitoring and measurement systems, technical means such as real-time monitoring and maintenance of network communication quality of energy monitoring and measurement systems, protection of information transmission in the network to prevent malicious attacks and steal from inside and outside the network, the timely response of network failures and quick network device recovery are realized. In addition, the development of information security technologies such as network firewall technologies, data encryption technologies, rights management and access control technologies and redundancy and backup technologies have also brought new development ideas for energy measurement information security protection strategies.

9.5.2 Information Security Countermeasures

(1) Energy Measurement Information Security Threats and Attack Means

The basic security goal is to realize the reliability, confidentiality, integrity and availability of energy measurement information and the legitimate use of energy monitoring and measurement system resources. A threat to energy measurement information security is a threat to these basic security goals. The main sources of energy measurement information security threats include internal personnel (including users of energy monitoring and measurement systems, managers and decision makers, maintenance operators and developers of energy monitoring and measurement systems, etc.), special status personnel (such as inspectors and auditors with special identities), external hackers, competitors, etc. Any threat may cause the system to be attacked by illegal intruders. Sensitive data in transmission may be leaked or modified. The transmitted information may be intercepted or changed by others. The main forms of information security threats are as follows:

Information leakage: Energy measurement information is inadvertently or intentionally leaked to an unauthorized person or entity. For example, interception or electromagnetic leakage is used to intercept information transmitted from the main station to a smart meter or collection terminal; illegally duplicate energy monitoring and measurement system sensitive information; and analyse information flow frequency, information length, information flow direction and information traffic to analyse useful information and cause information leakage and loss.

Integrity attacks: The energy measurement data are modified, inserted, deleted or retransmitted with important information by illegal means to achieve the expected response of the attacker, maliciously modify, or add data to interfere with the normal use of the user. The security threats to energy measurement systems mainly include malicious attacks on smart meters or collection terminals and other devices, which can prevent smart meters or collection terminals and other devices from working properly or modify parameters in the devices and destroy measurement accuracy.

Denial-of-service attacks: Attackers use the defects of the protocol stack of a protocol, operating system or network device. The system responds slowly or even reaches paralysis by sending illegal packets to the master system, leading to the user's legitimate access to the master system or visits being blocked unconditionally.

Illegal use: This refers to the use of energy monitoring and measurement system networks or computer resources without prior consent, such as intentionally circumventing the system access control mechanism, using equipment and resources abnormally, expanding authority without permission and accessing information beyond authority. It mainly includes counterfeiting, identity attacks, illegal users entering the network system to perform illegal operations, and legitimate users operating in unauthorized ways.

Logic bombs, worms, malicious code, Trojans and other virus threats: This refers to using energy monitoring and measurement system vulnerabilities to implant logic bombs, worms, malicious code, Trojans and other viruses to undermine the normal work of energy monitoring and measurement systems.

Backup data and storage media damaged or lost: Dedicated hardware such as disk arrays, tapes, tape libraries, CD-ROM towers, optical disk libraries and encryption U-disks, and appropriate backup software form backup and recovery systems. Even with perfect backup and recovery systems, there is still a risk of backup data and backup media being damaged or lost, such as the data in the storage medium cannot be read and the storage medium is lost or damaged.

Information security threats caused by management vulnerabilities: Information security solutions need to combine technology and management. In fact, 60% of information security threats come from inside the network, such as man-made violations, unintentional errors, malfeasance and so on. These behaviours are important causes of information insecurity. Therefore, strengthening information network management and improving employee safety awareness are key aspects of information security.

The main methods of threatening energy measurement information security include tampering, interception, denial-of-service attacks, unauthorized access/illegal use, counterfeit and forgery, replay attacks and so on. An attacker can destroy the confidentiality and integrity of information by changing the timing, order, content, form or flow direction; deleting all or part of the information; or

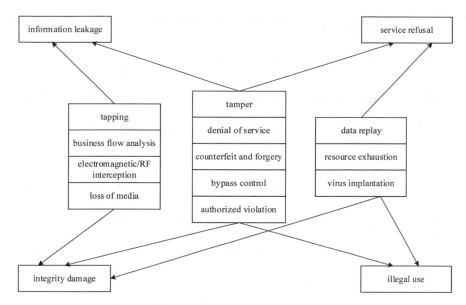

Fig. 9.16 Typical Information Security attacks and their relationships (Courtesy of [23])

inserting meaningless information or harmful information into a message. Typical information security attacks and their relationships are shown in Fig. 9.16.

(2) Energy measurement information security protection strategies
In terms of information security, for various attacks that energy monitoring and measurement systems may face, it is necessary to shift information security from passive defence to multi-level and active defence. To build an energy measurement information security system, security protection strategies need to include five aspects: physical security, communication link security, network security, system security and data security.

Physical security: The physical security of information includes medium security and device security. Medium security refers to data transmission and storage medium theft prevention, anti-destruction and anti-information leakage. In energy monitoring and measurement systems, data transmission and storage media include USB keys, CPU cards and cryptosystems. The storage and management of media must be formulated, and corresponding measures should be adopted to ensure the safety of the media. For example, important data in the system and data that play a key role in system operation and applications should be protected by encryption and other methods. The storage of important data and key data for various types of media should take effective measures, such as the storage or establishment of medium libraries in other locations to prevent theft, destruction and deterioration; important data that need to be destroyed and deleted should be centralized for destruction by a special officer to prevent illegal copying. Equipment safety refers to the related facilities in energy monitoring and measurement systems, such as computers, smart meters and terminals, which are protected against natural disasters, man-made damage and ride-through attacks.

The following protection measures are taken:

1. Use a standard computer room or computer field.
2. Adopt optical interfaces and optical cables to reduce the probability of a circuit being stolen.
3. The main network equipment should be placed in a special shielded room to prevent the circuit from being stolen.

Link security: To ensure that energy measurement information is not illegally accessed and used, we can adopt identity authentication technology and session key agreement technology. A secure communication link is established between a smart meter or terminal and the main station system. Identity authentication is the basis of the entire energy measurement information security system and can confirm the user's authority and responsibility. Identity authentication is used to confirm the identity of an operator in an information communication network and is divided into smart metering and identity authentication between a terminal and the main station. The most commonly used authentication method is "user + password". This authentication method has the drawback that the authentication process is not encrypted and that the password is easily cracked and monitored. Therefore, a variety of authentication methods have been proposed, such as smart card authentication, SMS password verification, dynamic port token verification, USB key verification and biometrics (such as fingerprint recognition and speech recognition). Session key agreement is an effective method to establish secure communication links. It provides a transparent network security transmission channel for network monitoring and the measurement system of the energy monitoring and measurement system through a secure communication protocol. It is generally divided into two phases: handshake and data transmission. During the handshake phase, the smart meter or terminal and the main station authenticate each other and establish the encryption key used to protect the data transmission. Generally, a public key is used, and the cryptosystem is implemented. After the handshake is completed, the data are converted into protected records for transmission.

Network security: In energy monitoring and measurement systems, a terminal and the main station need to use a public network for information transmission. Using VPN technology to establish an encrypted tunnel can ensure the safety of measurement information transmission on a public network. Additionally, we should leverage a firewall to limit unauthorized users' access to the internal network and internal network users' access to external systems. To reduce the pressure of network attacks on the firewall, it can prevent attacks that exploit operating system or protocol vulnerabilities by filtering packets entering the intranet or using network isolation techniques.

In the case of increasingly serious virus threats such as malicious code, bugs and Trojans, it is necessary to deploy vulnerability scanning systems, antivirus software, honeypot software and intrusion detection systems at key nodes of energy monitoring and measurement systems. To prevent attackers from exploiting system vulnerabilities to implant attacks, we can use human intrusion detection systems to monitor network data streams and discover external attacks in a timely manner.

In addition to applying the necessary technical means to ensure the safety of energy measurement information, the operation and maintenance of the intranet network is equally important. A rational means of preventing the leakage of internal information in energy monitoring and measurement systems is to deploy an intranet monitoring system.

System security: The information system can ensure its security through backups.

The backup of an energy measurement information system refers to backing up the core equipment and data information of the energy monitoring and measurement system so that the system can recover quickly and completely in the event of an accident. The backup of the core equipment mainly includes the backup of the core switch, the core router, the main station server and the power consumption server. The backup of data information includes the backup of configuration information and server data of the device. The data backup of energy measurement information systems is one of the important links in the routine maintenance of information systems. When related backup work is performed, the system can be restored promptly, quickly and effectively in the event of a system failure, ensuring the normal operation of the system. The operating system is connected to the hardware and other application software. Under network circumstances, the network security depends on the security and dependability of each host system. Without the security of the operating system, there will be no security of the host system and the network system.

Data Security: Data encryption technology is the most basic and core technology to ensure the safety of energy measurement information. Data encryption is a way of confusing information so that unauthorized people do not understand it. The data encryption process is implemented by a variety of encryption algorithms that can provide greater security at a lower cost.

The data encryption methods currently used can be classified into three types: soft encryption, hard encryption and network encryption. Soft encryption includes a more common cipher table encryption, software verification modes and license management methods. Hard encryption mainly includes encryption keys and encryption cards. Network encryption is different from software encryption methods and hardware encryption methods, and it performs encryption and decryption or verification work through other network-based computer or cryptographic devices. Network devices and clients communicate through secure channels. A typical data encryption process of an energy monitoring and measurement system is shown in Fig. 9.17.

As shown in the figure, the sender obtains the ciphertext C from the original information M by using the key K and the encryption algorithm and transmits it through the network. At the receiver's end, the ciphertext is interpreted. Illegal intruders can open passive or active attacks on the information transmission network, data may be modified on its path, or the entire system may send new data to the receiver. Passive attacks, although they are not a direct threat, are difficult to find. Active attacks are more destructive. They will change the ciphertext through some algorithm to form a new ciphertext C' and send it to the receiving end. The receiving end obtains the

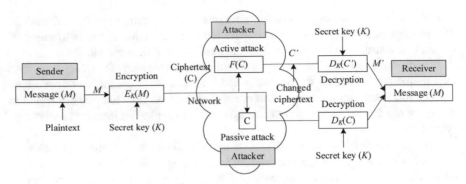

Fig. 9.17 Data encryption transmission process (Courtesy of [24])

changed original information M′ through the key and the decryption algorithm, but in most cases, it can be rapidly detected.

Conclusion

This section first introduces the basic concepts and requirements of information security in monitoring and measurement systems of the energy internet and then describes the security threats and attack means faced by energy measurement information systems. Finally, according to the analysis of security threats, it gives the corresponding defence strategies of energy measurement information systems.

9.6 Application Case Analysis

This section mainly describes existing typical energy internet projects, focuses on the composition and main technical features of different energy internet projects, and analyses the functions of the monitoring and measurement parts in the entire projects.

9.6.1 Energy Internet in a Smart Park

As a key part of the energy internet, the construction of a smart park energy internet project will speed up the development process of the energy internet. The park energy internet is based on local wind, photovoltaics, natural gas, and other distributed energy sources. It has a relatively high penetration of renewable energy. Through energy storage and an optimized configuration, local energy production can be achieved and energy loads can be supported in the smart park. In addition, it can flexibly connect to the public power grid according to the demand. The construction of an energy internet demonstration project in the park will promote the

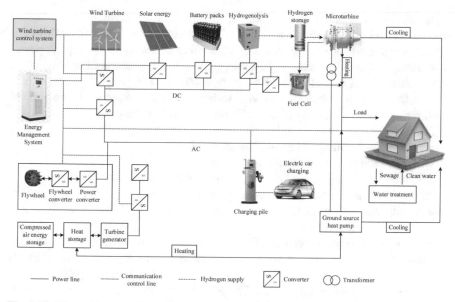

Fig. 9.18 The architecture of a Park Energy Internet System (Courtesy of [25])

development of a more dynamic electricity market and innovation and form a complete energy internet technology system and management system for the park. The structure of the park energy internet system is shown in Fig. 9.18.

The system mainly contains the following parts:

Photovoltaic power system: As an important part of the distributed renewable energy, a 400 kW solar photovoltaic power system is built in the park in the first step; in the second step, a 950 kW solar photovoltaic power system is built.

CCHP (combined cooling, heating and power) system: It serves as a standby power source for the park and relieves the pressure of the park peak period of electricity consumption. The park has two 1160 kW gas engine CCHP systems. The CCHP system uses natural gas as a fuel and uses a small-scale gas engine to generate electricity. Its power mainly meets the power load requirements for the park peak electricity charge. The residual heat generated by the prime mover can be absorbed by the heat exchanger or the absorption heat pump in the winter and can be cooled by driving the absorption chiller in the summer. Furthermore, the CCHP system can also supply domestic hot water. Its comprehensive energy utilization efficiency can reach 80–90%.

Air conditioning system: The park air conditioning system uses two chillers with a cooling capacity of 1,392 kW, two plate heat exchangers with a heat exchange capacity of 200 kW, and three sets of cooling water circulating pumps, a cold storage water pump, a cooling water pump and a water collection cooling tower with 2 low-noise groups. The park uses the company's mandatory fire-fighting pool (2000 m^3) to carry out secondary reforms such as thermal insulation and builds a "water storage" air conditioning system. By cooling water with electricity from the power grid at night, electricity is stored in the form of low-temperature cooling water. During the peak period of electricity consumption, cold water in the cold storage device is extracted and used for cooling to achieve the effect of transferring the peak power load, reducing the air conditioning operating costs, and improving the quality of air conditioning. The total cold storage capacity is 4000 RTH9.

Magnetic levitation vertical axis wind turbines: Five magnetic levitation vertical wind turbines with a capacity of 10 kW have been built in the park.

100 kW gas battery;

Electric vehicle charging and discharging station;

Uninterrupted power supply (UPS) system;

Micro energy management and service system; and cloud platform.

According to the overall development plan of the park energy internet, operation state monitoring, remote maintenance of equipment, energy efficiency analysis and prediction and user services of various distributed energy sources and smart devices in the park energy internet are realized. These functions need to use the cloud computing platform, big data technology, security protection technology and so on. Among them, the core cloud platform functional framework is shown in Fig. 9.19.

Operational state management: The park energy internet cloud platform will obtain operational data from various energy management systems in real time; provide system operation status monitoring, simulation analysis and operation management; and provide various forms of system interaction, such as large screens, control terminals and mobile terminals. The figure shows that the cloud platform is safe and reliable to operate, and the operating data are provided for support at the first time.

Remote monitoring: The system supports real-time remote monitoring and scheduling of various subsystems, such as energy storage, photovoltaics, gas and wind power, and is used in conjunction with the site's energy management system and production management system to coordinate and optimize the distributed power supply and load within the micro energy grid. Various regulating equipment is operated to ensure that each index of the subsystems is in a safe and stable state.

Remote maintenance of equipment: Remote maintenance, management and optimization of operating equipment through functions such as remote telescope monitoring, status monitoring and failure analysis of system equipment in the micro energy grid; once a subsystem appears unstable, it is fed back to the field manager and energy security experts on time. The daily maintenance and emergency maintenance of all equipment will be completed through expert remote technical guidance.

Energy Market Transactions: In conjunction with the main trading platform of the smart grid, the system realizes market trading functions such as user-level management, transaction price management, subsidy accounting and financial settlement and realizes multi-energy generation revenue for the company through the use of surplus electricity and grid connections to implement corporate energy operations.

Cloud platform system management: The core functions include resource allocation, service management, storage management, user management, rights management and security management, which ensure the stable and reliable operation of the cloud platform.

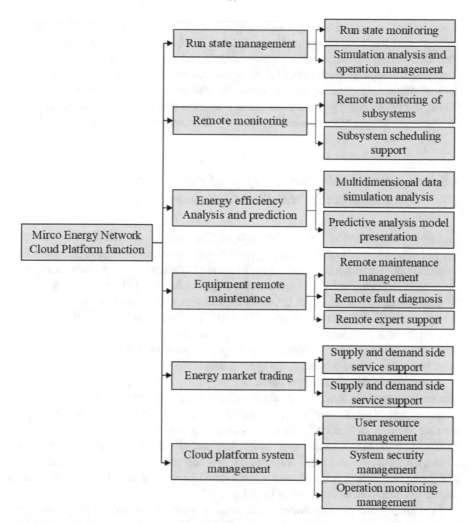

Fig. 9.19 Functional architecture of the Park Energy Internet cloud platform (Courtesy of [25])

   The park energy internet utilizes cloud platform technology and big data processing technology to make full use of energy monitoring and measurement data. These data include system parameters, park meteorological parameters and equipment operating status parameters. (1) For wind power generation, the monitoring data include wind speed, output voltage, output current, output power, grid-connected current, grid-connected power, power factor and voltage. (2) For photovoltaic power generation, the monitoring data include photovoltaic panels, temperature, light intensity, environmental temperature, grid-connected current, grid-connected power, grid voltage, power factor, output voltage and current, output power, and operating voltage of photovoltaic cells and arrays. (3) For the storage battery, the monitored data include indoor temperature and battery temperature, battery state of charge (SoC), battery voltage and battery input/output current. (4) For the load, the monitored data include user-side voltage, current, power, power factor and power consumption time. By analysing the measurement data of each measurement terminal and the monitoring data of smart equipment operation status, the efficient use of photovoltaics, wind power, energy storage, gas and other energy sources will be further enhanced. This provides an example of the promotion of energy internet demonstrations in the future.

9.6.2 Security Protection Analysis of Park Power Consumption Information Collection System

The power consumption information collection system is used as the key part of the park energy internet. It is mainly used for the collection and management of electricity consumption information for electricity-consuming enterprises, business users and residential users. Remote control is performed through the master system, acquisition terminal equipment, etc. With information collection, this system realizes the functions of power consumption information collection and cost control. It can be seen that the electricity information acquisition system is used as the collector of user information and the provider of basic information for the main station, and it must be well prepared to protect the system information.

According to the "Guidelines for rating the safety level of power industry information system protection", "Overall protection scheme for power secondary system security" and "Overall protection scheme for national grid company SG186 engineering security protection", the information acquisition system for power users' land will be deployed. In the information intranet of the State Grid Corporation of China, the information intranet is independently formed into a domain, and the security protection design is based on the three-level protection principle. The security protection architecture is shown in Fig. 9.20.

Specifically, the park electricity consumption information collection system can be divided into three parts according to the physical location: the main station, communication channel and collecting equipment. The physical devices of the main station network consist of the interface server [27, 28], disk array, application server, front server, crypto machines, firewall devices and other related devices. The communication channels represent the communication channels between the system's main station and all terminals, including fibre channels [29], GPRS/CDMA public network channels and 230 MHz wireless channels [30]. Acquisition equipment refers to terminals and metering equipment installed at the site, including dedicated terminals, concentrators, collectors and smart meters. The deployment of security protection equipment for the specific park electricity consumption information collection system is shown in Fig. 9.21.

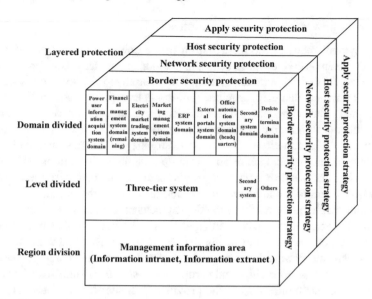

Fig. 9.20 The Security Protection Architecture of the Park Information Protection System (Courtesy of [26])

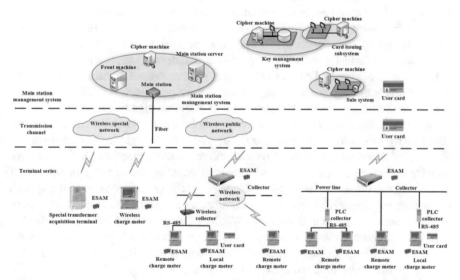

Fig. 9.21 Diagram of the deployment of security protection equipment in the park electricity consumption information collection system (Courtesy of [31])

(1) Security protection of the main station and channel

The deployment mode of the main station in the power consumption information collection system is divided into a centralized mode of the main station and a deployment mode of a distributed main station. The centralized deployment mode of the main station means that each province (autonomous region, municipality) deploys only one main station system and one unified communication access platform and directly collects all terminal and meter information within the province (municipality), focusing on information collection, data storage and business applications. The subordinate municipal companies do not set up separate main stations, and the users uniformly log into the provincial company's main station, authorize the use of data and perform operations and management functions within the region. A distributed main station deployment refers to the deployment of a main station system in the provinces (autonomous regions, municipalities) and cities to independently collect terminal and meter information within the region to achieve local area information collection, data storage and business applications. The provincial companies collect relevant data from various cities and complete the summary statistics of provincial companies and the applications of the province. Whether it is a distributed master deployment mode or a centralized master deployment mode, the overall framework of security protection is the same. Both need to protect borders, main stations, channels, collection equipment and application systems to meet the security protection needs of the entire system. To ensure the information security of the main station and related equipment, the communication channels of each equipment of the power consumption information collection system adopt independent private networks and independently deploy and adopt technologies such as firewalls, intrusion detection, vulnerability scanning and virus prevention at each main station [32]. A virus protection centre is deployed in the system, and antivirus clients are deployed on all computer terminals and servers to prevent malicious code, virus threats and hacker attacks. Two sets of firewall systems are deployed at the boundaries of each main station to implement boundary isolation and border policy protection: Deploy a server security enhancement system on the server to enhance the security of the server and ensure the security of server data. Deploy a security audit system in the system to monitor the operation behaviour of each server and terminal. Deploy a network intrusion detection system on the core switch of the system; deploy a full-hole vulnerability scanning system in the main station system of each provincial network; use a vulnerability scanning system to provide regular scanning services for terminal and server vulnerabilities; and apply patches in a timely manner.

The main station side deploys a useful information cryptographic service system and a front communication server, which are in the same local area network [32]. The electricity information cryptographic service system includes an electricity certificate service system and a crypto engine. The server/client working mode is adopted between the front communication server and the electricity information cryptographic service system, wherein the electricity information cryptographic service system is equivalent to the server and the front communication server is equivalent to the client. The front communication server initiates a request to the power information cryptographic service system, and the electricity information cryptographic service system is responsible for the functions such as signature verification, encryption and decryption processing. After processing, the result is sent to the front communication server to complete the identity authentication and data encryption and decryption functions.

The crypto machine mainly provides functions such as identity authentication, signature verification, key agreement, key update, encryption and decryption of key data, MAC calculation and data verification for the electricity information certificate service system. In the electricity information collection system, the asymmetric-key algorithm used in the content data interface is mainly used for terminal identity authentication, signature verification, key agreement and distribution and update of symmetric-key encryption algorithms. The symmetric key algorithm is mainly used for additional information fields specified in communication protocols such as setting parameters, controlling commands and data forwarding. China needs to carry cryptographic data processing of message authentication code field PW messages.

During the operation of the system, the front communication server should determine the data that needs to be encrypted according to the application layer function code and the requirements of Q/GDW3761-2012 and send the plaintext data that needs to be encrypted to the electricity information certification service system [33]. The electricity information certificate service system calls the crypto engine to encrypt the plaintext data and sends the encrypted ciphertext data and MAC to the front communication server. The front communication server packages the encrypted ciphertext data and then packages and issues the data. The data that do not need to be encrypted are processed by the cipher machine and are packaged and delivered directly by the pre-communication server. The packet encryption process is shown in Fig. 9.22.

(2)   Safety protection measures for collection equipment and measuring equipment

Collection equipment safety protection measures: In addition to the fact that acquisition equipment must have protection measures such as anti-theft, anti-damage, electrical safety and other protective measures, security modules should also be used to ensure communication data interface security within the system. Acquisition equipment includes dedicated terminals, collection concentrators and acquisition

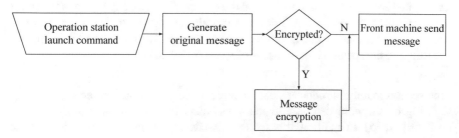

Fig. 9.22   Pre-message server encryption process (Courtesy of [31])

equipment. The collector does not need to install a security module. Other collection equipment must be equipped with a security module.

The security module is installed in the special collection terminal and concentrator for identification, security authentication, key information and secure transmission of sensitive information between the collection device and the main station and between the mining device and the smart energy meter. The encrypted data transmitted between the collecting device and the main station can be divided into management data and data transmission according to the functions. The transmission of management data implements identity authentication, key agreement, key distribution and update, MAC verification and data decryption as required by Q/GDW376.

The transmission data mainly uses the encrypted channel established between the main station and the collection device to transmit data. The main station and the smart meter can use this encrypted channel for data transmission. According to the requirements of the "power user electric consumption data acquisition system communication protocol of main station communication with data acquisition terminal", an encryption channel is established to transmit data in the form of data transfer. The forwarded data content is the data transmitted between the transmission data acquisition equipment and the smart meter to be carried out according to Q/GDW365-2012.

Metering equipment safety protection measures: In addition to the fact that metering equipment must have protection measures for burglary, vandalism, electrical safety and other protective measures, it should also adopt security modules to ensure the security of the communication data interface in the system. Metering equipment mainly refers to smart meters. The security module should be installed inside the smart meter and should at least have a symmetric-key algorithm.

The security modules installed in metering devices such as smart meters are mainly used for identification, key information and secure transmission of sensitive information between smart meters and main stations, between smart meters and collection devices, and between smart meters and infrared handheld devices.

The identification, key information and sensitive information communication between the smart meter and the main station mainly utilize the encrypted channel established between the main station and the collection device for data transmission. For example, the sales information ciphertext issued by the sales system should be transmitted between the smart meter and the collection device and between the smart meter and the infrared handheld device through the encrypted channel established between the main station and the collection equipment. The transmitted data interact according to the specific requirements of Q/GDW3652012.

Conclusion

This section mainly introduces existing typical energy internet projects, especially focusing on the composition and main technical features of the energy internet projects [34, 35]. In addition, this section also deeply analyses the functions of the monitoring and measurement parts in the whole projects [36].

Acknowledgements This research was funded by the National Key Research and Development Program of China (2017YFB0903000); National Natural Science Foundation of China (61571324); and Science and Technology Project of the Headquarters of the State Grid Corporation of China (5700-201946239A-0-0-00).

References

1. C. Shu-Yong, S. Shu-Fang, L.I. Lan-Xin, et al., Survey on Smart Grid Technology. Power System Technology 33(8):1–7 (2009)
2. D.G. Hart, Using AMI to realize the Smart Grid. Paper presented at the Power and Energy Society General Meeting-Conversion and Delivery of Electrical Energy in the 21st Century (2008)
3. Y. Liu, Wireless sensor network applications in smart grid: recent trends and challenges.International Journal of Distributed Sensor Networks 8(9): 492819
4. L. Wenpeng, W. Guan W, X. Daqing, Advanced metering infrastructure solution supporting multiple services and business integration. Proceedings of the CSEE 34(29): 5088–5095 (2014)
5. O.M. Longe, K. Ouahada, H.C. Ferreira et al., Wireless sensor networks and advanced metering infrastructure deployment in smart grid. Paper presented at International Conference on e-Infrastructure and e-Services for Developing Countries. Springer, Cham (2013)
6. Z. Dong, J. Zhao, F. Wen et al., From smart grid to energy internet: basic concept and research framework. Automation of electric power systems, 38(15):1–11 (2014)
7. A.G. Phadke, Synchronized phasor measurements-a historical overview. Paper presented at Transmission and Distribution Conference and Exhibition 2002, Asia Pacific. IEEE/PES (2002)
8. De La Ree J, Centeno V, Thorp J S et al (2010).Synchronized phasor measurement applications in power systems. IEEE Transactions on smart grid 1(1):20–27
9. Bian D, Kuzlu M, Pipattanasomporn M, et al (2014). Analysis of communication schemes for Advanced Metering Infrastructure. Paper presented at PES General Meeting Conference & Exposition, IEEE
10. A. Mahmood, N. Javaid, S. Razzaq, A review of wireless communications for smart grid. Renew. Sustain. Energy Rev. **41**, 248–260 (2015)
11. G. Mulligan, The 6LoWPAN architecture. Paper presented at Proceedings of the 4th workshop on Embedded networked sensors, ACM (2007)
12. Gungor VC, Sahin D, Kocak T et al (2011). Smart grid technologies: Communication technologies and standards. IEEE transactions on Industrial informatics 7(4):529–539
13. Cao J, Yang M(2013). Energy internet–towards smart grid 2.0.Paper presented at Networking and Distributed Computing (ICNDC),2013 Fourth International Conference on.IEEE,2013
14. Xu Y, Zhang J, Wang W et al (2011).Energy router: Architectures and functionalities toward Energy Internet, IEEE International Conference on Smart Grid Communications,IEEE, October 2011
15. C. Junwei, Y. Zhongda, M. Yangyang et al., Survey of Big Data Analysis Technology for Energy Internet. Southern Power System Technology **9**(11), 1–12 (2015)
16. L. Zhijian, W. Mingyu, The application of summingbird cloud computing platform in energy internet. Computer Science and Application. **5**(12), 464–471 (2015)
17. D. Wang, L. Yang, Stream Processing Method and Condition Monitoring Anomaly Detection for Big Data in Smart Grid. Automation of Electric Power Systems **40**(14), 122–128 (2016)
18. Y. Li, J. Wang, X. Wang, A Power System Network Loss Evaluation Method Based on Hybrid Clustering Analysis. Automation of Electric Power Systems **40**(1), 60–65 (2016)
19. K.Y. Lee, Y.T. Cha, J.H. Park, Short-term load forecasting using an artificial neural network. IEEE Trans. Power Syst. **7**(1), 124–132 (1992)

20. L. Li, J. Wei, C. Li et al., Prediction of Load Model Based on Artificial Neural Network. Transactions of China Electrotechnical Society **30**(8), 225–230 (2015)
21. W. Hu, L. Zheng, Y. Min et al., Research on Power System Transient Stability Assessment Based on Deep Learning of Big Data Technique. Power System Technology **41**(10), 3140–3146 (2017)
22. Hall David L, James Llinas(1997). An introduction to multisensor data fusion. Proceedings of the IEEE 85 (1): 6–23
23. L. Ying, *Safety protection technology for energy metering collection and billing* (China Electric Power Publishing House, China, 2014)
24. J.B. Ekanayake, Jenkins N, Liyanage K, et al(2012). Smart Grid: Technology and Applications. Wiley Publishing, p 69–76
25. Qingdong F(2015). Energy internet and smart energy. China Machine Press, China
26. Zhang Tao, Lin Weimin, Ma Yuanyuan, et al(2016) Power Information Network Security. Xi'an University of Electronic Science and Technology Press, Xi'an
27. Ting Yang, Young Choon Lee, Albert Zomaya, Collective Energy-Efficiency Approach to Data Center Networks Planning. IEEE Transactions on Cloud Computing **6**(3), 656–666 (2018)
28. Ting Yang, Haibo Pen, Albert Y. Zomaya, An Energy-efficient Storage Strategy for Cloud Datacenters based on Variable K-Coverage of a Hypergraph. IEEE Trans. Parallel Distrib. Syst. **28**(12), 3344–3355 (2017)
29. Ting Yang, Rui Zhao, Weixin Zhang et al., On the Modeling and Analysis of Communication Traffic in Intelligent Electric Power Substations. IEEE Trans. Power Delivery **32**(3), 1329–1338 (2017)
30. Ting Yang, Wenping Xiang, Linqi Ye, A Distributed Agents QoS Routing Algorithm to Transmit Electrical Power Measuring Information in Last Mile Access Wireless Sensor Networks,International Journal of Distributed Sensor Networks,2013.12.29,2013:1078–1086
31. Liu Ying(2014). Safety protection technology for energy metering collection and billing. China Electric Power Publishing House, China
32. Ting Yang, Haibo Pen, Albert Y. Zomaya, An Energy-Efficient Virtual Machine Placement and Route Scheduling Scheme in Data Center Networks. Future Generation Computer Systems **77**, 1–11 (2017)
33. Wei Li, Ting Yang, Flavia C. Delicato, Paulo F. Pires, Zahir Tari, Samee U. Khan, Albert Y. Zomaya, On Enabling Sustainable Edge Computing with Renewable Energy Resources. IEEE Commun. Mag. **56**(5), 94–101 (2018)
34. T. Yang, Y. Zhao, Z. Wang, Data Center Holistic Demand Response Algorithm to Smooth Microgrid Tie-line Power Fluctuation. Appl. Energy **231**, 277–287 (2018)
35. T. Yang, H. Pen, D. Wang et al., Harmonic analysis in integrated energy system based on compressed sensing. Appl. Energy **165**, 583–591 (2016)
36. T. Yang, Z. Wang, C. Chang, Feature knowledge based Fault Detection of Induction Motors Through the Analysis of Stator Current Data. IEEE Trans. Instrum. Meas. **65**(3), 549–558 (2016)

Chapter 10
Utilization of Big Data in Energy Internet Infrastructure

Songpu Ai, Chunming Rong, and Junwei Cao

Abstract With the maturation of technologies such as communication, sensors and networks, very large amounts of data are generated that are available for processing and analysis. With the popularizing process of the Internet of Things (IoT), available data resources will become even more plentiful and diversified in the near future. To provide feasible solutions in data science for the key features of the energy internet, such as energy interconnection and routing, a big data architecture could be utilized in the energy internet infrastructure to provide large-scale analysis of massive various types of data. In this chapter, the utilization of big data in the energy internet infrastructure is explored. A three-layer big data architecture for usage in the energy internet is presented. The characteristics of data utilized in the energy internet and the potential requirements of the energy internet for the big data architecture are studied. Then, analytics methods that could be executed in the energy internet big data infrastructure are introduced. Real-time and offline analyses, as two types of analysis modes for different requirements of application scenarios, are described. Several well-known open-source big data tools are discussed. In addition, the open challenges of utilizing big data in the energy internet are proposed.

S. Ai (✉) · C. Rong
Department of Electrical Engineering and Computer Science, University of Stavanger, CIPSI, IDE, UiS, 4036 Stavanger, Norway
e-mail: aisongpu@outlook.com

C. Rong
e-mail: chunming.rong@uis.no

S. Ai · J. Cao
Research Institute of Information Technology, Tsinghua University, 4-308, FIT, Beijing 100084, China
e-mail: jcao@tsinghua.edu.cn

© Springer Nature Switzerland AG 2020
A. F. Zobaa and J. Cao (eds.), *Energy Internet*,
https://doi.org/10.1007/978-3-030-45453-1_10

10.1   Introduction

Managing and analyzing data have always been challenges in the energy industry across infrastructures with different scales and distributed locations. Electricity companies on the generation side, network side and consumption side have always struggled to develop reliable approaches to capture and analyze information to support scientists and researchers in achieving a better understanding of their products, services and customers to offer advanced solutions and save cost. In the past forty years, collected data has been analyzed utilizing database technologies. In recent years, however, with the maturation of technologies such as communication, sensors and networks, a very large amount of data is generated that is available for processing and analysis. With the popularizing process of the Internet of Things (IoT), available data resources will become even more plentiful and diversified in the near future.

With this incoming surge of data, the generated data have the following notable characteristics: *Volume, Velocity, Variety* and *Variability*, which are the so-called "4 Vs" [49]. For the energy internet, specifically, the volume of generated data could be very large. Every device within the energy internet can generate logs with a regular interval or by events. Persons can also create records whenever required. Over time, the amount of data to be analyzed is considerable. Additionally, the velocity of data created in the energy internet can be especially rapid. To provide real-time analysis and execute further processing later on, the generated data flow needs to be stored, analyzed and visualized in a timely manner, which is the characteristic of velocity. In addition, data are generated from a variety of sources in- and outside of the energy internet with a variety of types, for instance, sensor readings, images, videos, ecommerce records, and social media streams. Each data source can be independently collected and analysed. In addition, utilizing multiple types of integrated heterogeneous data to explore their interaction and the insight between them to achieve the purpose of analysis is an interesting and required topic [64]. Simultaneously, the variability of the above three Vs, which refers to changes in data volume, velocity and variety over time or between different energy internet subnets, must be considered. The changes include data flow rate, format, quality, etc.

In addition, many additional Vs can be presented to summarize the characteristics of big data in the energy internet. For example, the *Veracity* [23] of each data resource, which refers to the noise, biases, missing and abnormality in the data or the unmeasurable certainty of truthfulness and trustworthiness of the data, is required to be considered; *Valence* [32] refers to the connectedness of big data. Data items always relate with each other directly or indirectly. To find novel relationships between diverse types of data, involving additional data into consideration for comprehensive analysis with the support of big data techniques is the work of researchers. Although we can keep listing some other Vs that are attributed to big data, we prefer to utilize the 4 Vs we discussed in the previous paragraph as the primary characteristics.

However, one special V, *Value*, that we should mention here is the fundamental motivation of the entire big data technology revolution [44] and is also an important reason that we should consider the utilization of big data in the establishment of the energy internet. Big data is important because we believe that there is a huge amount of untapped value among the collected data [27]. Even though the data are too large to process with existing tools, the data arrive too rapidly to store and index optimally by centralized database technologies, and the data are too heterogeneous to fit any rigid schema, there is still enough value to dedicate capital and time into the corresponding research and development to break through technical problems and achieve additional insight from the data.

Big data is not a new technique. It is an association of existing and novel technologies. It is a scalable architecture of efficient storage, manipulation and analysis. Traditional data architectures are no longer efficient enough to operate data through the architectures with the characteristics of volume, velocity, variety and variability simultaneously [37].

Big data has attracted attention in both academia and industry. Many ongoing and achieved innovations are engaged in or on big data technologies that obtain considerable results [28]. In business, enterprises utilize big data technology to understand and forecast consumer behaviour from all kinds of data sources they could collect. The quality, price and improvement of products or services rely on the results of big data analytics. Management departments are also able to utilize big data to optimize company operational efficiency and reduce personnel costs.

In health care, big data has been utilized to analyze and forecast patient condition and disease progression, for example, analyzing and comparing pathogenic characteristics integrating with patient physique through a very large number of medical cases. A more precise and customized treatment suggestion can be given by a big data system to assist the doctor in diagnosis and to reduce the incidence of patients.

Social media is another great known example of applying big data. Through analyzing user online behaviour, including but not limited to instant messages, published content, online social networking and sharing activities, platforms such as Facebook, Twitter and LinkedIn are able to understand user behaviour patterns and preferences to improve service quality and efficiency. For example, if LinkedIn is aware that Alice works on machine learning using big data analysis, the website background will then more likely tend to push news and advertisements related to machine learning to Alice. A notorious case of abuse of big data in social media is the Facebook scandal. Cambridge Analytica Ltd. (CA) acquired approximately 50 million Facebook users' personal data and utilized the analysis results in the 2016 US presidential election and UK Brexit referendum in a tendentious manner. Facebook lost users and popularity. In addition, CA has filed for bankruptcy. It is a warning for all researchers and companies that protecting the privacy of data is an important principle, including research and development in the energy field. Additionally, the General Data Protection Regulation (GDPR) (EU) became enforceable in May 2018. It gives all individuals within the EU control of their personal data. China and the US also have similar legislation being developed. We remind the reader here to comply with local laws while conducting research and development.

The utilization of big data in the energy field shows that existing studies generally focus on local and regional solutions with single functionality, as surveyed in [36]. More existing studies have been conducted to establish smart grid or smart city solutions. Project and research objectives are mostly limited to the microgrid level. Research on large-scale, multi-functional big data platforms operating in the energy internet is rare. However, as we discussed above, the capability of performing large-scale analysis on massive amounts of data is the critical advantage of big data technology. To provide feasible solutions in data science for the key features of the energy internet, such as energy interconnection and routing, big data architecture, analysis methods and platform building in the energy internet are introduced and analyzed in this chapter as a reference for energy internet researchers to better utilize big data as a powerful tool. In addition, the open challenges of utilizing big data in the energy internet are discussed. The remainder of this chapter is structured as follows. Section 10.2 introduces the architecture of big data specifically in the energy internet field. Big data analysis methods are summarized in Sect. 10.3. Section 10.4 provides an overview of big data platform building. The existing open challenges of utilizing big data in the energy field are presented in Sect. 10.5.

10.2 The Architecture of Big Data

If we focus on the technical level, big data in fact is an integrated term for a stack of technologies. The realization of big data requires an appropriate combination and cooperation among technologies from various disciplines. Relative technologies include data collection, storage, management, manipulation, analysis, results display, etc. Even though for each application scenario with specific conditions and requirements, a particular big data stack should be tailored for the implementation, a similar architecture is utilized in most of the implementations.

In this section, a layered reference architecture is presented to discuss the technologies of the big data stack for the energy internet, while the specific characteristics and requirements of the big data architecture targeting the energy internet scenario are presented.

10.2.1 The Architecture of Big Data

Big data is a flourishing technology stack. Novel components and functions are booming in the big data landscape, enriching and evolving the entire stack constantly. Diverse architectures are proposed and implemented in research and industry. Industry giants such as Google, Microsoft and IBM as well as academic leaders such as the University of Amsterdam and University of California, San Diego, all hold their own proposed architectures. Nevertheless, certain key components/common tasks are present as layer-like structures in the majority of architectures, which are data collection, the storage and management layer, the data analytics layer and the application layer, as shown in Fig. 10.1.

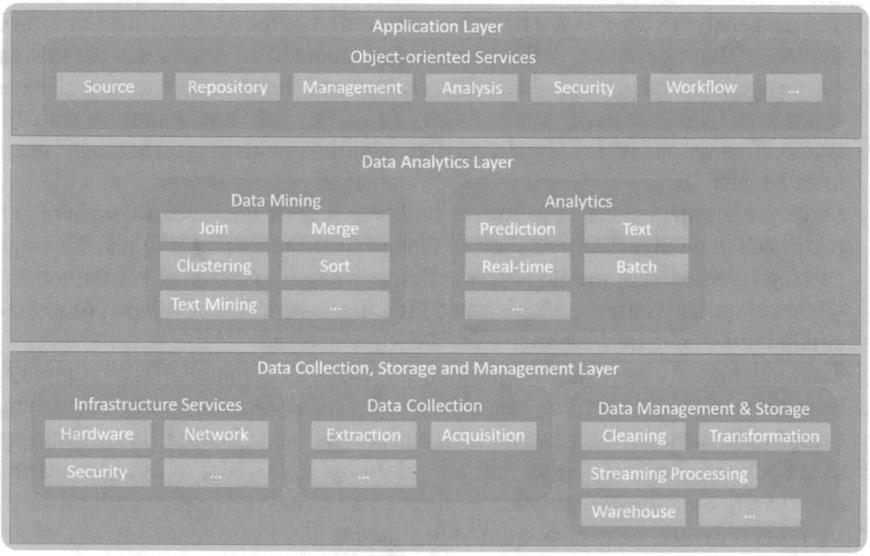

Fig. 10.1 Layer-like structure of big data

- Data Collection, Storage and Management Layer

  The data collection, storage and management layer is the basis of the entire big data architecture. As its name says, this layer first engages in collecting all types of data. The data can be structured, semi-structured or unstructured data records or data streaming from diverse data sources such as sensor readings, images, videos, ecommerce records and social media platforms.

  Then, considering scalability, reliability, security, stability and other reasons such as the data size and cost of communication, data are often stored in a distributed file system, such as the Hadoop Distributed File System (HDFS) [17] with duplications. The data can be stored as either SQL or NoSQL.

  Since the system is distributed, this layer also performs tasks to maintain the functionality of the entire system when node (a distributed server where data is stored) failure happens. Data should be able to be extracted, changed and deleted by the upper layers while nodes leave and return.

  Above the storage system, a global resource manager (such as YARN [10]) is allocated to manage data usage and computational power.

  To handle data with different formats, various data management tools are also engaged, such as Gephi [33] for graph data and MongoDB [45] for documents.

  In the big data architecture, the data collection, storage and management layer offers proper data material for further utilization by the analysis and application layer. Several open-source project-based big data services for the data collection, storage and management layer, such as HDFS and YARN, are introduced in Sect. 10.4.

- Data Analytics Layer

The data analytics layer is the intermediate portion of the architecture and is also the core of the entire stack. In this portion, data stored in the distributed file system can be operated and processed in real time, near real time or afterwards to provide diverse analytic results for applications. To realize data manipulations on data stored in numerous nodes simultaneously, parallel computing technology needs to be adopted [67].

Since the big data stack is normally built on at least hundreds of nodes [51], traditional parallel techniques such as Open Multi-Processing [50] and Message Passing Interface [46] are no longer efficient enough to fit the novel implementation condition. Thus, the MapReduce [18] framework was developed to realize parallel processing on a large-scale and with scalability.

The fundamental idea behind MapReduce is "moving computation is cheaper than moving data" [17]. Therefore, MapReduce puts computational power and activity towards the data side and just obtains the results back, instead of transmitting data back and forth. By following this philosophy, the cost of communication and the stress on the computation centre can be reduced. However, MapReduce supports only one type of programming model, the map-reduce model, which is introduced in more detail in Sect. 10.4.2.4. With continuous technological advancement, more flexible, object-oriented processing frameworks supporting more parallel programming models have been developed and released for both batch and stream processing of data mining and analysis, for instance, Spark [8] and Flink [6]. More details about the well-known processing framework Spark are discussed in Sect. 10.4.2.4.

By using batch and stream processing frameworks, big data analytical approaches such as deep learning can be achieved to study the association between variables and provide predictions/summaries to obtain the untapped value from big data. Several typical big data analytical approaches are presented in Sect. 10.3.

The data analytics layer in the big data architecture handles the data analysis requirements from the application layer and submits deep analytic, predictive analytic and/or summary analytical results for further utilization by top-layer applications.

- Application Layer
  The application layer is the topmost layer of the big data architecture. This layer offers object-oriented services.

  There are many tools that have been developed to realize different specific services. For instance, Hive [11] is a data warehouse infrastructure offering ad hoc querying. Moreover, we should note that there is no obligation that a tool should only belong to the data analytics layer or application layer. Depending on requirements, one tool, for example, Hive, can be either a portion of an integrated solution, as utilized in Facebook Messages [21], or an application itself, such as a data query platform [62], which is utilized by Facebook as well. Additionally, there is no restriction that a product/platform covers only one layer. As an example, the Spark [8] platform provides an in-memory big data processing framework for analyses and predictions as well as multiple practical tools that help it to be engaged in distinct types of applications. From this point of view, Spark covers both the intermediate and top layers. Furthermore, the Microsoft big data platform, Dryad [43], is capable of covering the development of an entire stack as utilized.

Moreover, tools can be adopted in combination to implement more complex functionalities. Furthermore, a stack should be able to run multiple services. Hence, a workflow management system is required in the big data stack to integrate, schedule, coordinate and/or monitor tools and services.

In addition, analysis results or extracted data items should be capable of interacting with other programmes such as visualization tools (for instance, Tableau [61]) and decision support mechanisms at this layer to assist people in understanding and handling the situation well.

Through the implementation of the three-layer architecture discussed above in the energy internet, this approach could be able to realize an entire workflow from big data acquisition, management and analysis to responses for a massive volume of various types of data that are generated within the energy internet or are captured from related resources with variability in data formats and generation velocities.

10.2.2 Implementing Big Data in the Energy Internet

In general, to build the energy internet through informatization and intellectualization to solve problems including improving equipment utilization, safety and reliability, power quality and access to renewable energy, the introduction of big data analytics into the energy internet appears to be an indispensable technology roadmap at present [41]. Determining how to integrate the software and hardware requirements of the big data stack together with the requirements of the energy internet and existing energy and communication infrastructure to create an efficient and affordable solution is a significant issue that energy internet practitioners need to solve jointly.

In this subsection, the characteristics of generated data in the energy internet and some potential requirements of the energy internet for big data architecture are discussed.

10.2.2.1 Characteristics of Energy Data

With the development of IoT technology, many devices in the energy internet are capable of generating and publishing data. Discussing and studying the characteristics of data generated in the energy internet is important for us to establish the big data stack and provide big data analyses and prediction services. From our perspective, data in the energy internet have the following characteristics:

- Data volume is very large and is generated with fast speed, including a large amount of streaming data.

  In the energy internet, each equipment can keep generating its status log, including but not limited to its U, I, P and Q records. The generation frequency can be quick enough by selecting appropriate configurations. Hence, collecting, processing, analyzing and giving responses through the big data stack with very large throughput is a challenge that needs to be solved, which is further discussed in Sect. 10.5.

- Considerable amounts of data generated in the energy internet are monitoring data, which are not critical for energy internet operations. However, once a situation fluctuates, the useful information is very dense.
  Data generated by equipment are mostly in a regular situation when the power prediction, scheduling and management procedures of the energy internet work well. Hence, in many circumstances, the data collected in the energy internet are used to confirm the prediction and management results and to monitor the situation. However, if an unexpected situation appears, the density of useful information in the data is high. Real-time responses need to be given through the analysis results derived from the data.
- The main portion of the data includes the real-time status information throughout the grid (such as U, I, P and Q) as well as communication data about demand and supply. Additionally, there are various types of data generated outside the energy internet that should also be captured, stored and analyzed by the energy internet big data stack, such as weather predictions and holiday pattern announcements.

10.2.2.2 Potential Requirements in Energy Internet Big Data Architecture

Combining the expectations of the energy internet to increase equipment utilization, promote safety and reliability, boost power quality and advance renewable energy access, with the characteristics of energy internet data, we consider that the energy internet big data architecture should have the following requirements:

- Capability to process and monitor large-scale mass streaming data in real time.
  To collect, process and manage data generated for the energy internet, the big data stack should be able to handle large-scale mass streaming data in real time. However, a certain amount of real-time data is not critical for energy internet operations and does not have a high information density for big data analyses. Therefore, the stack should be capable of "just" monitoring these data streams but not paying "too much" attention to them unless they are called by some applications (in real time or as historical records).
- Capability to process and analyze in parallel large-scale nodes whenever necessary.
  When a demand or supply fluctuation appears, the energy internet big data stack should be capable of providing auto-decision making or decision support through processing and analyzing the streaming data as well as historical data. If the fluctuation affects the load/supply scheduling in a wide area, parallel processing and analysis of the large-scale streaming data are demanded in the energy internet big data architecture.

- Low-latency real-time feedback, response and decision making/support.
  The requirement of energy internet big data analysis on timeliness is sensitive. Real-time treatments, such as transient state balancing, are always important for power quality, grid stability, etc. In addition, response and decision making/support here do not refer to a centralized mechanism but to a distributed decision making/support mechanism or framework. By implementing treatments at different locations and on various infrastructures simultaneously, even lower latency is expected to be achieved.
- Scalability, stability and robustness in data analysis and prediction.
  The energy internet big data stack should be scalable along with the development of urban/rural areas, access to renewable energy, etc. In addition, the analysis and prediction should be robust and stable during issues such as network failure and grid damage. Furthermore, the stability and robustness of the stack are also crucial since the services that the stack offers relate to the most important terminal energy in modern society—electricity.
- Security of data, communication and decision making, as well as protection of user privacy.
  Historical energy data are required to be stored and manipulated safely. Similarly, the corresponding communication and decision making both need to be performed with security guarantees. In addition, energy data are relevant to privacy, not only for individuals but also for enterprises. Thus, the protection of user privacy is also needed for the big data stack.

10.3 Big Data Analytics Methods

In the energy internet field, massive various types of data are collected and required to be analyzed [36]. With the development and release of new generations of big data processing frameworks and service platforms, an increasing number of database transformation operations, such as add, join and filter, are supported. Most traditional analytical methods are able to be implemented in a big data stack easily and efficiently. There are tools or tool combinations that can provide similar data operations, interfaces and language environments as database management systems for a big data stack, such as Apache Impala [14], to make their users easily switch from a database to a big data stack.

The analytical capability of a big stack is much greater than this. Along with the iterative operations that are supported by many analytical frameworks, machine learning as an intense user of iterative operations is widely utilized in big data analysis. Analytical applications include regression, classification, clustering, etc. Regarding the type of learning, the tasks can be classified into two main categories: supervised learning and unsupervised learning.

In this section, the two categories of big data analysis methods are introduced. Several commonly used analysis methods in the two categories are discussed. Moreover, deep learning and ensemble learning are presented as two notable methods that have promising potential in the energy internet field.

10.3.1 Supervised Learning

Supervised learning is a category of machine learning that learns the mapping between an input data set and the output data set (target). The objective of supervised learning is to precisely forecast the target for a given input. Frequently utilized supervised learning models include regression, Random Forest (RF), adaptive boosting (AdaBoost), Naive Bayes, Artificial Neural Networks (ANNs), K-Nearest Neighbours (KNN), Support Vector Machine (SVM), etc.

Regression analysis, including linear regression, logistic regression, etc., is used to find the most likely mathematical explanation between the dependent variable and independent variables. The advantage of this approach is that after the learning process, we obtain the mathematical relationship of the problem. By checking the residuals, the accuracy of the mathematical model can be verified. However, regression analysis needs researchers to assert the type of mathematical expression manually, which is an empirical task. More details on regression analysis can be found in [65].

RFs use decision trees as base prediction models to classify input data with different labels [22]. A decision tree is a tree-like graph used to model the different classifications among sets of input features. An RF first trains multiple individual decision trees constructed with a certain number of randomly selected features. In these trees, leaves correspond to the target classes, and branches represent the different variables of features selected by the tree that lead to those class labels. To classify a test input vector, it is passed through the trees in the forest. The forecast results are voted on by all trees with their own decisions.

Adaptive boosting (AdaBoost) is another type of learning method that uses a sequence of decision trees as base prediction models to produce a forward stagewise additive model. In AdaBoost, an additional weight feature is allocated to each training sample [57]. The weight of a sample is a response to the importance of forecasting the particular sample correctly. In each boosting iteration, according to the forecast result, a vote coefficient of the iteration is calculated, and the weight of each sample is updated. Incorrect training samples get higher weights to receive more attention in the next iteration of operation. A better iteration yields a higher vote coefficient in the final vote. More details on AdaBoost can be found in [24].

A Naive Bayes algorithm is an efficient probabilistic classifier that applies Bayes' probability theorem with the assumption that the input features are independent of each other [38]. According to the distribution of input features (for instance, Gaussian or multinomial), disparate Naive Bayes classifiers have been developed. More details on Naïve Bayes can be found in [47].

Support Vector Machine (SVM) is a machine learning model that attempts to find the optimal hyperplane to separate a dataset into two groups [60]. The direction of optimization is to find the hyperplane that has the maximum margins with the two groups. Margin here is the distance between the hyperplane and the data points (one or multiple) that are the closest to the hyperplane. The so-called support vector refers to the vector from the hyperplane directed towards the closest points. SVM was adopted in [52] to forecast the energy market and utilized in [26] to assess the solar radiation of a day to assist renewable energy generation.

K-Nearest Neighbours (KNN) is a widely used instance-based classification algorithm. It provides class forecasting based on the "distance" relationship between a testing sample and training samples. The "distance" can be decided by any distance functions (e.g. Manhattan, Euclidean, Minkowski and Hamming) [63]. The classification is computed through a simple/weighted vote within a certain number of nearest neighbours of the testing sample. It was utilized in [1] as a component algorithm in an ensemble learning method to recognize the activities of ageing people in smart homes.

Artificial Neural Networks (ANNs) are widely utilized in smart home solutions, such as electrical appliance power profile classification in residential buildings [59] and human activity recognition in smart homes [20]. These networks model a biological neural network as an interconnected group of nodes called neurons. In each neuron, the weighted summation of an input vector is sent to the activation function to attain the output of the node. Normally, nodes in the network are structured as feedforward layers. Feedforward here refers to the fact that the outputs of these layer nodes are used as the inputs of the next layer. The data flow from the first (input) layer to the last (output) layer without looping back. When an input row is processed through the network by edges, a network output is obtained. An ANN learns by updating the neurons' weights after each training iteration. The weights are updated in the direction of minimizing the cost function of the problem.

10.3.2   Unsupervised Learning

Unsupervised learning uses unlabelled data as input to let the machine study the "structure" of the data. Unsupervised learning is normally utilized on problems that do not have a determined solution, which is also the reason why the data are unlabelled. Hence, it is hard to evaluate the results, which is an important difference between unsupervised and supervised learning. Commonly adopted unsupervised learning approaches include k-means clustering and ANNs. Within the energy internet big data stack, unsupervised learning could be useful for tasks such as auto-classifying sudden events to learn manual treatments to provide auto-decision making and decision support services.

K-means clustering uses the distances between each data item and the cluster centroids in the dataset vector space to classify data items into a certain number of clusters. The number of clusters should be assigned manually. The initial centroids are randomly selected from the vectors where the data items are held. The clusters are updated after each iteration of study by renewing the centroids of the clusters. The learning process finishes when the centroids no longer change or the changes remain within a certain range.

The learning process of an ANN in unsupervised learning is similar to that in supervised learning. The cost function is decided depending on the task, for example, the objective of unsupervised learning, as well as a priori assumptions such as the implicit properties of the model and observed variables.

10.3.3 Deep Learning

Deep learning attempts to use a multi-layer structured learning model to study the data, which can be both supervised and unsupervised learning. Neural networks, as an important approach to structure multi-layer architectures, have been widely discussed and adopted in deep learning studies. Neural network-based deep learning architectures, such as deep neural networks, recurrent neural networks and convolutional deep neural networks, have become representative deep learning approaches.

In general, a Deep Neural Network (DNN) refers to an ANN with multiple feedforward hidden layers. Hidden layers are the intermediate layers between the input layer and output layer. However, in simple ANNs, only one hidden layer is involved. By introducing multiple hidden layers into the network, DNNs have the potential to model complex data with fewer neurons than single hidden layer networks [30].

A Convolutional Neural Network (CNN) is another type of feedforward neural network that consists of one or multiple convolutional layers, pooling layers, fully connected layers, etc. CNNs have achieved better performance in image recognition studies [42].

A Recurrent Neural Network (RNN) connects neurons by a directed graph along with a sequence, which enables the so-called "memory" in the network. It is capable of exhibiting dynamic behaviour on time series. Long Short-Term Memory (LSTM) [34] and Gated Recurrent Unit (GRU) [29] are two types of complex recurrent units for RNNs to promote the advantage of memory by gated state [3].

10.3.4 Ensemble Learning

Ensemble learning works to integrate multiple learning methods to obtain better analytical performance rather than focusing on a specific algorithm [55]. It was developed based on evaluation results of different machine learning models for historical data. A common realization of ensemble learning is taking a vote among multiple promising models to obtain the forecast result [54].

In addition, many other model integration architectures have been adopted in studies to adapt to realistic requirements, which is called hybrid ensemble learning

[2]. Prediction of device status in smart homes was proposed by integrating random forest and gradient boosting as the basis of an ensemble method [19]. Using a hybrid ensemble learning model to predict time series data with a weighted mean of the forecast results of several algorithms was discussed in [66]. In [2], a two-layer hybrid stacking ensemble learning model was employed to forecast EV charging demand.

Moreover, one additional objective or benefit of utilizing ensemble learning could be to improve the stability of the analysis [3], which is particularly required by the predictions in the energy internet.

10.4 Big Data Platforms

This section provides an overview of big data platform building in the energy internet. Real-time and offline analysis platforms are discussed. Some well-known open-source big data tools that could be used in energy internet big data platforms are introduced as well.

10.4.1 Real-Time and Offline Analysis

When we consider establishing a big data platform, or stack based on its structural characteristics, the three-layer architecture discussed in Sect. 10.2 or a similar architecture is often followed. There are numerous tools for each aspect of the stack. The tools provide us with diverse kinds of services with different features. When we choose from them to build our stack, speed and scale are the two essential aspects that we need to consider. Due to the requirements of different application scenarios, two types of analysis modes, real-time and offline analysis, should be available in the energy internet big data stack based on the trade-off between speed and scale.

10.4.1.1 Real-Time Analysis

Real-time analysis, just as its name suggests, tends to access, process, analyze data and to give responses as quickly as possible [39]. It is normally utilized in situations where the situation constantly changes, immediate analyses are required, and the response should be executed with very short latency.

Implementations of real-time analysis mainly use two types of structures. One uses a traditional relational database in a parallel processing cluster, which is not capable enough to satisfy the growing requirements of speed and scale. The other structure is integrating in-memory analytics platforms with a distributed file system [40], which is very good for performance.

Existing tools include portions or the entire platform of Spark [8], Storm [9], Beam [4] from Apache, Greenplum [31] from EMC and HANA [56] from SAP, etc.

In the energy internet, potential utilization areas include real-time electricity demand prediction, pricing adjustment, load auto-balancing/scheduling, etc.

10.4.1.2 Offline Analysis

In contrast to real-time analysis, offline analysis focuses on more comprehensive data processing and analysis. It is able to make use of larger amounts of and more complex data in more sophisticated analysis methods.

For offline analysis implementations, data are normally already acquired and stored in the stack in advance, or the incoming data rate and response requirement are not high. Then, more sophisticated and, at the same time, more time-consuming operations are able to be utilized during the data processing. For example, we can realize iterative operations with fewer restrictions in offline analysis, which is one of the fundamental necessities of employing deep learning.

It should be mentioned that to promote processing speed considerably, real-time analysis technologies as well as the frameworks that we mentioned in Sect. 10.4.1.1 could be adopted for offline analyses as well.

In summary, real-time analysis has novel requirements for big data analysis. Hence, techniques for instance in-memory processing are developed to fill the gap. However, currently, the gap has not been fully filled. We are in a situation in which some relatively simple operations have been successfully achieved while some complex operations have yet to be realized. Since well-known platforms such as Spark and Storm [9] have supported real-time analysis functionality at a certain level, we can utilize real-time analysis operations together with complex operations that still currently belong to offline analysis to find the sweet spot between speed and scale.

10.4.2 Open-Source Big Data Tools

Some well-known open-source big data tools are introduced in this subsection. Along with big data workflows, namely, acquisition, storage, management, analysis and response, we introduce one or two tools for each workflow pivot.

10.4.2.1 Acquisition Tool: Kafka

Data should be acquired by a big data stack before any further manipulation. To improve data acquisition efficiency and reduce the cost of format conversion during usage, acquisition tools have been developed based on distributed file systems, such as HDFS.

Existing tools include Kafka [7] developed by LinkedIn and Apache, Chukwa [5] developed by Apache Hadoop, Scribe [58] from Facebook, TimeTunnel [35] from Taobao, etc. Here, we introduce Kafka as an example of an acquisition tool.

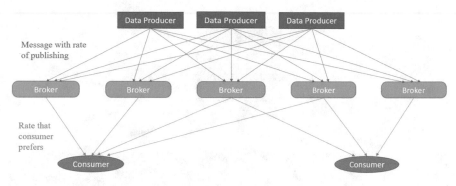

Fig. 10.2 Architecture of Kafka

Kafka is a high-throughput, low-latency, fault-tolerant stream processing platform that subscribes streaming data and distributed republishes the data in a proper way to feed the storage portion or for other purposes. It is able to collect and transmit hundreds of MB of data in each second.

In the Kafka architecture, it acts as an intermedia role between data generators/publishers and data consumers/subscribers. Data records (messages) are pushed to Kafka as key-value pairs. Kafka runs one or multiple servers (brokers) to receive the messages. The pairs are grouped by keys (topics) in each broker. Consumers can poll brokers to obtain messages from Kafka, as shown in Fig. 10.2. As the key components, the brokers store the published messages and prepare them plainly for consumers to pull their required data at the rate they prefer.

10.4.2.2 Storage Tool: Hadoop Distributed File System (HDFS)

The HDFS [17] is the most famous and widely used distributed file system. It uses one or multiple servers to store split large files in blocks.

The HDFS architecture is a master and slave structure that consists of a cluster of nodes. The communication between nodes is based on TCP/IP. Each node (both the master and slaves) has a DataNode to store blocks that are allocated by the master. The master contains an additional NameNode to manage the namespace of the file system and to regulate file access from clients of the system. A file is first split into blocks with a standard block size, which is decided by the NameNode. Then, the blocks are stored in a set of DataNodes, as described in Fig. 10.3. To promote reliability, each block is replicated a certain number of times, and the replications are stored in different nodes.

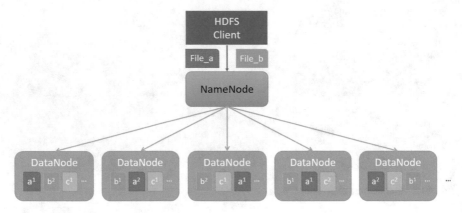

Fig. 10.3 Data storage in HDFS

10.4.2.3 Resource Management Tool: Yet Another Resource Negotiator (YARN)

YARN [10] is a global resource manager on top of the HDFS used to schedule and optimize cluster utilization. Optimization can be practised in different criteria, such as capacity and fairness.

In the architecture of YARN, there is a ResourceManager (RM) to arbitrate the resource usage (CPU, memory, disk, network, etc.) among all applications running on the file system. Each application (one or multiple jobs) is assigned a master manager to negotiate resources from the ResourceManager and execute the application. Moreover, each node is assigned a NodeManager to report the resource and resource usage regularly to the ResourceManager. A brief overview of how an RM obtains information from other YARN key components is presented in Fig. 10.4.

YARN supports many types of processing models operating on different execution engines, such as Tez [15], Spark [8], Flink [6] and Dryad [43], as well as big data applications, such as HBase [13].

10.4.2.4 Analytics Tool: MapReduce, Spark

To analyze massive data stored in a distributed system, a parallel programming framework is required to achieve efficient data processing and computation. Hadoop MapReduce [18] is an open-source implementation used to execute the MapReduce processing model on the HDFS.

The processing model of MapReduce only has two sequential stages, i.e. map and reduce. In the first stage, a map programme is mapped towards certain nodes that store the blocks of input data. The map programme is executed within the nodes in parallel. Then, in the second stage, the executed intermediate results are shuttled towards a small number of nodes to merge the intermediate results as a smaller

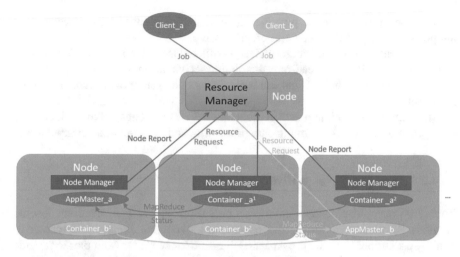

Fig. 10.4 Operations of an RM obtaining information within YARN

set. Through several rounds of merging, the final output is achieved. The process of merging intermediate results towards the output is called "reduce". The entire procedure is illustrated for a word count process example in Fig. 10.5.

MapReduce is a simple but powerful processing model. The open-source MapReduce released by Apache, Hadoop MapReduce, simplifies the programming procedure of research and development on big data. Only two programmes are needed to develop a MapReduce processing application.

However, Hadoop MapReduce still has several shortcomings as an early big data execution engine. A fatal shortcoming is that Hadoop MapReduce supports only one processing model, and no other DAGs or iterations are supported.

Spark [8] is an open-source unified analytics platform first developed by AMPLab at the University of California, Berkeley, to support iterative and interactive big data processing pipelines. In addition to map and reduce operations, Spark supports a range of transformation operations, such as add, join, and filter. Moreover, there

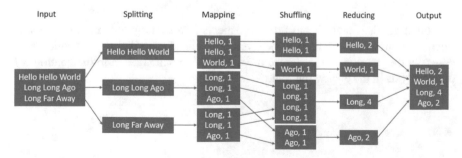

Fig. 10.5 MapReduce process of a word count program

is no sequence restriction on pipeline implementation. Hence, Spark is capable of supporting high-performance iteration processing.

A Resilient Distributed Dataset (RDD) is utilized to enable in-memory computation in Spark. An RDD is a read-only distributed set of data items. A parallel operation is not able to change an RDD but can build a new one to store the intermediate/final result set.

Spark includes multiple practical tools that help it to be engaged in distinct types of applications, such as streaming data processing and graph-parallel computation.

10.4.2.5 Application Tool: Hive

Tools for the application layer are mostly object-oriented to realize certain specific services. For distinct scenarios, various tools or tool combinations can be adopted to achieve a solution. Here, we use Apache Hive and Zookeeper as examples to provide the reader a preliminary impression of application layer tools.

Hive [11] was initially developed by Facebook and joined Apache later for big data querying. It is a data warehouse infrastructure providing data summarization and ad hoc querying. Users can use SQL-like language (HiveQL) to access data stored in various databases with different storage types efficiently. Hive or its further developed commercial versions are utilized by companies such as Netflix and Amazon as part of their big data implementations.

Apache ZooKeeper [16] provides centralized services to coordinate distributed big data stacks. It stores configuration values hierarchically and provides distributed stack group services, such as maintaining and synchronizing configurations and naming registries among nodes. This kind of maintenance and synchronization service is necessary for large distributed systems to enable highly reliable distributed coordination. Zookeeper is utilized by companies such as Yahoo!, eBay, and Reddit.

10.5 Open Challenges for Utilizing Big Data in the Energy Internet

Although big data analysis has broad prospects in the energy internet field, several open challenges are required to be overcome ahead of utilizing it beyond demo cases. In this section, we introduce the main open challenges in implementing big data analysis in the energy internet.

10.5.1   Data Throughput

In the practice of the energy industry, a very large amount of data can be generated from infrastructure allocated at various places with a fast speed. For instance, records of energy receiving and forwarding between energy routers are created from each energy router continuously. The records can change quickly depending on situations such as transient state change. This method is required to handle the situation through processing, analyzing the records and providing suggestions as soon as possible. Moreover, to obtain a big data stack with better performance, it is necessary to collect, store, manage, process and analyze as much data as the system can, as rapidly as the system can. Hence, increasing the data throughput of the entire big data procedure, including data collection, storage, management and analysis, is a continuing open challenge. This challenge remains along with the development of related technologies, which are able to generate and transmit more data with faster rates, such as the IoT.

One possible way to tackle the challenge stands with the IoT as well. Fog computing and edge computing can push the data processing even further towards the sensor itself or at the local sensor network level to reduce the cost of data transmission and central computing to promote the throughput of the big data stack.

In addition, integrating the big data stack with the energy internet is another aspect of this challenge because the processing requirements for the energy internet functionalities overlap with the big data services. If we design the physical, transport and application layers of the energy internet with consideration of big data analytical requirements, then we are able to achieve a better solution for utilizing big data in energy internet circumstances.

10.5.2   Data Privacy and Security

For big data analytics, obtaining as much data as possible is a perpetual tendency. This is because a better performance analysis is more likely to be obtained by using enough data for analysis. However, in practice, it is not easy to obtain enough data to feed a big data stack. For researchers, electricity generation and consumption companies do not desire to let streaming data or huge amounts of records be taken outside their companies. It is even more difficult when negotiating with residential users to share their usage data. The situation has become worse after the Facebook scandal. Companies and individuals have become more conservative on data privacy and security issues. For companies in the power supply chain, obtaining data from upstream and downstream companies is also very difficult. Therefore, providing an efficient approach to obtain data with guaranteed privacy and security protection is an open challenge for both the research and industrial communities.

The enforcement of the GDPR can be a positive signal of normative data usage related to business. The GDPR regulates business behaviour around data, especially on data security and privacy protection. The severe sanctions can, to some extent, curb the occurrence of data abuse like that in the case of Cambridge Analytica Ltd.

Technically, blockchain [53] is a promising mechanism to protect data privacy and security. The data owner, which can be a company or an individual, can publish the authorizations of utilizing the data on the chain. An authorization includes the data abstract as well as the personnel of access, location of access, times of access, etc. A data user must obtain an authorization to access the data. Then, the utilization of data is easier to monitor by the owner and public. Thus, data privacy and security are able to be protected more efficiently.

10.5.3 Data Storage

At present, large amounts of data are generated with fast speed, but the progress of data storage capacity is not able to follow the increasing requirements on storage. The requirements generally regard two aspects, scale and speed, which are also the essentials of big data analysis.

For the challenge of increasing the speed of storage, in-memory databases [48] can be a potential approach to tackle this challenge. Through storing in-memory data, performance such as the reading and writing speed is faster than that for data stored on disks or on flash drives. However, the disadvantages of utilizing in-memory technology are still obvious at present: data should fit in-memory processing; in-memory databases have difficulty remaining persistent for long periods; databases should be loaded from/to disk images before/after usage and data communication between in-memory databases and other databases is not straightforward.

For the challenge of scale, web giants such as Google and Facebook keep building big data stacks with a larger scale of nodes for specific utilization by themselves. However, the increasing number inspires our expectation of a large data storage scale in the energy internet. Public cloud service providers, such as Amazon and Microsoft, are also developing cloud storage-based big data stacks continuously for usage by industrial users.

10.5.4 Data Stream Processing

Although data acquisition frameworks such as Kafka and Flume [12] are capable of feeding the analytics layer hundreds of MB of streaming data per second, processing all valuable data in time, in other words, the timeliness of large-scale real-time processing, is still a challenge. For instance, despite the fact that Spark supports both in-memory iterative processing and streaming data applications, its technical implementation uses micro-batching to handle streaming, which allows stream granulation. It is not native stream processing, and the latency can be on the order of seconds with the configuration of the size of a micro-batch.

However, timeliness is a crucial requirement for utilization in the energy internet, including consumption pattern recognition, real-time power adjustment, scheduling, management, etc. Moreover, historical data should be capable of being extracted from the file system and interacting with new incoming data.

To overcome this challenge, big data processing frameworks, such as Apache Flink [6] and Beam [4] have been natively developed for streaming processing. These streaming processing frameworks just passed through their preview versions and entered the public view in 2018. The development of these frameworks could provide a promising solution for the timeliness challenge of large-scale real-time processing.

10.5.5   Data Opening

Data opening is a common challenge that researchers and public service providers face in many fields. For the purpose of data reuse and the social good, anonymous data should be stored for future usage. Data records and streams are commonly difficult to catch but easily disappear and are always valuable for research as well as public product development. These features are significant in energy data records.

To open up data records and data streams, financial support is necessary. Data with better quality, for example, more accurate, complete and consistent data, often means more investment. In addition, better data quality also brings more accurate analysis, prediction and management. The trade-off regarding data quality is a challenge of data opening.

Furthermore, managing the opened data is also a challenge. Before data are opened, anonymization should be performed. However, determining the edge of privacy is difficult. Several levels of anonymization exist. If we anonymize all possible factors, the value of the data will also be lost. Another point that can be considered together with anonymization is who can access the open data. In general, if the data owner wants to check the data, it is not necessary to be anonymized. If it is a person or company without any credit background and confidentiality technology certification, he or it should only be able to access the data after the strictest anonymization process. Blockchain is also a potential solution to this challenge [25].

10.6   Summary

In this chapter, the utilization of big data in the energy internet infrastructure is explored. A three-layer big data architecture of usage in the energy internet is presented. The characteristics of data utilized in the energy internet and the potential requirements of the energy internet for the big data architecture are studied. Then, analytics methods that could be executed in the energy internet big data infrastructure are introduced. Real-time and offline analyses, as two types of analysis modes for different requirements of application scenarios, are described. Several well-known open-source big data tools are discussed. In addition, the open challenges of utilizing big data in the energy internet are proposed.

References

1. B. Agarwal, A. Chakravorty, T. Wiktorski, C. Rong, Enrichment of machine learning based activity classification in smart homes using ensemble learning, in *2016 IEEE/ACM 9th International Conference on Utility and Cloud Computing (UCC)*, 6–9 December 2016, pp. 196–201
2. S. Ai, A. Chakravorty, C. Rong, E.V. Household, Charging demand prediction using machine and ensemble learning, in *2018 IEEE International Conference on Energy Internet (ICEI)*, 21–25 May 2018, pp. 163–168
3. S. Ai, A. Chakravorty, C. Rong, Household power demand prediction using evolutionary ensemble neural network pool with multiple network structures. Sensors **19**(3) (2019)
4. Apache Beam (2018), https://beam.apache.org/
5. Apache Chukwa (2018), http://chukwa.apache.org/
6. Apache Flink (2018), http://flink.apache.org/
7. Apache Kafka (2018), https://kafka.apache.org/
8. Apache Spark (2018), http://spark.apache.org/
9. Apache Storm (2018), http://storm.apache.org/
10. Apache Hadoop YARN (2019), https://hadoop.apache.org/docs/current/hadoop-yarn/hadoop-yarn-site/YARN.html
11. Apache HIVE (2019), https://hive.apache.org
12. Apache Flume (2019), http://flume.apache.org/
13. Apache HBase (2019), http://hbase.apache.org/
14. Apache Impala (2019), https://impala.apache.org/
15. Apache Tez (2019), https://tez.apache.org/
16. Apache (2019) ZooKeeper. https://zookeeper.apache.org/
17. Apache Hadoop, HDFS Architecture Guide (2019), https://hadoop.apache.org/docs/r1.2.1/hdfs_design.html
18. Apache Hadoop, MapReduce Tutorial (2019), https://hadoop.apache.org/docs/r1.2.1/mapred_tutorial.html
19. M. Bhole, K.Phull, A. Jose, V. Lakkundi, Delivering analytics services for smart homes, in *2015 IEEE Conference on Wireless Sensors (ICWiSe)*, 24–26 August 2015, pp. 28–33
20. T. Bier, D.O. Abdeslam, J. Merckle, D. Benyoucef, Smart meter systems detection & classification using artificial neural networks, in *IECON 2012–38th Annual Conference on IEEE Industrial Electronics Society* (IEEE, 2012), pp. 3324–3329
21. D. Borthakur, J. Gray, J.S. Sarma, K. Muthukkaruppan, N. Spiegelberg, H. Kuang, K. Ranganathan, D. Molkov, A. Menon, S. Rash, R. Schmidt, A. Aiyer, Apache Hadoop goes real-time at Facebook, in *Proceedings of the 2011 ACM SIGMOD International Conference on Management of data, Athens, Greece* (ACM, 2011), pp. 1071–1080
22. L. Breiman, Random forests. Mach. Learn. **45**(1), 5–32 (2001)
23. V. Bureva, S. Popov, E. Sotirova, K.T. Atanassov, Generalized net of MapReduce computational model, in *Uncertainty and Imprecision in Decision Making and Decision Support: Cross-Fertilization, New Models and Applications* (Springer International Publishing, Cham, 2018), pp. 305–315
24. Y. Cao, Q.-G. Miao, J.-C. Liu, L. Gao, Advance and prospects of AdaBoost algorithm. Acta Autom. Sinica **39**(6), 745–758 (2013)

25. A. Chakravorty, C. Rong, Ushare: user controlled social media based on blockchain, in *Proceedings of the 11th International Conference on Ubiquitous Information Management and Communication, Beppu, Japan* (ACM, 2017), 3022325, pp. 1–6
26. J.-L. Chen, G.-S. Li, S.-J. Wu, Assessing the potential of support vector machine for estimating daily solar radiation using sunshine duration. Energy Convers. Manag. **75**, 311–318 (2013)
27. J. Chen, Y. Chen, X. Du, C. Li, J. Lu, S. Zhao, X. Zhou, Big data challenge: a data management perspective. Front. Comput. Sci. **7**(2), 157–164 (2013)
28. M. Chen, S. Mao, Y. Liu, Big data: a survey. Mob. Netw. Appl. **19**(2), 171–209 (2014)
29. K. Cho, B. Van Merriënboer, C. Gulcehre, D. Bahdanau, F. Bougares, H. Schwenk, Y. Bengio, Learning phrase representations using RNN encoder-decoder for statistical machine translation (2014). arXiv preprint arXiv:14061078
30. L. Deng, D. Yu, Deep learning: methods and applications. Found. Trends® Signal Process. **7**(3–4), 197–387 (2014)
31. EMC, Greenplum Database (2018), https://greenplum.org/
32. B. Fiore-Gartland, G. Neff, Communication, mediation, and the expectations of data: data valences across health and wellness communities. Int. J. Commun. **9**, 1466–1484 (2015)
33. Graph, The Open Graph Viz Platform (2019), https://gephi.org/
34. S. Hochreiter, J. Schmidhuber, Long short-term memory. Neural Comput. **9**(8), 1735–1780 (1997)
35. Huolong, TimeTunnel (2018), http://code.taobao.org/p/TimeTunnel/wiki/index/
36. H. Jiang, K. Wang, Y. Wang, M. Gao, Y. Zhang, Energy big data: a survey. IEEE Access **4**, 3844–3861 (2016)
37. A. Katal, M. Wazid, R. Goudar, Big data: issues, challenges, tools and good practices, in *2013 Sixth International Conference on Contemporary Computing (IC3)* (IEEE, 2013), pp. 404–409
38. J.D. Kelleher, B.M. Namee, A. D'Arcy, *Fundamentals of Machine Learning for Predictive Data Analytics: Algorithms, Worked Examples, and Case Studies* (The MIT Press, London, 2015)
39. P.A. Laplante, *Real-time Systems Design and Analysis* (Wiley, New York, 2004)
40. J. Lee, Y.S. Kwon, F. Färber, M. Muehle, C. Lee, C. Bensberg, J.Y. Lee, A.H. Lee, W. Lehner, SAP HANA distributed in-memory database system: transaction, session, and metadata management, in *2013 IEEE 29th International Conference on Data Engineering (ICDE)*, 8–12 April 2013, pp. 1165–1173
41. S. Liu, D. Zhang, C. Zhu, W. Li, W. Lu, M. Zhang, A view on big data in energy internet. Autom. Electric Power Syst. **40**(8), 14–21 (2016)
42. W. Liu, Z. Wang, X. Liu, N. Zeng, Y. Liu, F.E. Alsaadi, A survey of deep neural network architectures and their applications. Neurocomputing **234**, 11–26 (2017)
43. Microsoft Dryad (2019), https://www.microsoft.com/en-us/research/project/dryad/
44. H.G. Miller, P. Mork, From data to decisions: a value chain for big data. IT Profess. **15**(1), 57–59 (2013)
45. MongoDB (2019), https://www.mongodb.com/
46. MPI Forum (2018), https://www.mpi-forum.org/
47. K.P. Murphy, Naive bayes classifiers. University of British Columbia (2006), pp. 1–8
48. J. Najajreh, F. Khamayseh, Contemporary improvements of in-memory databases: a survey, in *2017 8th International Conference on Information Technology (ICIT)*, 17–18 May 2017, pp. 559–567
49. NIST Big Data Public Working Group (NBD-PWG), The NIST Big Data Interoperability Framework, Vol 1, Definitions. USA Patent (2015)
50. Openmp (2019) Openmp, https://www.openmp.org/
51. A. Oussous, F.-Z. Benjelloun, A. Ait Lahcen, S. Belfkih, Big data technologies: a survey. J. King Saud Univ. Comput. Inf. Sci. **30**(4), 431–448 (2018)
52. T. Papadimitriou, P. Gogas, E. Stathakis, Forecasting energy markets using support vector machines. Energy Econ. **44**, 135–142 (2014)
53. M. Pilkington, Blockchain technology: principles and applications, *Research Handbook on Digital Transformations* (Edward Elgar, UK, 2016), pp. 225–253

54. R. Polikar, Ensemble based systems in decision making. IEEE Circ. Syst. Mag. **6**(3), 21–45 (2006)
55. L. Rokach, Ensemble-based classifiers. Artif. Intell. Rev. **33**(1), 1–39 (2010)
56. SAP HANA (2018), https://www.sap.com/sea/products/hana.html
57. R.E. Schapire, Y. Freund, *Boosting: foundations and algorithms* (MIT Press, London, 2012)
58. Scribe (2018), https://www.scribesoft.com/
59. A. Songpu M.L. Kolhe, L. Jiao, External parameters contribution in domestic load forecasting using neural network, in *International Conference on Renewable Power Generation (RPG 2015)*, 17–18 Oct. 2015, pp. 1–6
60. J.A.K. Suykens, J. Vandewalle, Least squares support vector machine classifiers. Neural Process. Lett. **9**(3), 293–300 (1999)
61. Tableau (2019), https://www.tableau.com/
62. A. Thusoo, J.S. Sarma, N. Jain, Z. Shao, P. Chakka, S. Anthony, H. Liu, P. Wyckoff, R. Murthy, Hive: a warehousing solution over a map-reduce framework. Proc. VLDB Endow. **2**(2), 1626–1629 (2009)
63. J. Walters-Williams, Y. Li, Comparative study of distance functions for nearest neighbors, in *Advanced Techniques in Computing Sciences and Software Engineering*, edited by Elleithy, Dordrecht, 2010. Springer Netherlands, pp 79–84
64. Y. Wang, Q. Shi, H. Song, Z. Li, X. Chen, Multi-source heterogeneous data integration technology and its development. Data Management and Three-dimensional Visualization of Global Velocity Mode l Crust **20**, 133 (2016)
65. S. Weisberg, *Applied Linear Regression*, vol. 528 (John Wiley & Sons, New Jersey, 2005)
66. J.D. Wichard, An Adaptive forecasting strategy with hybrid ensemble models, in *2016 International Joint Conference on Neural Networks (IJCNN)* (IEEE, 2016), pp. 1495–1498
67. C. Yang, Q. Huang, Z. Li, K. Liu, F. Hu, Big data and cloud computing: innovation opportunities and challenges. Int. J. Digit. Earth **10**(1), 13–53 (2017)

Chapter 11
Artificial Intelligence Models Used for Prediction in the Energy Internet

Cristina Heghedus and Chunming Rong

Abstract Decision-making in the energy field, especially in recent years, is highly data-driven. The crucial information extracted from historical data tells a lot about future behaviour and expected events. The extraction of this information is traditionally performed by statistical analysis and predicting future trends. However, in recent years, the amount of collected data has increased substantially and has become harder to handle by traditional means. The processing power and storage capacity of computers have also increased and have opened new horizons for data analysis and prediction tasks. One of the great advantages of modern technology is the possibility for automation in the execution of some tasks; thus, less human interaction and higher efficiency can be achieved. Therefore, in this chapter, the energy field is combined with the field of computer science that contributes to automation, that is, artificial intelligence (AI). The subject of artificial intelligence for the Energy Internet is analysed in this chapter mainly from the point of view of future consumption prediction (electricity, wind speed and solar radiation), and prediction models, efficiency measurements and implementation techniques are described in detail. Models are analyzed in terms of accuracy and design (stand-alone and hybrid models). Implementations include descriptions of different libraries used for AI models and characteristics of programming languages. Recently developed data processing and prediction models are numerous and highly efficient; thus, applying these models to data from the energy field has various advantages, including cost-effective resource management, asset management and energy efficiency.

C. Heghedus (✉) · C. Rong
Department of Electrical Engineering and Computer Science, University of Stavanger, Stavanger, Norway
e-mail: kaveczkicristina@gmail.com

C. Rong
e-mail: chunming.rong@uis.no

A. F. Zobaa and J. Cao (eds.), *Energy Internet*,
https://doi.org/10.1007/978-3-030-45453-1_11

11.1 Introduction

Artificial intelligence (AI) is a field of computer science that refers to the ability of machines (computers) to work and make decisions in a similar manner to humans. AI models are designed to gain knowledge from the given information and/or their previous action. In other words, computers can be programmed to learn based on real-world scenarios. Thus, with precise implementation, computers are capable of solving problems, predicting future behaviour, planning, reasoning and even moving objects.

AI is a complex and still developing area, although it is already highly efficient in multiple domains. There are many ways to categorize AI models; however, based on the scope for which these models are designed, the following two types are mentioned: task-specific AI and memory-based AI. A task-specific AI does not have experience from the past, as it has no memory, so it can only perform a strategic move based on given information (for instance, Deep Blue, the IBM chess program, and Apple's Siri). On the other hand, a memory-based AI makes decisions using past experience and can be trained in a way that if it receives an unfamiliar task, it will have the intelligence to solve it.

A core part of AI that has received increasing attention in the recent past is machine learning (ML). It represents the computer's ability to learn to extract hidden patterns in a series of data, classify and cluster objects with high accuracy or even make predictions based on past behaviour. This learning can be supervised (the expected output is given, and the model uses its own errors to improve) or unsupervised (the output is not known, and the model has to identify specific patterns and characteristics of the dataset). ML has applicability in multiple domains, such as education, business, robotics, healthcare, finance and even law. Combining ML with the very large amount of data that is generated all over the world leads to efficient data processing and improved future decision-making.

A tremendous amount of data is being generated every day by humans and even by computers and other devices (video cameras, smart metres, etc.). This amount has been rapidly increasing over the years; therefore, there is a need for new and efficient technology to process it and gain valuable knowledge. In addition, the rapid evolution of technology has led to advanced computers with high processing capabilities. The attempt to deal with such data using strong computers revolutionized ML, creating a new branch, that is, deep learning (DL). The depth in DL is given by the complex and sometimes complicated structure of models similar to ML. In the composition of DL models, there are multiple functions interacting in a specially defined way. To ensure high accuracy for these models, they need to "learn" from massive datasets.

Examples of ML and DL models, together with their applicability in the Energy Internet, are presented in this chapter. These include models and results of different experiments from the past years, from the point of view of efficiency, implementation, data preparation and other aspects. For the reader to better understand the presented performance, four measures of effectiveness are introduced that are used in most scientific papers to analyze how close the results produced by an AI model are to the true values that are expected from the dataset.

Moreover, certain issues and challenges related to the data collection process are detailed (as the high quality of historical data is crucial for data analysis and future behaviour forecasting), along with various factors that can influence the forecasting process (such as weather aspects since many scientific papers include this as input in their experiments; details about buildings and houses; and details about inhabitants, which are often not available for research purposes or non-existent). The scope of this chapter is to offer the reader an overview of all AI models that are highly used in the Energy Internet and perhaps a hierarchy of these models, so that the reader is able to group them based on the tasks for which they are applied.

11.2   Motivation for Applying AI in the Energy Internet

World modernization is a rapid process that implies expected and unexpected consequences. One of the phenomena closely related to it is urbanization. The causes are multiple, including job opportunities, education possibilities, transportation and diversity in general. The population trend is presented in Fig. 11.1, which evidently highlights the increasing tendency towards urban areas.

The urban and rural populations represented less than one-third, 30%, and more than two-thirds, 70%, respectively, in 1950. It is notable that in an interval of 100 years, these numbers are expected to change in the opposite direction, 66% and 34% for urban and rural, respectively. Equality was reached in 2007, and in 2014, the urban population was already more than half, 54%, of the total, and the rural population was 46%, according to World Urbanization Prospects [65]. With increasing population comes increasing consumption in many fields, including the

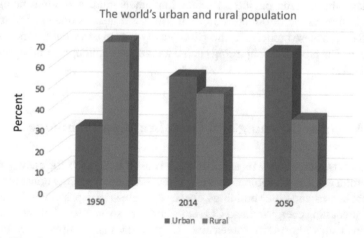

Fig. 11.1 Historical global population trend, indicating urbanization

energy field. Increasing energy consumption brings new challenges that suppliers need to face and solve efficiently.

Figure 11.1 provides an indication of the changing ratio of urban and rural populations. Considering this situation, there are several consequences of urbanization and of evolving technology. An important fact to note is that in the past, humanity was less dependent on electricity or energy in general, as house appliances were very limited or even non-existent. People did not depend much on electricity, compared to modern times, when most of appliances work with electricity. Additionally, not only is the urban/rural ratio changing but also the population size is as well. The population is increasing, and the Earth's resources are limited; therefore, it is crucial to efficiently manage usage and assets.

As mentioned previously, the consequences of urbanization are also multiple, and it comes with increasing demand for housing, cars, televisions, kitchen appliances and smart devices. The majority of these goods are electricity-dependent. Moreover, there is a noticeable difference in the number of electric cars used back in the day (when most cars were run on petrol or gasoline) and in recent years (when electric cars are ready to take over and chargers are placed around every second corner in a city). Thus, one major goal of the field of the Energy Internet is the efficient management of finite resources and efficient energy storage and distribution.

One way of improving energy storage and distribution schemes is to ensure a stable energy supply and demand ratio. Consequently, this implies that suppliers can forecast future electricity demand with high accuracy and that they have enough information on consumption patterns. For companies and researchers to forecast future electricity consumption, they need two indispensable factors: a decent amount of historical data on past consumption and a high-performance forecasting technique. Plenty of work has been done in this direction, and each work aims to bring improvement to the field.

Section 11.3 focuses on ML and DL models, presenting their main characteristics and applicability. Different AI models come with different advantages, some being faster than others, some having an easier implementation, and some being more accurate in forecasting future consumption. However, there is one general truth for each of them, as these models require good quality data to carry out the most precise forecasting. The process of capturing such data is often challenging and is described in this chapter.

11.3 Machine Learning and Deep Learning Techniques

Over the years, numerous ML and DL models have been used for energy market-related problems. These problems include forecasting future energy consumption for individual houses and other buildings, forecasting electricity price, wind speed prediction, forecasting energy-related CO_2 emissions and so on. Electricity consumption forecasting can be short-term forecasting, from minutes ahead up to days; medium-term forecasting, from days ahead up to several months or a year; and long-term forecasting, for a few years ahead. In this section, some of the most highly used models are described, with corresponding references, so that the reader will be able to follow the exact results.

11.3.1 Machine Learning Models

Some of the most widely used ML models in the energy field are introduced here, and their performance is presented. These models are present in numerous scientific papers, having many advantages such as high accuracy, simple implementation and rapidity. Used in their regular form, some of the models are quite fast in training and testing; however, adding complexity to their structure (using them in a hybrid form) can cause these models to become more powerful and accurate (even if it implies losing computation speed). To analyze the models based on their performance, first, four methods for accuracy evaluation are introduced in Sect. 11.3.1.1. These are present in the majority of scientific papers that analyze AI models in the Energy Internet.

11.3.1.1 Accuracy Evaluation

ML models are frequently used in the energy field because they are suitable for time series problems. The good results obtained by ML models are attested by the small differences between the actual known values and the forecasted values. There are multiple ways to compute these differences and evaluate the models. Measuring the performance of AI models is quite complex, as it involves numerous factors such as computation speed and memory usage. However, in the majority of cases, accuracy (the smallest error) is the most determining factor. There are four most frequently used measures described as follows.

Mean absolute percentage error (MAPE) is a measure of accuracy used for many ML and DL models. It shows how far the forecasted values are from the actual known values, also called labels. Mean square error (MSE, also called L2 loss) is another measure of effectiveness used in prediction problems; however, MAPE is more often found in such problems, as it outputs the error in percentage. The formulas for both measures are found in Eqs. (11.1) and (11.2):

$$MAPE = 100 \cdot \frac{1}{n} \cdot \sum_{i=1}^{n} \left| \frac{A_i - P_i}{A_i} \right| \tag{11.1}$$

$$MSE = \frac{1}{n} \cdot \sum_{i=1}^{n} (A_i - P_i)^2 \tag{11.2}$$

where A_i are the true values, P_i are the predicted values and n is the size of the data. In other words, these measure the differences between the predicted values and the actual values (labels).

Other loss measures, similar to the first two, are mean absolute error (MAE, also called L1 loss) and root mean squared error (RMSE), presented in Eqs. (11.3) and (11.4).

$$MAE = \frac{1}{n} \cdot \sum_{i=1}^{n} |A_i - P_i| \qquad (11.3)$$

$$RMSE = \sqrt{\frac{1}{n} \cdot \sum_{i=1}^{n} (A_i - P_i)^2} \qquad (11.4)$$

Choosing one type of loss function over another is a matter of the problem one is trying to solve and the desired results. There are publications stating the advantages of MSE [16] since it is more robust to outliers and other publications in which RMSE is more preferred; however, its sensitivity to outliers remains a concern [62]. Moreover, in addition to these measures of performance, it is also important to compare models' running times and implementation techniques for a clear view of the advantages of the models. In general, the implementation of AI models is relatively easy (varying for different programming languages), and if the dataset is not extremely large (or there are not too many iterations), the running time is also satisfying.

11.3.1.2   Support Vector Machine

Support vector machine (SVM) is an ML model based on supervised learning and is used for classification and regression problems. Thus, it receives a set of data for training and a new set of data for testing. SVM is memory efficient and performs well in high-dimensional spaces; however, its performance decreases when the dataset is too large or noisy. It has been used over the years in the energy field for forecasting energy consumption [8, 25, 34, 54, 61], wind power prediction [46, 70] and forecasting energy-related CO_2 emissions [17]. SVM has been used in the Energy Internet in regular form and in combination. Algorithms combined with SVM include simulated annealing algorithm (SA), particle swarm optimization (PSO), harmony search algorithm (HS), fruit fly optimization algorithm (FOA), firefly algorithm (FA) and so on [23].

For instance, in [8], SVM was tested in 2005 for energy load forecasting for four buildings. These buildings were all for commercial use. To assess the performance of SVM, the authors calculated the MSE for the forecast produced for each building, which varied between 0.14 and 0.73. They concluded that the model outperformed other models in related research; however, their dataset consisted of only monthly utility bills for a period of four years.

In a more recent work, in 2015 [34], a variation of SVM was used, again for building energy consumption forecasting, namely, least squares SVM (LSSVM). This is a real-coded genetic algorithm (RCGA)-based model that was used in three forms: a hybrid model of the direct search optimization (DSO) algorithm and RCGA-LSSVM, a regular RCGA-LSSVM, and a differential evolution algorithm (DEA)-LSSVM. Experiments were conducted for three different cases, and the models were evaluated in terms of RMSE, average convergence iteration, average calculation time and average convergence time. In the first case, the RMSE was approximately 7.64, 7.70 and 7.59 for RCGA-LSSVM, DEA-LSSVM and DSORCGA-LSVM (which was the proposed model), respectively. The proposed model was also proven to be superior to the other two in terms of convergence speed and computation time.

Hybrid models of SVM have been proven to be extremely efficient, as stated in [23]; however, all have advantages and disadvantages as well. A single SVM has the advantage of structural risk minimization (SRM), which means it can deal with overfitting problems but not for very large datasets. On the other hand, the hybrid HS-SVM and FOA-SVM are much faster in terms of computational speed and searching algorithms, but both have complicated structures [23]. Other similar hybrid models were presented in [35], a recent work from 2016, where the authors implemented PSO-SVM in comparison with a simple artificial neural network (ANN), a genetic algorithm (GA)-based SVM and AutoRegressive Integrated Moving Average (ARIMA). The results were in favour of the proposed model, which achieved an MAPE of 2.53%, in comparison with ARIMA (11.21%), the GA-SVM (5.27%) and the ANN (6.62%).

As mentioned previously, SVM is a supervised learning-based model, and the main concept is that it creates a line in a 2-dimensional space and a hyperplane in a multidimensional space to separate different classes in a dataset. The concept is presented in Fig. 11.2. To achieve better performance, the parameters of the model, such as the regularization parameter, kernel and margins, can be tuned in the experimental process. Since electricity consumption forecasting is, in fact, a regression problem (not a class separation problem), a version of SVM named support vector regression (SVR) [20] better fits the forecasting task. SVR is based on similar steps as SVM,

Fig. 11.2 Classes separation in a 2-dimensional space using SVM

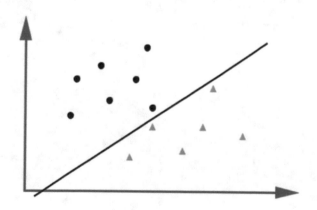

with a few minor changes. Here, the output must be a real number (not a class); therefore, the number of possibilities is infinite. This also means that regression is a slightly harder task to perform than classification.

In a slightly different context, SVM and SVR were presented in [66] as models for forecasting solar radiation. They were compared with other ML models in terms of accuracy. Although SVM and SVR have successfully been used in time series problems, the experimentation on solar radiation forecasting was quite limited. As a conclusion of that work [66], SVM and SVR produced similar results to k-means in solar radiation forecasting and presented promising applicability in future work.

11.3.1.3   Artificial Neural Networks

An artificial neural network (ANN) is a biological neural network-inspired ML model. It learns based on a set of examples (training) and then makes predictions or classifies objects in the testing dataset. It is formed by an input layer, one hidden layer and an output layer, each containing fully connected nodes (neurons) to perform mathematical operations. The connections are assigned *weights*, which help optimize the learning process. These weights are updated at each iteration as needed based on the error obtained at the output, which is carried back to the input.

The structure of an ANN is presented in Fig. 11.3. The number of nodes in the input layer is equal to the number of inputs for the specific problem, and the number of hidden nodes is generally calculated based on the input and hidden nodes [50] (or established based on trial and error). The output nodes always match the number of outputs in the problem.

To achieve better performance, which means minimizing the error at the output, the parameters of the neural network can be modified (tuned) during experiments. Thus, the activation function can be changed based on the type of data, the number

Fig. 11.3 Structure of ANN, with 2 input nodes, multiple hidden nodes, 1 output node, weights (*full black arrows*) and errors (*dotted arrows*)

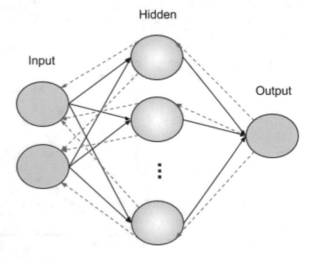

of iterations can be varied by keeping in mind to minimize the error but avoid over-fitting [60], and the number of hidden nodes can be changed. The goal is to obtain predicted values as close as possible to the actual consumption values as the output. The error that we obtain in an ANN shows how far the predicted values are from the actual consumption values. This error can be minimized by sending the values back to the network and adjusting the weights between all nodes. This process is called backpropagation [27], and this step is performed as many times as necessary (iterations) to obtain the smallest error.

ANNs have been successfully used in many energy forecasting problems for households and industrial buildings, renewable energy systems and user demand-based forecasting [4, 15, 36, 57–59, 74]. Moreover, ANNs presented compara-ble results with the traditional AutoRegressive Integrated Moving Average model (ARIMA) in [66] in a solar radiation forecasting task, where the prediction accura-cies of an ANN and ARIMA were fairly similar; however, the ANN was presented as carrying the advantage of flexibility in implementation and structure.

A simple ANN model was built in 2008, in [4], for forecasting the energy consump-tion of a university building. This model was compared with a traditional physical model, EnergyPlus, which allows the user to introduce details about the building, such as the geometry of the building and construction material, and then forecast the energy consumption. The two models were separately described in the paper, together with the characteristics of the dataset and building. The building is gener-ally populated between 8 AM and 6 PM by 1,000 employees and has a manually controlled air conditioning system. The authors also considered outside temperature data collected from the campus meteorological station in the warmest period of the year. The EnergyPlus model produced close forecasting to the actual measured val-ues in 80% of the testing period. The inputs for this model included the building's geometry, lighting, material of the walls and windows, and details about the period of occupancy and the equipment. On the other hand, the ANN was implemented in two different ways. First, the ANN took only the minimum and maximum external dry-bulb temperatures as inputs and produced the energy consumption in kWh as the output. The second type of ANN was based on temperature (daily average), humidity and solar radiation. The same testing period was used for the ANN as for Energy-Plus. Additionally, the data were divided into three sets: all days, working days and weekends (+ holidays). The results of the ANN were slightly better than those of EnergyPlus, with an error range of approximately 13% and 10% for EnergyPlus and the ANN, respectively. Moreover, the authors concluded that the humidity and solar radiation might not be of high importance since they did not affect the forecasting much [4].

In a more recent work [59], in 2014, a similar problem was treated, that is, estimat-ing a university building's energy consumption. The studied spaces were a library and an amphitheatre. The models compared were linear regression, fuzzy models and an ANN. Similar to the previous described work, here, the output was the future energy consumption, and the inputs were as follows: the space's occupancy (people/hour), length of the day, solar radiation for the current and previous day and the heating and cooling degree days. The electricity consumption data were gathered for periods of 4 months for the amphitheatre and the library, respectively. The amount of data used for training was 60% and the amount used for testing was 40% for the linear regression and fuzzy models. The ratio of training and testing data was not known for the neural network. Two fuzzy models were implemented in the paper, with two different clustering methods.

One was subtracting clustering (SC), and the other was a fuzzy c-means algorithm (FCM). In regard to the ANN, in the experiments, the authors varied the numbers of hidden layers, hidden neurons and epochs. The results were presented in comprehensive tables for the amphitheatre and for the cold and warm seasons in the library. In terms of MSE, the ANN produced 3.5, 576 and 456 for the amphitheatre, library warm season and library cold season, respectively, whereas the linear regression produced MSE of 119.8, 925 and 607, respectively; the results varied for the different types of fuzzy models and different types of clustering, but all were higher than the ANN errors. MAE was the smallest for the ANN, with values of 1.4, 17.5 and 12.9 for the amphitheatre, library warm season and library cold season, respectively. Although the results produced by the ANN were proven to be much more promising than those produced by the linear regression and fuzzy models, there was space for improvement in the following work, and according to the authors, some drawbacks of the experiments were the following: the range of the dataset was quite low and the representation of the space's occupancy might have been less accurate.

Forecasting energy consumption for commercial buildings was also treated in [74] for short-term (sub-hour) forecasting. Here, a feature extraction technique was first implemented with an ANN for consumption forecasting. The work included three neighbouring commercial buildings with varying numbers of floors (from 2 to 5). The database contained weather data, building management system data (air conditioning and others) and electricity consumption data; however, for the model's input, the authors used outdoor temperature, outdoor humidity and the schedule of air conditioning operation. To assess the ANN's performance, it was compared to eight other ML models, and in terms of the coefficient of variance of RMSE, it was proven to be the most efficient. The errors varied from 0.08 for the proposed ANN model to 0.18. Based on the obtained results, the authors claimed that, for short-term energy usage forecasting with 15-minute resolution, for the analyzed buildings, their proposed ANN model with feature selection was most suitable.

A large part of the applicability of ANNs related to the energy field was presented in [57, 58]. First, an entire section presented the concept of ANNs, starting from the biological neuron and continuing to the artificial neuron and the parameters of the network. In these reviews, multiple applications of ANNs in the Energy Internet were described, including forecasting the energy consumption in commercial buildings, evaluating and optimizing the energy consumption in buildings, forecasting of solar radiation and wind speed, and management of different air conditioning and ventilation systems. While there is no clear answer on which model is most suitable for forecasting or optimization in the Energy Internet, it was concluded that ANNs hold numerous advantages that increase their applicability in the energy field as follows:

- ANNs are models that learn from real-world scenarios and examples
- Their capability of dealing with complex (noisy and incomplete) datasets is well known.
- Unlike linear regression, ANNs can solve non-linear problems
- ANNs are able to "learn" hidden patterns in a dataset that are not always obvious for other models
- Depending on the size of the dataset, they can carry out computations relatively quickly.

In addition to these advantages, there is one crucial advantage related to implementation. ANNs are relatively easy to implement even on systems with GPUs or multicore processors. The implementation is easily readable, and adding layers and hidden nodes is possible at any time.

Various studies regarding energy load forecasting have been performed using hybrid ANNs, such as an ANN with the fruit fly optimization algorithm (FOA-NN) [33, 43], ANN with particle swarm optimization (PSO-NN) [40, 44], ANN trained by a hybrid artificial immune system (AIS-NN) [47, 73] and wavelet transform ANN (WT-NN) [31, 41]. The aim of all these works was to obtain forecasting as accurately as possible. The measure of effectiveness for these models was mean absolute percentage error (MAPE), and the obtained values were in a small range from 1.25 to 2.95%.

11.3.1.4 Other ML Models for the Energy Internet

In 2017, a review work [72] thoroughly presented a number of traditional methods and ML models that have been used over the years for forecasting electricity loads for commercial buildings. Moreover, a set of factors that can influence the forecasting of electricity loads related to weather conditions and day-type aspects were presented. Two traditional models were described in the work, namely, a thermal model and an autoregressive model. The thermal model can calculate the energy behaviour and heat transfer in a building. Usually, historical data are not required for these models; however, they often need a high number of inputs. These inputs are often details about the building, about the surrounding area, about the number of people in the building and so on. Intuitively, such data are not always available or are quite limited; thus, this is a disadvantage of the models.

More often, two autoregressive models are being used in the Energy Internet (and other fields as well), namely, AutoRegressive Moving Average (ARMA) and AutoRegressive Integrated Moving Average (ARIMA). Both models fit electricity time series data well and are used for consumption forecasting. In the components of both models, we find the autoregressive part (AR) and the moving average part (MA); and in ARIMA, we see an additional integrated part (I), which means that we integrate (sum) over the *differenced* (making stationary data from nonstationary data is called differencing) values of the values that we want to predict. For ARIMA, in [6], the results for hourly forecasting were promising, producing an MAPE between 2.21 and 4.29% and between 1.45 and 1.99% for a modified forecasting that included temperature data.

When a dataset contains multiple features and the interaction among them happens in a non-linear way, linear regression does not always offer satisfactory performance. In such cases, a regression tree can be used, which is another regression model that divides the task onto its nodes and produces an output at the leaves. An extension of a regression tree is a random forest (RF) model, used in [24], which can offer better performance than a simple regression tree. An RF is, in fact, running a regression tree model multiple times, on multiple variations of the dataset, which has been resampled, and takes the average of all trees to output the forecasted values. In [24], the authors compared the RF to seven other models for two problems: forecasting the electricity consumption and peak power demand. They computed the RMSE, MAE and MAPE as measures of performance for each model in both cases. While the RF and SVR performed well, with an MAPE equal to 3.17% and 3.11%, respectively, for the consumption task and 3.63% and 3.34%, respectively, for the peak demand, the best performance was achieved by an ensemble [24] model with an MAPE of 2.32% and 2.85% for the consumption and peak tasks, respectively.

The performance of all ML models has been analyzed in a multitude of research papers, and the advantages of each have been presented extensively. While many of the models offer extremely good performance in different cases for different problems, there is still a general truth about forecasting future behaviour, which is that it always depends on the dataset. It depends on the set of features, the quality of the data, the information that is contained in the data and the data preprocessing. Additionally, there are always methods to improve a model's performance, which include the data preprocessing and parameter tuning steps.

Although ML models have been successfully used over the years in the Energy Internet (and other fields as well), their performance is weaker for very large datasets. Often, ML models are not very suitable for very large datasets for one or more of the following reasons: they are not able to capture long-term dependencies, they are not able to manage too many computations, computing operations is extremely time consuming, the accuracy of forecasting drops, or the patterns in the dataset are not well captured. Depending on the type of data, these models still exhibit remarkable performance; however, DL models (a class of ML) have started to outperform ML models, including on energy (and other types of) datasets containing time dependencies.

11.3.2 Deep Learning Models

The emergence of Big Data created numerous challenges for scientists and researchers. Not only the amount of produced data but also the complexity, inconsistent patterns, and other factors made it difficult to manage datasets with the techniques mentioned above. Therefore, researchers started to build models in a different way, adding complexity and power to their structures [28]. By adding more hidden layers to a simple ML model, an artificial neural network (ANN), we say that its depth is increased, and it becomes a deep neural network (DNN). An ANN with two or more hidden layers can be considered a DNN, called a multi-layer perceptron (MLP), although it must be applied to larger datasets to avoid overfitting [60] (overfitting is the case when a network "gets used" to the training data, so it has great performance on the training data but weak performance on the testing data).

On the other hand, more powerful and complex DNNs have been created than the multi-layer ANN called multi-layer perceptron (MLP). Since energy consumption data are a time series, a network with *memory* is needed to handle current and past information. Therefore, recurrent neural networks (RNNs) are a type that better fits energy forecasting. An RNN has a chainlike structure in which the information flows in a directed cycle. Each cell contains a group of activation functions interacting in a special way. In contrast with simple ML models, there are two advantages of DNNs: the capability to handle extremely large amounts of data and automatically performing feature selection inside the model (therefore, there is no special need for a pre-applied feature selection technique). Models with good performance used for energy load forecasting and energy price forecasting problems include Long-Short Term Memory (LSTM); gated recurrent unit (GRU), which is a sibling of LSTM with a small difference; non-linear autoregressive with exogenous inputs (NARX); and conditional restricted Boltzmann machine (CRBM) [18, 21, 30, 37, 38, 56].

11.3.2.1 LSTM and GRU

Long-Short Term Memory (LSTM) and gated recurrent unit (GRU) are designed to capture different time scale dependencies; however, they have minor differences in structure. Both approaches use a combination of functions, such as the sigmoid and hyperbolic tangent, to decide the output based on the input and, in some cases, the previous output. However, the main difference is that, compared to GRU, LSTM uses a memory cell to compute the output. This makes the GRU cell less complex; therefore, the training time is less than that for LSTM. Their performance is comparable, both being highly used in successful forecasting tasks; therefore, the user has the freedom to select the model most suited for the given specific scenario.

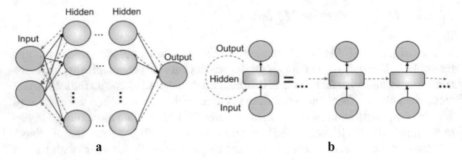

Fig. 11.4 **a** Layers and nodes in MLP, **b** LSTM and GRU unrolled chain structure

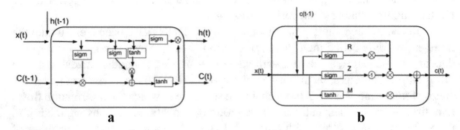

Fig. 11.5 **a** Functions in the LSTM cell, **b** (same) functions in the GRU cell [12]

The chainlike structure of LSTM and GRU cells is presented in Fig. 11.4b, in comparison with the structure of an MLP, shown in Fig. 11.4a, where we see multiple layers with fully connected nodes. Figure 11.4b represents the unrolled structure of LSTM/GRU, and it is notable that the information is passed on over time. "Remembering" information through time is one of the key characteristics that makes LSTM/GRU neural networks efficient and accurate. The layers in the LSTM cell and in the GRU cell interact in a special way. For a better understanding of how both cells work, their structure is presented in Fig. 11.5a and b. In Fig. 11.5a, we observe the structure of LSTM with four interactive layers (forget gate layer, store information, update the cell and decide the output). For each layer, we have either a sigmoid or hyperbolic tangent activation function and at least one computation. Figure 11.5b shows sigmoid and hyperbolic tangent functions interacting in a specific way in the GRU cell, where we have three layers (a reset gate R, an update gate Z and a new value M that is passed to the next state).

LSTM was also used in solar power forecasting in 2016 [26], in combination with a deep belief network and autoencoder, for 21 solar power plants. For results assessment, these models were compared to an MLP and showed better performance in forecasting. The authors used the hyperbolic tangent as the activation function in all ANNs and DNNs, except in the last output layer, where they used a rectified linear unit (ReLu). To measure the performance, the authors used five measures, including RMSE and MAE, concluding that Auto-LSTM (the combination of an autoencoder with LSTM) performed the best, with an average RMSE = 0.0713 and an average MAE = 0.0366.

In the same year, 2016, LSTM, and a variation of it, was implemented in [18] for forecasting building energy loads. Two models were designed, standard LSTM and sequence-to-sequence (S2S) LSTM, both trained and tested for one-hour and one-minute resolution energy data. The S2S model outperformed the standard LSTM; hence, the authors used the S2S model with different numbers of layers (from one to three) and different numbers of units in the layers (from five up to 100) for comparison. Computing the RMSE for all scenarios, the conclusion was that a two-layer S2S LSTM with 10 units performed best, with RMSE = 0.625, for the one-hour resolution data, and a two-layer S2S LSTM with 50 units presented the smallest RMSE = 0.667 for the one-minute resolution data. The authors claimed that the results are comparable to those obtained in [21] with a factored conditional RBM.

In 2018, LSTM was compared to ML models in [9] for medium- to long-term electricity consumption forecasting. The idea was to use wrapper and embedded feature selection methods and then genetic algorithms (GAs) to determine the best number of layers for LSTM and the optimal time lags. In this way, the forecasting was more accurate than that of the tested traditional ML models (ridge regression, k-nearest neighbours, random forest, gradient boosting, ANNs and extra trees; results are shown in [9]). The coefficient of variation of RMSE for LSTM with 30 time lags was 0.643, in comparison with that of, for instance, the extra tree model, which was 0.78.

Because LSTM and GRU have such similar architectures, both are equally likely to be implemented in forecasting tasks in the Energy Internet. Their performance was first compared in 2014 [37], for a different task of music and speech signal modelling, and the results obtained from both were in the same range. The equations presented in the respective paper in addition to Fig. 11.5a and b provide a better understanding of how the layers interact with each other. Both models outperformed a single tanh unit; however, there was no clear statement on which of the two units was better.

For the Energy Internet, GRU was used in 2018 [12] to forecast short-term electricity demand. Before feeding the data to the neural network, the authors performed autocorrelation to quantify the strength of the relationship between the data points. Then, GRU was designed with varying numbers of hidden layers (from one to three) and varying numbers of units (from ten to 30). Then, each model was run, and the comparison of their performance was made by computing the MAPE. The results showed that, for the respective dataset, a GRU with 2 hidden layers and 20 units offered the best results, and there was no need to increase the number of hidden layers for that amount of electricity consumption data. In a different scenario, the number of inputs was changed from one to two (meaning that two time stamps were fed to the network as input) to test the same model. According to the results, the model with only one input at a time performed better than the one with two inputs for all numbers of layers, and in the case of the GRU with 2 layers and 20 units, the MAPE was equal to 0.79%.

11.3.2.2 Non-linear Autoregressive Model

NARX is another recurrent neural network that uses a combination of input and past output values to compute the current output. Therefore, unlike a regular RNN, the communication in NARX between the hidden and output layers is bidirectional [64]. This is illustrated in Fig. 11.6. NARX takes the linear combination of the inputs and previous outputs, adds biases and applies an activation function. Then, it applies weights and biases again and runs the values through the function to map the output.

For a better understanding of the communication between the layers in an NARX neural network, Fig. 11.7 is presented, where x(t) represents the inputs (input layer) that are being passed to the hidden layer, y(t) is the current output and "steps" are all the previous outputs that are being passed back to the hidden layer and contribute to the current output.

In 2018, NARX was used in the Energy Internet, in a work published in the International Journal of Forecasting [30], for electricity price forecasting of a day-ahead market. Here, the authors also aimed to show the importance of the seasonal component in the long term. For a long time, it was believed that the seasonal component in the long term does not add much value to electricity price study; thus, it was ignored in most works. Instead, the daily and weekly seasonal components were used in many cases for day-ahead electricity price forecasting. In [30], the authors were inspired by the seasonal component autoregressive model, which was recently introduced. They treated four benchmarks separately, first the so-called naïve method, then the expert benchmark, then the seasonal component autoregressive benchmark (SCAR) and finally the proposed NARX neural network, which they aimed to prove outperforms the previous benchmarks. To compare the models, the authors computed the weekly weighted mean absolute error (WMAE), which is similar to MAPE but normalizes the absolute error by the mean weekly price. They concluded that the seasonal component (SC) plays a huge role in forecasting the day-ahead electricity price with NARX and that the SC is more beneficial in a non-linear context than

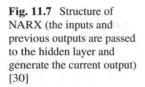

Fig. 11.6 Communication in an RNN (one direction) and NARX (two directions) [64]

Fig. 11.7 Structure of NARX (the inputs and previous outputs are passed to the hidden layer and generate the current output) [30]

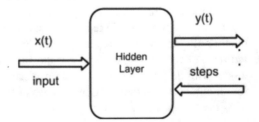

in linear modelling. Finally, the non-linear models used with the SC were able to outperform the previously proposed SCAR model.

Similar work has been conducted for electricity price forecasting, in [39, 49, 69], using ML models such as simple feed-forward ANNs, SVM, and DL models such as recurrent neural networks (RNNs).

A very recent work, [10], in 2018, applied NARX to a different problem, that is, forecasting daily solar radiation. For this problem, ANNs had already produced remarkable results and proved their performance. However, in [10], the authors proposed a NARX model that is designed to help supply a sailboat with electricity, with the ultimate goal being estimation of solar energy availability.

The model took as inputs the theoretical direct solar radiation, the predicted cloud cover, and the global solar radiation and produced as output the direct solar radiation on a horizontal surface. To measure the model's effectiveness, two validation methods were implemented, the well-known MSE and the Daily Mean of the Power Error (DMPE).

The latter was used to ensure that error with opposite signs neutralize each other in the end. In the experiments, the number of nodes in the input and hidden layers was varied from 10 to 22 (number of output neurons being 1 at any time), and it was noticeable that the MSE varied between 0.0041 and 0.01438 and that the DMPE varied between 30.4164 and 73.4646. However, the best performance was produced when the weights were not initialized (were randomly initialized), so the MSE was 0.00279 and the DMPE was 24.0584.

Although the NARX neural network started to be implemented quite recently in the Energy Internet, its performance is extremely promising and has the potential to become a highly used model for energy-related tasks, especially forecasting.

11.3.2.3 Restricted Boltzmann Machine

A restricted Boltzmann machine (RBM) is a useful algorithm for many problems, such as feature learning, classification, regression and dimensionality reduction. It has a simple structure consisting of two layers, one visible (input) and one hidden layer, with fully connected nodes, as presented in Fig. 11.8. The inputs are multiplied by the weights, and biases are added to them in the hidden layer; then, the results are

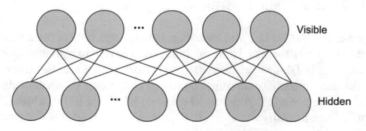

Fig. 11.8 Structure of an RBM with 2 layers, one visible and one hidden [55]

passed through the activation function. In this way, an output is produced for each node in the hidden layer. An RBM finds patterns in the data (in an unsupervised way) by reconstructing the input. It passes the activations back from the hidden layer to the visible layer by adding biases, and these steps are repeated until the inputs and the approximations are as close to each other as possible.

RBMs have been successfully applied in electricity load forecasting, such as in [32], where it outperformed a simple neural network.

In another work [55], an RBM was used, together with a rectified linear unit (ReLu), for electricity consumption forecasting of individual customers' consumption data and weather elements. For results assessment, the RBM was here compared to a simple ANN and ARIMA models, and the conclusions were drawn based on MAPE and relative RMSE efficiency measures. In that work, detailed descriptions of the ANN, RBM, ARIMA and ReLu activation functions were provided before the experimental section. Computational experiments were conducted for three different cases, first with an individual customer and second with a data combination of eight industrial categories ("public administration, retail business, R&D services, networking business, healthcare, vehicle and trailer manufacturing industry, electronic component and computer manufacturing industry and other manufacturing industries" [55]). The DNN here with a ReLu activation function was easy to train and presented an MAPE and a relative RMSE of 17% and 22% less than those of the simple ANN, respectively.

As for work [21], mentioned in Sect. 11.3.2.1, energy load forecasting was conducted on the same one-minute resolution data for a period of four years. The authors implemented two DL models, conditional restricted Boltzmann machine (CRBM) and factored conditional restricted Boltzmann machine (FCRBM), claiming that FCRBM outperformed CRBM, a recurrent neural network (RNN), SVM and an ANN. Both proposed models were carefully described, with supporting pictures and equations.

The dataset and the implementation were meticulously illustrated before drawing the conclusion. Again, RMSE was computed for results assessment, together with the correlation coefficient (R). For the three different scenarios (forecasting for 15 min, 1 h and 1 day ahead), the RMSE was 0.6211, 0.6663 and 0.8286, respectively. Multiple tables support their claims for all different scenarios and models.

11.3.2.4 Rectified Linear Unit (ReLu)

A rectified linear unit (ReLu) is an activation function that is most often used in hidden layers of neural networks. However, it has been proven to significantly improve the performance of DL models. One example is work [55], where ReLu contributed to an RBM in an electricity consumption forecasting problem. Another study [48] presented a more powerful hybrid model using an RBM and ReLu, which can naturally handle large intensity variations. In [2], the authors showed how they changed a ReLu from being an activation function for the neural network's hidden layers to contributing to predictions. They used the network on a classification problem, and the twist was that they implemented a DNN by using ReLu in the prediction units. They claimed that DL using ReLu is at least comparable or even better than conventional DL models.

11.3.2.5 Other DL Models Used in the Energy Internet

DL models are widely used in the field of the Energy Internet for different tasks and in different forms. As mentioned previously, the models are either used alone or in combination with other structures. A good example for this is work [42], in which the authors used a stacked autoencoder (SAE) with an extreme learning machine (ELM) for forecasting the energy consumption of buildings. They compared the proposed model with other highly used models, such as a simple back propagation ANN, SVM, and multiple linear regression (MLR). The SAE was used to extract relevant features from the building's energy consumption data, whereas the ELM was used for forecasting. Measuring the effectiveness with MAE, RMSE and mean relative error (MRE), the authors concluded in the paper [42] that their model outperformed several traditional models due to its complexity and learning strategy.

An MLP was also designed in [7] for forecasting a daily electric load profile. The data consisted of not only electricity consumption but also weather information. An unsupervised neural network was first applied to cluster the different load profiles before actually forecasting the demand. A detailed description of the MLP was given in the paper together with a description of PCA, which was also applied to the data in the processing. PCA is described in Sect. 11.3.2.6. The work underlines the suitability of the proposed model for energy load forecasting in urban areas.

In 2016, another DL model, a deep belief network (DBN), was used for the first time for forecasting wind speed, an aspect that plays a huge role in the field of the Energy Internet, in what concerns planning and operation of energy systems [67]. In this work, a DBN was precisely described as having a RBM used for learning and a logistic regression layer used for forecasting. Moreover, the model applied here is a hybrid form of DBN, wavelet transform (WT) and quantile regression (QR). The authors tested their approach in many different scenarios, which involved seasonal variation (spring, summer, fall, winter and in addition average). Clear and comprehensive tables in work [67] presented comparisons of the proposed model with other ML models, such as autoregressive moving average (ARMA), a simple back propagation neural network (BPNN) and a Morlet wavelet neural network (MWNN). For this purpose, three measures of effectiveness were computed, namely, MAE, RMSE and MAPE, and for instance, for one-hour-ahead spring wind speed forecasting at one of the farms, these values were the smallest for their proposed model, 0.4097, 0.5510 and 6.15%, for MAE, RMSE and MAPE, respectively. The hybrid DBN approach outperformed the rest of the tested models and presented high stability during simulations.

A deep belief network (DBN) was also used for electricity load forecasting in [19] compared to a traditional ANN and to the forecast provided by the local system operator (MEPSO) in terms of MAPE. The authors presented the importance of using as input variables other than only the previous electricity consumption, such as hour of the day, day in the week (with respect to holidays), previous day's average load, and temperature, gathering a total of 8 inputs. As a result, for a model of 24-hour-ahead consumption forecasting, the MAPE improved by 8.6% compared to MEPSO, and for the peak consumption forecasting for the next day, the MAPE improved by 21% compared to the same MEPSO. The MAPE for forecasting the consumption was just below 3.7% for the proposed DBN, just below 3.8% for the traditional ANN, and just above 4% for the MEPSO. The authors also concluded that their model was able to efficiently indicate the most critical period in a day and the most critical day in a week regarding electricity consumption. Moreover, there was a difference in accuracy between forecasting in a section composed of mainly households and a section with industrial buildings, with higher accuracy in the household section.

As previously mentioned, ML models are often used in combination with other models, and a number of *hybrid* SVMs and ANNs were presented. In the case of DL, there is an amount of work in the literature, where we find DL models combined with other models, and these are called *hybrid* or ensemble deep learning (EDL) models, and often, using only neural networks, ensemble neural networks (ENNs).

A good example of an EDL is presented in [52], a work from 2014, where the authors implemented the DBN mentioned before, in combination with an SVR. The main idea was to train the DBN and produce 20 outputs, which were later fed to an SVR as inputs, and other new outputs were produced. To assess their model's efficiency, two measures of effectiveness were computed, the RMSE and MAPE, while testing on different datasets. Moreover, the proposed model was compared to 4 other models, SVR, a traditional feed-forward ANN named FNN, a DBN and an ENN with only feed-forward networks, on all of their datasets (a Mackey-Glass dataset, 3 electricity load datasets and 3 regression datasets). In terms of RMSE and MAPE, the proposed EDL (DBN + SVR) outperformed the other 4 models on both time series and regression datasets. Some examples are given as follows: for California housing, RMSE was 0.1637, 0.1803, 0.1773, 0.1615 and 0.1508 for the SVR, FNN, DBN, ENN and proposed EDL, respectively, and the MAPE was 28.36%, 33.31%, 32.26%, 29.44% and 27.33%, respectively. Other datasets show in the same way that the proposed ENN in [52] offered a higher performance than other ML and DL models, which were used as comparisons.

Three years later, in 2017, the same authors as in [52] proposed another approach for EDL in [53] for energy load forecasting. Here, the ensemble model consists of empirical mode decomposition (EMD) and a DBN. The proposed model was again compared to 8 other ML/DL models on 3 different datasets, in terms of RMSE and MAPE. As observed, the proposed model outperformed the other models with single structures. Additionally, the advantage of DL handling non-linear datasets and the effectiveness and rapidity in computation of the random forest (RF) model were pointed out. The proposed ensemble model was able to efficiently forecast the electricity demand for a day ahead, with a MAPE varying between 2.08 and 10.46%.

The applicability of EDL is demonstrated not only on electricity demand forecasting but also on wind speed-related tasks.

In the same year, 2017, an EDL approach was proposed in [68] for probabilistic wind power forecasting. Here, a convolutional neural network (CNN) was implemented for forecasting probabilistic wind power (in a hybrid form with wavelet transform), and an ensemble technique was then applied to deal with the errors. CNNs are a class of DNNs that are implemented in a multilevel way, with convolution and pooling layers, similar to an MLP. These are most highly used in image classification and recognition problems; however, their applicability started to extend in the Energy Internet. Therefore, the proposed method for wind power forecasting, based on the WT, CNN and ensemble techniques, was compared to other benchmark models, such as the persistence method, ANN and SVM. On the dataset collected from a wind farm containing various seasons and time resolutions, the proposed model demonstrated the highest reliability.

In addition to these DL models and their ensemble variations, work has been conducted on pooling-based forecasting in the Energy Internet. Examples of these methods are found in [3, 56]. In [56], the pooling system was used to generate inputs for an RNN on a problem of forecasting household energy consumption. The model demonstrated the best results (6.5) in terms of RMSE when compared to ARIMA and SVR (19.5 and 13.1, respectively). Another approach of pooling was presented in [3] on forecasting the household electric vehicle (EV) charging demand. Here, a two-layer ensemble model was used, combining several methods such as random forest and naïve Bayes. The proposed two-layer model achieved the highest accuracy of 0.724.

Based on the analysis of all ML and DL models above, it is visible that AI has a huge applicability in the field of Energy Internet, and it still continues to grow.

While in the beginning of the ML era, the models were designed mainly for forecasting electricity demand or price, we can observe that in recent years, these models are used in wind speed forecasting, EV power demand, household and commercial building power consumption optimization, etc. Moreover, higher accuracy on many of the tasks is achieved by combining neural networks with other ML models, DL models with other ML and DL models, or pooling-based techniques. The obtained hybrid or ensemble methods demonstrate very large benefits for the majority of the problems that they are applied on in the Energy Internet.

11.3.2.6 Dimensionality Reduction Techniques

One major difference between ML and DL models is the way in which we feed the data to the input layer. Gathering the data is a complicated and hard process, and researchers usually obtain data with missing values, with different shapes or that are not clean. In this case, data must be run through a preprocessing step. Even if the data are clean and with no missing values, the preprocessing is indicated to obtain better results in forecasting. If in ML learning many preprocessing steps need to be done by human interaction, in DL some of the steps can be neglected, as DL models learn features of the dataset on their own [45, 71].

As the data sometime come in unexpected or not needed dimensions, in ML, it is indicated to perform a dimensionality reduction before feeding the inputs. The advantages of dimensionality reductions include better visualization, computing and space efficiency, avoiding the "curse of dimensionality" and of course relevant feature extraction. Methods for this are multiple; however, principal component analysis (PCA), linear discriminant analysis (LDA), and t-distributed stochastic neighbour embedding (T-SNE) are among the most popular. PCA and LDA work similarly, as both compute eigenvectors and eigenvalues, and their performance is comparable [5]; however, the major difference is that PCA finds the directions of the highest variance, and LDA is designed to find the set of features that help maximize the class separability.

Consequently, to implement a DNN, we need two important elements: large data and more computational power. Deep learning is, in fact, a result of technology evolution, as today's computers are able to process tremendous (increased over years) amounts of data in a much faster and more elegant manner.

11.3.3 Advantages of AI in the Energy Internet

Data analysis, energy consumption optimization, and future energy consumption prediction were treated over the years according to emerging technologies. Before the existence of AI models, data scientists and energy suppliers worked with statistical, or so-called traditional models, that often included a database composed of details about the studied building (such as construction material, occupancy period, dimensions and others), outside dry-bulb temperature, outside humidity, solar radiation, time indicators (such as week day or weekend, holyday, seasonality and others). The availability of such data is most often questionable, and the only dataset that is accessible for research is energy consumption. Here, AI plays a huge role for the Energy Internet because AI models are well known for learning patterns in consumption data and estimating future behaviour. In Table 11.1, the advantages of the AI for the Energy Internet over statistical models are illustrated, as found in most of the scientific work presented in this chapter. In what concerns the choice between an ML and a DL model, there are some characteristics of the dataset that make the decision possible: the size of the dataset, the number of features, the complexity and the noise of the data, and the computational power. If any of these should be high, a DL model will offer a more satisfactory performance.

Table 11.1 Advantages of using AI over statistical approaches in the Energy Internet

Statistical models	AI models
Lower accuracy	Simple structure and easy implementation
Sensitivity to outliers (which are quite common in electricity consumption data)	Suitable for time series (with special attention to RNNs)
Possibly inadequate features in the data, due to, not always accurate, human assumptions	Fast/accurate computation
More human interaction, thus more subjectivity	"Learned" pattern based on historical consumption data
Time consuming	"Learning" and improving based error
	Efficient feature selection
	Handling noisy and non-linear data
	Reduced human interaction in the analysis and forecasting process
	Flexibility in modifying the model's structure (especially when adding layers and nodes to a neural network)

11.4 Electricity Prediction Challenges and Influential Factors

In the field of the Energy Internet, predicting future electricity consumption is a challenging and complex task for companies and researchers. As the presented related works show, it can be treated in many different ways, including several inputs. It consists of multiple steps and depends on a multitude of factors that can influence the process. In this section, the detailed process of collecting electricity consumption data, processing, predicting and other steps is presented for a better understanding of the field. Moreover, a description of the set of factors that can influence electricity consumption and the prediction process is given as well. Influencing factors can contribute immensely to electricity forecasting and are therefore crucial when designing the prediction model.

11.4.1 Challenges

Figure 11.9 shows the process of collecting raw consumption data from the real world (this step being done by an authorized party—e.g. an electricity supplier), and the processing and cleaning part of the data (which is done by the same party or an authorized third party). After this step, a conclusion can be drawn already by performing some statistical analysis regarding the trend and consumption pattern.

Next, using artificial intelligence techniques (DL was presented in this case), they can forecast future behaviour, or in other words, future electricity consumption.

In this way, the supply and demand ratio can be stabilized, and decision-making is automatically supported.

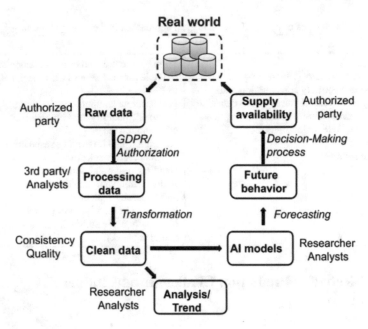

Fig. 11.9 Illustration of data collection, processing, analysis, forecasting and future decisions using Deep Learning [13]

One important aspect of the Energy Internet is data protection. It is important that the data are anonymized and no unauthorized person receives access to personal information from clients. Therefore, while researching in the energy field, it is crucial to consider the GDPR [1] regulations, as also shown in Fig. 11.9.

11.4.2 Factors Affecting Prediction

The Energy Internet is an extremely complex field, considering the multitude of methods that can be used to predict future electricity consumption, as well as the multitude of factors that influence the process of data analysis and prediction. First, we need to take into consideration what type of prediction is needed, then weather characteristics and house (or building) characteristics. A few of these are mentioned below:

- Short-term prediction
- Medium-term prediction
- Long-term prediction
- Seasonality/Temperature
- Humidity
- Solar gain [72]

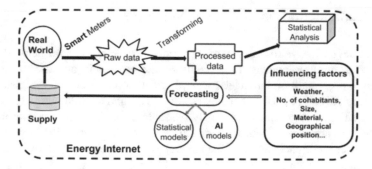

Fig. 11.10 Visual representation of steps and influential factors in the Energy Internet field [14]

- House size
- Material of construction
- Number and age of inhabitants
- Geographical position
- Nature and number of appliances
- Electric cars, and so on.

On multiple occasions, many of these aspects are not included in the data that are provided to use for research. These aspects are either unavailable or extremely difficult to provide. In general, it is most common for researchers to include weather data in their analysis, as this is quite easy to collect, time of the day, day of the week, holiday season and peak intervals. When including temperature data in the set of inputs, the trend, seasonality and outliers are easily readable. Moreover, predicting future behaviour strongly relies on future weather conditions (see Fig. 11.10).

A work relevant to this topic was conducted in [32], where the authors presented four of these features, related to electricity load, as well as a comparison of a deep neural network, namely, a RBM with a simple backpropagation neural network for electricity load forecasting. By computing the MAPE (%) as a measure of performance, it was visible that RBM (MAPE = 1.98) carried out more accurate results than the simple back propagation neural network (MAPE = 2.07). The most challenging electricity consumption prediction type is the granular prediction, which includes separate consumption data from all house appliances. This type of data is, in most cases, unavailable, as it involves the existence of smart metres in each home/building. In addition, the number and the age of cohabitants would have a huge effect on the prediction, as each of them presents different, but extremely specific, consumption patterns over a longer period of time. Figure 11.10 shows a visual representation of data collection, including some of the presented aspects and the consumption prediction process.

11.5 Implementation of Neural Networks

There are several ways to implement AI models, especially neural networks, and several programming languages that facilitate the process in one way or another. Choosing a programming language for implementation depends on the goal of the users for their algorithm. This goal may be centred around the following aspects: fast implementation, fast computation, easily readable code, short versus long instructions (as there are several libraries for each language that can facilitate writing) or any other user preference. Some of the languages that have been used over the years for implementing neural networks are as follows: C, C++, C#, Lisp, Java, MATLAB, R, and Python. Table 11.2 briefly presents some of the advantages of different programming languages, so it raises the interest of the reader for investigating the most suitable one. From this list, in recent years, Python began to stand out due to its readability and fast and efficient implementation. Python supports procedural programming, object-oriented and functional programming.

Implementing neural networks has become easier and more time-efficient over the years with the emergence of different libraries. In the beginning, neural networks were implemented from scratch, programming each layer line by line, activation function, error computing function, optimizer and so on. However, these days, programming languages, especially Python, include a multitude of dependencies and libraries, such as NumPy, Scipy, Sklearn and Matplotlib, that help implementing matrix multiplications, additions, functions and plots for visualization. In addition, we find a multitude of libraries and packages that facilitate code writing for neural networks, such as Keras, Theano, TensorFlow, Tflearn, Mxnet (Amazon's DL framework), Microsoft Cognitive Toolkit (CNTK) and Pytorch, some of which will be described (with their advantages) in the following paragraphs.

The implementation of ML and DL models is multiple; therefore, users choose the right implementation based on their scope and preferences. One of the most frequently used programming languages for implementing neural networks is Python. As mentioned in the previous paragraph, two of the most important reasons for choosing Python are readability (it is an easy to learn and to read language, less complex than other programming languages) and the variety of packages. Here, the focus is on open-source, high-level libraries/packages, Keras, TensorFlow and Theano, and Pytorch efficiently and frequently used for building neural networks. In addition, as mentioned in the previous paragraph, there are different packages used for implementing mathematical operations, scientific computing and plotting, such

Table 11.2 Characteristics of five different popular programming languages

Python	Java	R	MATLAB	Lisp
Fast and readable implementation Packages	Packages Easy debugging	Suitable for data, statistical analysis	Immediate testing of the implemented methods	Suitable for AI development (one of the oldest)

as NumPy, Scikit, Matplotlib and SciPy. Keras is a high-level API that is written in Python and simplifies the design and experimentation of neural networks. It can run effectively on both a CPU and GPU, on top of TensorFlow and Theano. It is easy to implement due to imported dependencies such as different layers, optimizers, cost and activation functions and initialization schemes [22].

TensorFlow is an open-source library that allows computations on one or more CPUs or GPUs. It was developed by the Google Brain team and has a high performance due to the way the operations are evaluated in the computational graph. In this graph, the nodes contain mathematical operations executed according to the dependencies between them [29].

Theano is an open-source library also written in Python, which efficiently defines, optimizes and evaluates complex mathematical expressions. Elementary computations are performed and expressed NumPy techniques. It has a familiar Python syntax and transparent set of instructions, as well as high performance on both a CPU and GPU [63].

On the other hand, Pytorch [51] is a newer library, implemented as a relative of Torch, which was designed in the Lua programming language. Pytorch was designed by Facebook, as opposed to TensorFlow, designed by Google. One of the main differences between TensorFlow and Pytorch is the way in which the computational graph is designed. TensorFlow supports static computational graph, whereas Pytorch dynamic computational graph (in which the graph is built while running the network). Some of the advantages are as follows: easy for beginner users to learn, due to its simple implementation (which also means that it is easier to implement neural networks from scratch), slightly faster than TensorFlow, considering less complexity, debugging while coding, data parallelism. On the downside, because it is a relatively new library, the community behind it is smaller than the TensorFlow community, which consequently means less documentation and less support for problems encountered during coding. Pytorch can be used on GPUs that support CUDA [51]. Regarding operation systems, TensorFlow and Pytorch run as follows:

- **TensorFlow**: Ubuntu (16.04 and later), macOS (Sierra and later), Windows 7 (and later), Raspbian (9.0 and later),
- **Pytorch**: Linux (glibc $\geq$ v2.17, Arch Linux, CentOS, Debian, Fedora, Mint, OpenSUSE, PCLinuxOS, Slackware, Ubuntu), macOS (Yosemite and later), Windows 7 (and later, Windows Server 2008 r2 and greater).

Most of today's neural network implementation libraries support GPU programming or data parallelism, which means that a large amount of computations can be carried out in reduced time on parallel devices. This is extremely useful in the case of very large datasets and DL models and can reduce training time from days (or even weeks) to hours. TensorFlow and Pytorch are both implemented with CUDA (parallel computing platform) extensions; therefore, both support parallel training.

In conclusion, there are plenty of possibilities for the implementation of neural networks in Python (and other languages); however, the choice of the implementation style is greatly dependent on the user's needs and preferences, the characteristics of the task, and the computational power.

11.6 Conclusion

The Energy Internet is a complex field that, treated meticulously, can open new horizons for existing and future energy suppliers and even academia. It represents a careful combination of existing data and knowledge with performant emerging technology. A huge advantage of the world's modernization is represented by the extremely fast evolution of technology.

Over the years, methods have been applied in the energy field, from statistical traditional methods to artificial intelligence (AI) techniques. These AI techniques have become increasingly powerful due to increasing computational power. Today's performant computers are able to handle a large amount of data and perform millions of computations in a shorter time; therefore, we are able to see a transition from ML models to DL models. These DL models require a large amount of training data (from consumption to weather and house characteristics, house occupancy, etc.), which, as mentioned previously, are unavailable or non-existent in multiple cases.

In this chapter, we learned about aspects regarding the Energy Internet, general artificial intelligence notions and how these affect data analysis and prediction processes. Moreover, we presented in detail different ML/DL techniques, including visual representations of some models, as well as the advantages and strengths of these methods. All of the abovementioned AI models present remarkable performance in the energy field, as well as other fields, such as cancer detection, drug discovery, cybersecurity, land and surface changes researched by NASA and weather forecasting. As a conclusion of this chapter, we note that for the question of which AI model is the most performant, the answer is debatable since estimating energy consumption is a complex task that involves a multitude of components, and the models are highly dependent on the quality of the dataset. Although the answer is debatable, it is clear that DL brings a very large contribution to the Energy Internet and carries out the most accurate forecasting on large datasets. The results may improve in the future by including multiple influential factors as inputs in the mentioned AI models to support consumption prediction.

References

1. M.C. Addis, M. Kutar, The General Data Protection Regulation (GDPR), Emerging Technologies and UK Organisations: Awareness, Implementation and Readiness (2018)
2. A. F. Agarap, Deep learning using rectified linear units (relu) (2018). *arXiv preprint* arXiv: 1803.08375
3. S. Ai, A. Chakravorty, C. Rong, Household EV charging demand prediction using machine and ensemble learning, in *2018 IEEE International Conference on Energy Internet (ICEI)* (IEEE, 2018), pp. 163–168
4. A.H. Neto, F.A.S. Fiorelli, Comparison between detailed model simulation and artificial neural network for forecasting building energy consumption. Energy Build. **40**(12), 2169–2176 (2008)
5. A.M. Martínez, A.C. Kak, Pca versus lda. IEEE Trans. Pattern Anal. Mach. Intell. **23**(2), 228–233 (2001)

6. N. Amjady, Short-term hourly load forecasting using time-series modeling with peak load estimation capability. IEEE Trans. Power Syst. **16**(3), 498–505 (2001)
7. M.A.R.C.O. Beccali et al., Forecasting daily urban electric load profiles using artificial neural networks. Energy Convers. Manag. **45**(18-1), 2879–2900 (2004)
8. B. Dong, C. Cao, S.E. Lee, Applying support vector machines to predict building energy consumption in tropical region. Energy Build. **37**(5), 545–553 (2005)
9. Salah Bouktif et al., Optimal deep learning lstm model for electric load forecasting using feature selection and genetic algorithm: Comparison with machine learning approaches. Energies **11**(7), 1636 (2018)
10. Z. Boussaada, O. Curea, A. Remaci, H. Camblong, N.A. Mrabet Bellaaj, Nonlinear Autoregressive Exogenous (NARX) neural network model for the prediction of the daily direct solar radiation. Energies **11**, 620 (2018)
11. K. Greff, R.K. Srivastava, J. Koutník, B.R. Steunebrink, J. Schmidhuber, LSTM: a search space odyssey. IEEE Trans. Neural Netw. Learn. Syst. **28**(10), 2222–2232 (2017)
12. C. Heghedus, A. Chakravorty, C. Rong, Energy load forecasting using deep learning. In *2018 IEEE International Conference on Energy Internet (ICEI)* (IEEE, 2018), pp. 146–151
13. C. Heghedus, A. Chakravorty, C. Rong, Energy informatics applicability; machine learning and deep learning, in *2018 IEEE International Conference on Big Data, Cloud Computing, Data Science & Engineering (BCD)* (IEEE, 2018), pp. 97–101
14. C. Heghedus, A. Chakravorty, C. Rong, Deep learning for short-term energy load forecasting using influential factors, in *The 13th World Congress on Energy Asset Management (WCEAM)* (2018)
15. C. Deb, L.S. Eang, J. Yang, M. Santamouris, Forecasting diurnal cooling energy load for institutional buildings using Artificial Neural Networks. Energy Build. **121**, 284–297 (2016)
16. C.J. Willmott, K. Matsuura, Advantages of the mean absolute error (MAE) over the root mean square error (RMSE) in assessing average model performance. Clim. Res. **30**(1), 79–82 (2005)
17. S. Dai, D. Niu, Y. Han, Forecasting of energy-related CO_2 emissions in china based on $GM(1,1)$ and least squares support vector machine optimized by modified shuffled frog leaping algorithm for sustainability. Sustainability **10**, 958 (2018)
18. D. L. Marino, K. Amarasinghe, M. Manic, Building energy load forecasting using deep neural networks, in *IECON 2016-42nd Annual Conference of the IEEE Industrial Electronics Society* (IEEE, 2016), pp. 7046–7051
19. A. Dedinec, S. Filiposka, A. Dedinec, L. Kocarev, Deep belief network based electricity load forecasting: an analysis of Macedonian case. Energy **115**, 1688–1700 (2016)
20. H. Drucker, C. J. Burges, L. Kaufman, A. J. Smola, V. Vapnik, Support vector regression machines, in *Advances in Neural Information Processing Systems* (1997), pp. 155–161
21. E. Mocanu, P.H. Nguyen, M. Gibescu, W.L. Kling, Deep learning for estimating building energy consumption. Sustain. Energy Grids Netw. **6**, 91–99 (2016)
22. F. Chollet, Keras, Github, (2015). https://keras.io
23. S.N. Fallah, R.C. Deo, M. Shojafar, M. Conti, S. Shamshirband, Computational intelligence approaches for energy load forecasting in smart energy management grids: state of the art, future challenges, and research directions. Energies **11**, 596 (2018)
24. C. Fan, F. Xiao, S. Wang, Development of prediction models for next-day building energy consumption and peak power demand using data mining techniques. Appl. Energy **127**, 1–10 (2014)
25. F. Kaytez, M.C. Taplamacioglu, E. Cam, F. Hardalac, Forecasting electricity consumption: a comparison of regression analysis, neural networks and least squares support vector machines. Int. J. Electr. Power Energy Syst. **67**, 431–438 (2015)
26. A. Gensler, J. Henze, B. Sick, N. Raabe, Deep learning for solar power forecasting— an approach using AutoEncoder and LSTM Neural Networks, in *2016 IEEE International Conference on Systems, Man, and Cybernetics (SMC)*, pp. 002858–002865
27. A.T. Goh, Back-propagation neural networks for modeling complex systems. Artif. Intell. Eng. **9**(3), 143–151 (1995)

28. I. Goodfellow, Y. Bengio, A. Courville, Y. Bengio, *Deep learning*, vol. 1 (MIT Press, Cambridge, 2016)
29. Google Brain Team, TensorFlow: a system for large-scale, machine learning, in *12th USENIX Symposium on Operating Systems Design and Implementation (OSDI 16)*
30. G. Marcjasz, B. Uniejewski, R. Weron, On the importance of the long-term seasonal component in day-ahead electricity price forecasting with NARX neural networks. Int. J. Forecast.
31. C. Guan, P.B. Luh, L.D. Michel, Y. Wang, P.B. Friedland, Very short-term load forecasting: wavelet neural networks with data pre-filtering. IEEE Trans. Power Syst. **28**, 30–41 (2013)
32. W. He, Deep neural network based load forecast. Comput. Model. New Technol. **18**(3), 258–262 (2014)
33. R. Hu, S. Wen, Z. Zeng, T. Huang, A short-term power load forecasting model based on the generalized regression neural network with decreasing step fruit fly optimization algorithm. Neurocomputing **221**, 24–31 (2017)
34. H.C. Jung, J.S. Kim, H. Heo, Prediction of building energy consumption using an improved real coded genetic algorithm based least squares support vector machine approach. Energy Build. **90**, 76–84 (2015)
35. H. Jiang, Y. Zhang, E. Muljadi, J. Zhang, W. Gao, A short-term and high-resolution distribution system load forecasting approach using support vector regression with hybrid parameters optimization. IEEE Trans. Smart Grid (2016)
36. J. Ahn, S. Cho, D.H. Chung, Analysis of energy and control efficiencies of fuzzy logic and artificial neural network technologies in the heating energy supply system responding to the changes of user demands. Appl. Energy **190**, 222–231 (2017)
37. J. Chung, C. Gulcehre, K. Cho, Y. Bengio, Empirical evaluation of gated recurrent neural networks on sequence modeling (2014). arXiv preprint arXiv:1412.3555
38. W. Kong, Z.Y. Dong, D.J. Hill, F. Luo, Y. Xu, Short-term residential load forecasting based on resident behaviour learning. IEEE Trans. Power Syst. **33**, 1087–1088 (2018)
39. J. Lago, F. De Ridder, P. Vrancx, B. De Schutter, Forecasting day-ahead electricity prices in Europe: the importance of considering market integration. Appl. Energy **211**, 890–903 (2018)
40. C.M. Lee, C.N. Ko, Time series prediction using RBF neural networks with a nonlinear time-varying evolution PSO algorithm. Neurocomputing **73**(1–3), 449–460 (2009)
41. H.-Z. Li, S. Guo, C.-J. Li, J.-Q. Sun, A hybrid annual power load forecasting model based on generalized regression neural network with fruit fly optimization algorithm. Knowl.-Based Syst. **37**, 378–387 (2013)
42. S. Li, P. Wang, L. Goel, Short-term load forecasting by wavelet transform and evolutionary extreme learning machine. Electr. Power Syst. Res. **122**, 96–103 (2015)
43. C. Li, Z. Ding, D. Zhao, J. Yi, G. Zhang, Building energy consumption prediction: an extreme deep learning approach. Energies **10**(10), 1525 (2017)
44. N. Liu, Q. Tang, J. Zhang, W. Fan, J. Liu, A hybrid forecasting model with parameter optimization for short-term load forecasting of micro-grids. Appl. Energy **129**, 336–345 (2014)
45. M.M. Najafabadi, F. Villanustre, T.M. Khoshgoftaar, N. Seliya, R. Wald, E. Muharemagic, Deep learning applications and challenges in big data analytics. J. Big Data **2**(1), 1 (2015)
46. M.A. Mohandes, T.O. Halawani, S. Rehman, A.A. Hussain, Support vector machines for wind speed prediction. Renew. Energy **29**(6), 939–947 (2004)
47. S. Mishra, S.K. Patra, Short term load forecasting using a neural network trained by a hybrid artificial immune system, in *Proceedings of the IEEE Region 10 and the Third international Conference on Industrial and Information Systems, ICIIS 2008, Kharagpur, India* (2008)
48. V. Nair, G. E. Hinton, Rectified linear units improve restricted boltzmann machines, in *Proceedings of the 27th International Conference on Machine Learning (ICML-10)* (2010), pp. 807–814
49. J. Nowotarski, R. Weron, Recent advances in electricity price forecasting: a review of probabilistic forecasting. Renew. Sustain. Energy Rev. **81**, 1548–1568 (2018)
50. G. Panchal, Mahesh Panchal, Review on methods of selecting number of hidden nodes in artificial neural network. Int. J. Comput. Sci. Mob. Comput. **3**(11), 455–464 (2014)
51. Pytorch Documentation, http://www.pytorch.org

52. X. Qiu, L. Zhang, Y. Ren, P. N. Suganthan, G. Amaratunga, Ensemble deep learning for regression and time series forecasting, in *2014 IEEE Symposium on Computational Intelligence in Ensemble Learning (CIEL)* (IEEE, 2014), pp. 1–6

53. X. Qiu, Y. Ren, P.N. Suganthan, G.A. Amaratunga, Empirical mode decomposition based ensemble deep learning for load demand time series forecasting. Appl. Soft Comput. **54**, 246–255 (2017)

54. R.K. Jain, K.M. Smith, P.J. Culligan, J.E. Taylor, Forecasting energy consumption of multi-family residential buildings using support vector regression: Investigating the impact of temporal and spatial monitoring granularity on performance accuracy. Appl. Energy **123**, 168–178 (2014)

55. S. Ryu, J. Noh, H. Kim, Deep neural network based demand side short term load forecasting. Energies **10**(1), 3 (2016)

56. H. Shi, M. Xu, R. Li, Deep learning for household load forecasting—a novel pooling deep RNN. IEEE Trans. Smart Grid **9**(5), 5271–5280 (2018)

57. S. A. Kalogirou, Applications of artificial neural-networks for energy systems, in *Energy Systems*, ed. by T. Ohta (Elsevier, Oxford, 2000), pp. 17–35, ISBN 9780080438771

58. S.A. Kalogirou, Artificial neural networks in renewable energy systems applications: a review. Renew. Sustain. Energy Rev. **5**(4), 373–401, ISSN 1364-0321

59. H.R. Pombeiro, C.A. Silva, Linear, fuzzy and neural networks models for definition of baseline consumption: early findings from two test beds in a University Campus in Portugal, in *Science and Information Conference (SAI)* (IEEE, 2014), pp. 481–487

60. N. Srivastava, G. Hinton, A. Krizhevsky, I. Sutskever, R. Salakhutdinov, Dropout: a simple way to prevent neural networks from overfitting. J. Mach. Learn. Res. **15**(1), 1929–1958 (2014)

61. S. Paudel, M. Elmitri, S. Couturier, P.H. Nguyen, R. Kamphuis, B. Lacarrière, O. Le Corre, A relevant data selection method for energy consumption prediction of low energy building based on support vector machine. Energy Build. **138**, 240–256 (2017)

62. T. Chai, R.R. Draxler, Root mean square error (RMSE) or mean absolute error (MAE)?–Arguments against avoiding RMSE in the literature. Geosci. Model Dev. **7**(3), 1247–1250 (2014)

63. The Theano Development Team, Theano: a Python framework for fast computation of mathematical expressions (2016), arXiv:1605.02688v1 [cs.SC]

64. C. Tian, R.N. Horne, Recurrent neural networks for permanent downhole gauge data analysis. Soc. Petrol. Eng. (2017). https://doi.org/10.2118/187181-MS

65. United Nations, Department of Economic and Social Affairs, Population division. World Urbanization Prospects: The 2014 Revision, CD-ROM Edition

66. C. Voyant, G. Notton, S. Kalogirou, M.L. Nivet, C. Paoli, F. Motte, A. Fouilloy, Machine learning methods for solar radiation forecasting: a review. Renew. Energy **105**, 569–582 (2017)

67. H.Z. Wang, G.B. Wang, G.Q. Li, J.C. Peng, Y.T. Liu, Deep belief network based deterministic and probabilistic wind speed forecasting approach. Appl. Energy **182**, 80–93 (2016)

68. H.Z. Wang, G.Q. Li, G.B. Wang, J.C. Peng, H. Jiang, Y.T. Liu, Deep learning based ensemble approach for probabilistic wind power forecasting. Appl. Energy **188**, 56–70 (2017)

69. R. Weron, Electricity price forecasting: a review of the state-of-the-art with a look into the future. Int. J. Forecast. **30**(4), 1030–1081 (2014)

70. X. Yuan, Q. Tan, X. Lei, Y. Yuan, X. Wu, Wind power prediction using hybrid autoregressive fractionally integrated moving average and least square support vector machine. Energy **129**, 122–137 (2017)

71. X.W. Chen, X. Lin, Big data deep learning: challenges and perspectives. IEEE Access **2**, 514–525 (2014)

72. B. Yildiz, J.I. Bilbao, A.B. Sproul, A review and analysis of regression and machine learning models on commercial building electricity load forecasting. Renew. Sustain. Energy Rev. **73**, 1104–1122 (2017)
73. Y. Yong, W. Sun'an, S. Wanxing, Short-term load forecasting using artificial immune network, in *Proceedings of the International Conference on Power System Technology, PowerCon 2002*, Kunming, China (2002)
74. Y.T. Chae, R. Horesh, Y. Hwang, Y.M. Lee, Artificial neural network model for forecasting sub-hourly electricity usage in commercial buildings. Energy Build. **111**, 184–194 (2016)

Part IV
Energy Management Systems for Energy Internet

Chapter 12
Multiple Source-Load-Storage Cooperative Optimization of Energy Management for Energy Local Area Network Systems

Tao Zhang, Fuxing Zhang, Hongtao Lei, Rui Wang, Kaiwen Li, Yang Chen, and Yonghua Gui

Abstract This chapter proposes an effective model and algorithm for energy management within an energy local area network (ELAN), considering the cooperative optimization of resources, loads and battery energy storage systems (BESSs). First, the operation conditions of various resources and their interactions for the management of the ELAN are introduced. Second, an energy management model is proposed that aims to maximize the utilization of renewable energies and minimize the overall system cost. Additionally, the optimal scheduling strategies are provided for cases of productive capacity and shortage. Finally, three different optimization schemes considering the demand response effect and direct interaction effect are proposed in which the Lagrangian multiplier method and an evolutionary large-scale global optimization algorithm (self-adaptive neighbourhood search differential evolution) are used to find the optimal solution. Experimental results indicate that the developed model and algorithm are efficient and also present the realization of the goal of "intranet autonomy, network synergy and global optimization."

12.1 Introduction

With the rapid development of renewable energy, energy utilization and consumption have changed significantly [1–3], and related research is introduced as follows.

The research in [4] reviewed regional renewable energy planning; introduced the present situation, problems and future development trends of domestic and foreign classic energy models (such as the national energy model); and pointed out that regional renewable energy planning and effective utilization can contribute to

T. Zhang (✉) · H. Lei · R. Wang · K. Li · Y. Chen
College of Systems Engineering, National University of Defense Technology, Changsha, People's Republic of China
e-mail: zhangtao@nudt.edu.cn

F. Zhang · Y. Gui
G. Hardware R&D Department, HNAC Technology Co. Ltd., Changsha 410073, People's Republic of China

© Springer Nature Switzerland AG 2020
A. F. Zobaa and J. Cao (eds.), *Energy Internet*,
https://doi.org/10.1007/978-3-030-45453-1_12

energy structure adjustment, economic and social development, environmental protection and mining the potential value of renewable energy. Considering the negative influence on the grid structure and the real-time supply and demand balance caused by access of large-scale renewable energy [5] proposed a bundled scheduling and optimization strategy that combined traditional power generators and distributed renewable energy. This method can help large-scale renewable energy to connect to the utility grid and realize the optimization goal of environmental protection as well as peak load shifting. Three kinds of cooperative control abilities were proposed and discussed in [6], in which an economic dispatch model was constructed and a frequency regulation mechanism was introduced during the optimization process. Experimental results showed that the proposed method could eliminate the adverse effects of non-controllable factors, and the reliability and validity of the proposed method were examined; [7] pointed out that traditional adjustment and control methods have difficulty meeting the requirements of peak load shifting with the access of large-scale renewable energy. Therefore, a scheduling and optimization model considering source-load interactions and predicting the output of the source was built, and an interior-point method was used to solve and analyze four different scenarios. A demand response mechanism was regarded as a feasible and cost-effective way to access the utility grid for renewable energy in [8], and a combinatorial optimization model considering the demand response mechanism was proposed, which can contribute to saving investment in renewable energy [9] proposed the architecture, optimization model and scheduling strategy of an active power distribution network system that included renewable energy generators, battery energy storage systems and flexible loads. The collaborative optimization of scheduling resources within the active distribution network had important effects on realizing the renewable energy consumed locally and improving the satisfaction of terminal users. The architecture of an active distribution network was proposed in [10], which considered the collaborative scheduling and optimization of multiple resources and regarded renewable energy generators, battery energy storage systems and flexible loads as scheduling resources. The related key technologies, optimization scheduling strategy and multiple application scenarios were also discussed. In addition [11] discussed the combination optimization of large-scale electric vehicles and wind power generators and generated typical application scenarios by Monte Carlo simulation. To solve renewable energy location and sizing issues, [12] modelled the goal of minimizing network loss and solved it by evolutionary programming.

However, deficiencies in renewable energies, such as geographical dispersion, low capacity density and stochastic nature, create great challenges for the use of renewable energy [12]. The energy Internet (EI), which integrates advanced concepts and technologies of both energy and Internet networks, appeared in [13–15]. Internal and external collaborative optimization and joint scheduling of different levels and regions based on energy local area networks (ELANs) is a new and hot research topic in the field of energy management for EI systems. By adjusting the interactive relationship of schedulable units, such as internal generation-load-storage units and external adjacent energy local area networks, as well as the utility grid, the task is to maximize the use of renewable energies, minimize the total operating cost and maintain the balance between supply and demand.

To achieve this aim, this study considers the energy management of ELANs as large-scale optimization problems and proposes a modelling and collaborative optimization method suitable for solving such problems. This method can effectively meet point-to-point energy production, consumption and interaction requirements. With the advantages of existing information and communication networks, an ELAN can integrate its internal and external units located in different regions by a distributed energy network. Additionally, with the aim of maximizing the utilization of renewable energy and minimizing the difference between operating cost and revenue, a demand response mechanism is introduced to build an energy management and optimization model of time of use (TOU) for comparative analysis of local energy networks under different application scenarios.

The remainder of this chapter is organized as follows: The system optimization model and optimization strategies are introduced and discussed in Sect. 12.2. Section 12.3 presents the model solution process and analyzes the optimization results. Section 12.4 concludes the chapter.

12.2   System Optimization Model

12.2.1   Structure Model

As the basic optimization unit of the energy Internet, an ELAN consists of many interconnected schedulable/non-schedulable energy production, energy consumption and energy storage units, which can independently accomplish the process of energy exchange and maintain the balance among various schedulable resources in the network. Within each energy local area network, all schedulable resources have the ability to interact and share information with their local energy controller (such as a local scheduling and optimization centre) and can decide whether to participate in a certain period of energy trading activity. During this period, the local energy controller is in charge of overall management. By predicting the input/output parameters and/or relations of these schedulable resources, it is possible to coordinate the user's local interests with the overall benefits of the system within a certain range, as shown in Fig. 12.1.

An ELAN can run reliably both in and out of the network [16]. In the grid-connected mode [17], each ELAN participates in the energy scheduling and optimization process of the transmission and distribution system by which it can realize the bottom-up requirement transfer, top-down instruction execution, supply–demand response and energy exchange among its neighbouring and/or non-neighbouring ELANs at the same level. In the structural model of energy local area networks, the independent system operator (ISO) and distribution system operator (DSO) constitute two main managers of the transmission and distribution network system. The former

Fig. 12.1 Structural model
of energy local area networks

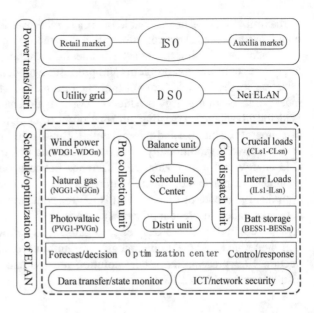

introduces more incentives to facilitate wider, more flexible and more diversified transactions within the jurisdiction of the retail and auxiliary market. The latter, on the one hand, is responsible for collecting the demand information from all ELANs in its jurisdiction and uploading the aggregated demand information to the ISO and, on the other hand, receives ISO scheduling instructions and accordingly coordinates the energy interactions between different ELANs to achieve a real-time supply and demand balance.

The scheduling and optimization management of ELANs (dotted line in Fig. 12.1) include two main centres: a scheduling centre and optimization centre, whose main work processes are shown as follows:

First, the predicted and response information of terminal equipment (such as sources, networks, loads and BESSs) are monitored and transmitted by the communication network. After analysis and decision-making, all the information is changed into executable dispatching instructions. Second, the practical requirement information of production units (including wind power, solar energy and natural gas), consumption units (including crucial loads and interruptible loads) and adjustable units (such as BESSs) is collected separately. Finally, all the predicted and practical information is compared through the distribution unit and the balance unit, which can help to match different requirements and achieve the real-time balance between the production and consumption demand. Note that each ELAN uses the centralized control and scheduling strategy to achieve internal management, and different ELANs can achieve the state of regional autonomy and mutual synergy.

12.2.2   Mathematical Model

To highlight the focus of this chapter, a simplified ELAN system structure is shown in Fig. 12.2, which is divided into two main parts: an internal part and an external part. Among them, the internal part of the network includes various energy production units that are composed of various types of natural gas generation, wind generation and photovoltaic generation; energy storage units; and different consumption units, including crucial loads and interruptible loads. The external part of the network refers to the natural gas grid used for natural gas power generation, the neighbouring/non-neighbouring ELANs and its directly connected high-voltage utility grid. Additionally, the energy management system (EMS) is responsible for the collection, analysis, optimization and decision-making of various types of information inside and outside of the ELAN. Then, the determined information is formed into executable dispatching instructions, which can guide the energy router to coordinate and optimize the dispatchable resources inside/outside of the ELAN accordingly. The main optimization goal of the ELAN is to meet the real-time supply and demand balance, utilize the distributed renewable energy and realize the economic dispatch of internal and external schedulable units.

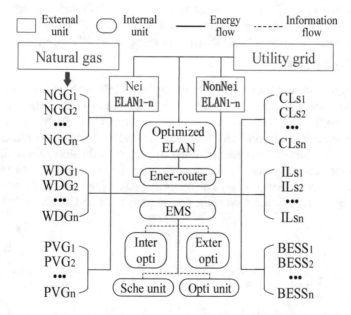

Fig. 12.2 Mathematical model of energy local area networks

12.2.2.1 Objective Function

In the energy management process of an ELAN system considering multiple source-load-storage units, the optimization goal during period T is to minimize the difference between the system operation cost and revenue, shown as follows:

$$
\begin{aligned}
C_{\text{Total}}(T) &= \sum_{t=1}^{T} C_{\text{Exp}}(t) - E_{\text{Inc}}(t) \\
&= \sum_{t=1}^{T} \left(C_{\text{PRO}}(t) + C_{\text{ILs}}^{\text{Offset}}(t) + C_{\text{BESS}}(t) + C_{\text{ExGrid}}^{\text{Input}}(t) - E_{\text{CON}}(t) - E_{\text{ExGrid}}^{\text{Output}}(t) \right)
\end{aligned}
$$

$$(12.1)$$

where $C_{\text{Exp}}(t)$ and $E_{\text{Inc}}(t)$ represent the total cost and revenue generated in period t, respectively, and $C_{\text{Total}}(T)$ means the difference between the total revenue and all costs during the entire period T. Among them, the total cost $C_{\text{Exp}}(t)$ includes the operating cost of production units $C_{\text{PRO}}(t)$ (such as natural gas generators), the compensation cost for the interrupted part of the interruptible load $C_{\text{ILs}}^{\text{Offset}}(t)$, the operating cost of BESS units $C_{\text{BESS}}(t)$, the cost used for purchasing electricity from external networks $C_{\text{ExGrid}}^{\text{Input}}(t)$, etc. The total revenue $\text{Inc}(T)$ includes the revenue from selling electricity power to the crucial and interruptible loads $E_{\text{CON}}(t)$, the revenue from selling external power to the external power grid $E_{\text{ExGrid}}^{\text{Output}}(t)$, etc.

(1) Production units

The production units considered in this chapter mainly include controllable production units and uncontrollable production units. Among them, the controllable production units are all kinds of natural gas generators (NGGs); the uncontrollable production units include wind-driven generators (WDGs) and photovoltaic generators (PVGs). The operating costs generated by all the production units mainly refer to the natural gas consumed by the natural gas generators and the losses caused by the energy conversion process. The operating cost of the gas production units is presented by a quadratic function of the output power [18, 19], as shown in Eq. 12.2:

$$
\begin{aligned}
C_{\text{PRO}}(t) &= C_{\text{NGG}}(t) \\
&= \sum_{\text{NGG}i=1}^{\text{NGG}n} \left(\alpha_{\text{NGG}i} \times (P_{\text{NGG}i}(t))^2 + \beta_{\text{NGG}i} \times P_{\text{NGG}i}(t) + \gamma_{\text{NGG}i} \right) \times \Delta t \times \text{Pr}_{i\text{Gas}}
\end{aligned}
$$

$$(12.2)$$

where $C_{\text{NGG}}(t)$ represents the operating cost of the gas production units during the t-th period; NGGn is the number of natural gas production units; $\alpha_{\text{NGG}i}$, $\beta_{\text{NGG}i}$ and $\gamma_{\text{NGG}i}$ are the quadratic term, primary term and constant term of the relationship equation between the operating costs of the NGGi natural gas production unit and its output power, respectively; $P_{\text{NGG}i}(t)$ represents the output power of the NGGi natural gas production unit during period t; Pri_{Gas} represents the price per unit of natural gas; and Δt means the optimization time step.

(2)   Consumption units

The consumption units of the ELAN are divided into crucial loads (CLs) and interruptible loads (ILs), which depend on their schedulable availability. The operating profit of a consumption unit depends on its energy usage during each period t, the market clearing price and the corresponding tariff. Specifically, the revenue (negative value) of the crucial load in a certain period is equal to the product of the energy used, the market clearing price and the corresponding tariff rate. The interruptible load consists of the uninterrupted portion and the interrupted portion, in which the revenue of the uninterrupted portion (negative value) during period t is equal to the product of the actual energy consumption, the market clearing price and the corresponding tariff, and the compensation cost of the interrupted portion is equal to the product of the revenue of the actual interrupted power, the market clearing price and the corresponding compensation rate. The specific information is shown as follows:

$$E_{\text{CON}} = E_{\text{CLs}}(t) + E_{\text{ILs}}(t) \tag{12.3}$$

$$E_{\text{CLs}}(t) = \sum_{\text{CL}si=1}^{\text{CL}sn} (P_{\text{CL}si}(t) \times \Delta t \times pri_{\text{CL}si}(t)) \tag{12.4}$$

$$E_{\text{ILs}}(t) = \sum_{\text{IL}si=1}^{\text{IL}sn} (P_{\text{IL}si}(t) \times (1 - R_{\text{IL}si}(t)) \times pri(t)) \times \Delta t \tag{12.5}$$

$$C_{\text{ILs}}^{\text{Offset}}(t) = \sum_{\text{IL}si=1}^{\text{IL}sn} \left(P_{\text{IL}si}(t) \times R_{\text{IL}si}(t) \times pri_{\text{IL}si}^{\text{offset}}(t) \right) \times \Delta t \tag{12.6}$$

where $E_{\text{CLs}}(t)$ and $E_{\text{ILs}}(t)$ stand for the income from the crucial loads and uninterrupted part of the interruptible loads during period t, respectively; $C_{\text{ILs}}^{\text{Offset}}(t)$ represents the compensation cost of the interrupted part of the interruptible loads during period t; CLsn and ILsn represent the amounts of the critical loads and interruptible loads, respectively; $P_{\text{BESS}i}^{\text{Char max}}(t)$ represents the market clearing price, such as the time-of-use price; $pri_{\text{CL}si}(t)$ and $pri_{\text{IL}si}^{\text{offset}}(t)$ represent the payable rates for the critical loads and interrupted part of the interruptible loads during the t-th period, respectively, which are expressed by the following formulas:

$$pri_{\text{CL}si}(t) = \eta_{\text{CL}si} \times pri(t) \tag{12.7}$$

$$pri_{\text{IL}si}^{\text{offset}}(t) = \begin{cases} \eta_{\text{IL}si}^{H} \times pri(t) \ if \ R_{\text{IL}si}(t) > \theta_{\text{IL}si}^{H} \times R_{\text{IL}si}^{\max}(t) \\ \eta_{\text{IL}si}^{M} \times pri(t) \ if \ \begin{pmatrix} \theta_{\text{IL}si}^{L} \times R_{\text{IL}si}^{\max}(t) < R_{\text{IL}si}(t) \\ R_{\text{IL}si}(t) \leq \theta_{\text{IL}si}^{H} \times R_{\text{IL}si}^{\max}(t) \end{pmatrix} \\ \eta_{\text{IL}si}^{L} \times pri(t) \ if \ R_{\text{IL}si}(t) \leq \theta_{\text{IL}si}^{L} \times R_{\text{IL}si}^{\max}(t) \end{cases} \tag{12.8}$$

where $\eta_{\text{CL}si}$ represents the payable rate of the CLsith crucial load; $\eta_{\text{IL}si}^{H}$, $\eta_{\text{IL}si}^{M}$ and $\eta_{\text{IL}si}^{L}$ represent different corresponding compensation rates when the interrupted amount of the ILsi-th interruptible load remains high, medium and low grade, respectively; and $\theta_{\text{IL}si}^{H}$ and $\theta_{\text{IL}si}^{L}$ represent the classification coefficients of the ILsi-th interruptible load, which divides the interrupt ratio of each interruptible load $R_{\text{IL}si}(t)$ into three types together with the maximum interruptible proportion $R_{\text{IL}si}^{\max}(t)$. Generally, the larger the interrupt proportion is, the higher the corresponding compensation rate will be.

(3) Battery energy storage system

The operating cost of the battery energy storage system includes three parts: the loss cost for the charging/discharging process, the state transition cost and the loss cost for self-discharging, shown as follows:

$$
C_{\text{BESS}i}(T) = \sum_{t=1}^{T}(C_{\text{BESS}i}(t)) + C_{\text{BESS}i}^{\text{State Change}}(T) \tag{12.9}
$$

$$
C_{\text{BESS}i}^{\text{StateChange}}(T) = \left(\text{Num}_{\text{BESS}i}^{\text{Char 2 DisChar}}(T) + \text{Num}_{\text{BESS}i}^{\text{DisChar 2 Char}}(T)\right) \times C_{\text{BESS}i}^{\text{Change Cost}} \tag{12.10}
$$

$$
C_{\text{BESS}i}(t) =
\begin{cases}
\left(P_{\text{BESS}i}^{\text{Char}}(t) \times \left(1/\eta_{\text{BESS}i}^{\text{Char}} - 1\right) + P_{\text{BESS}i}^{\text{Self DisChar}}(t)\right) \times pri(t), & P_{BESSi}^{Char}(t) > 0 \\
\left(P_{\text{BESS}i}^{\text{DisChar}}(t) \times \left(1 - \eta_{\text{BESS}i}^{\text{DisChar}}\right) + P_{\text{BESS}i}^{\text{Self DisChar}}(t)\right) \times pri(t), & P_{BESSi}^{DisChar}(t) > 0 \\
P_{\text{BESS}i}^{\text{Self DisChar}}(t) \times pri(t), & \text{else}
\end{cases}
\tag{12.11}
$$

In the formula, $C_{BESSi}(T)$ and $C_{BESSi}(t)$ represent the total cost of the $BESSi$-th battery energy storage system in the whole period T and the sum of the charging/discharging loss and the self-discharge loss in the single period t, respectively. $C_{\text{BESS}i}^{\text{StateChange}}(T)$ represents the state transform cost of the $BESSi$-th battery energy storage system. $\text{Num}_{\text{BESS}i}^{\text{Char2DisChar}}(T)$ and $\text{Num}_{\text{BESS}i}^{\text{DisChar2Char}}(T)$, respectively, represent the numbers of state transitions during the entire period T, in which the former represents the transitions from charging to discharging and the latter stands for the transitions from discharging to charging. $P_{\text{BESS}i}^{\text{Char}}(t)$ and $P_{\text{BESS}i}^{\text{DisChar}}(t)$, respectively, indicate the charging and discharging energy of the $BESSi$-th battery energy storage system during period t. $\eta_{\text{BESS}i}^{\text{Char}}$ and $\eta_{\text{BESS}i}^{\text{DisChar}}$, respectively, represent the corresponding charging and discharging efficiencies. $P_{\text{BESS}i}^{\text{SelfDisChar}}(t)$ represents the self-discharging power of the $BESSi$-th battery energy storage system during period t. $pri(t)$ presents the market clearing price during period t, such as time-of-use. Δt is the time step of the optimization process.

(4) Interaction with the external grid

The cost of energy exchange between the ELAN and its external power grid includes two parts: one is the cost for the ELAN to purchase electricity from the external grid, and the other is the revenue of selling the surplus electricity power to the external power grid, which is presented as follows:

$$C_{\text{ExGrid}}^{\text{Input}}(t) = C_{\text{Neighbour}}^{\text{Input}}(t) + C_{\text{NonNeighbour}}^{\text{Input}}(t) + C_{\text{ExGrid}}^{\text{Input}}(t)$$

$$= \sum_{\text{Neighbour}i}^{\text{Neighbour}n} \left(P_{\text{Neighbour}i}^{\text{Input}}(t) \times R_{\text{Neighbour}i}^{\text{Input}}(t) \right) \times \Delta t \times pri(t)$$

$$+ \sum_{\text{NonNeighbour}i}^{\text{NonNeighbour}n} \left(P_{\text{NonNeighbour}i}^{\text{Input}}(t) \times R_{\text{NonNeighbour}i}^{\text{Input}}(t) \right) \times \Delta t \times pri(t)$$

$$+ P_{\text{UGrid}}^{\text{Input}}(t) \times R_{\text{UGrid}}^{\text{Input}}(t) \times \Delta t \times pri(t) \tag{12.12}$$

$$dE_{\text{ExGrid}}^{\text{Output}}(t) = E_{\text{Neighbour}}^{\text{Output}}(t) + E_{\text{NonNeighbour}}^{\text{Output}}(t) + E_{\text{ExGrid}}^{\text{Output}}(t)$$

$$= \sum_{\text{Neighbour}i}^{\text{Neighbour}n} \left(P_{\text{Neighbour}i}^{\text{Output}}(t) \times R_{\text{Neighbour}i}^{\text{Output}}(t) \right) \times \Delta t \times pri(t)$$

$$+ \sum_{\text{NonNeighbour}i}^{\text{NonNeighbour}n} \left(P_{\text{NonNeighbour}i}^{\text{Output}}(t) \times R_{\text{NonNeighbour}i}^{\text{Output}}(t) \right) \times \Delta t \times pri(t)$$

$$+ P_{\text{UGrid}}^{\text{Output}}(t) \times R_{\text{UGrid}}^{\text{Output}}(t) \times \Delta t \times pri(t) \tag{12.13}$$

In the formula, $C_{\text{ExGrid}}^{\text{Input}}(t)$ and $E_{\text{ExGrid}}^{\text{Output}}(t)$ represent the total cost and the total benefit generated by the energy exchange between the ELAN and its external power grid during the t-th period, respectively. $C_{\text{Neighbour}}^{\text{Input}}(t)$, $C_{\text{NonNeighbour}}^{\text{Input}}(t)$ and $C_{\text{ExGrid}}^{\text{Input}}(t)$, respectively, represent the cost generated by the ELAN purchasing power from its neighbouring energy local area network, non-neighbouring energy local energy network and the utility grid. $E_{\text{Neighbour}}^{\text{Output}}(t)$, $E_{\text{NonNeighbour}}^{\text{Output}}(t)$ and $E_{\text{ExGrid}}^{\text{Output}}(t)$, respectively, represent the benefits generated by the ELAN selling surplus power to its neighbouring energy local area network, non-neighbouring energy local energy network and the utility grid. Neighbour$n$ and NonNeighbour$n$, respectively, represent the number of neighbour and non-neighbour energy local area networks. $R_{\text{Neighbour}i}^{\text{Input}}(t)$, $R_{\text{Neighbour}i}^{\text{Output}}(t)$, $R_{\text{NonNeighbour}i}^{\text{Input}}(t)$, $R_{\text{UGrid}}^{\text{Input}}(t)$ and $R_{\text{UGrid}}^{\text{Output}}(t)$, respectively, represent the purchase/sell rate for the ELAN interacting with its Neighbour$i$-th neighbour ELAN, NonNeighbour$i$-th non-neighbour ELAN and the utility grid during the t-th period, respectively.

12.2.2.2 Constraints

(1) Energy production unit constraints

According to the principle of "maximizing the utilization of renewable energy," the power generated by the wind power and photovoltaic generators in the production units is preferentially given to internal energy consumption and energy storage units. During the period of operation, the natural gas production units need to adjust their output according to external factors (such as demand and supply changes) and satisfy the constraints (such as the output power range of the generation units), as shown in Eq. 12.14:

$$P_{\text{NGG}i}^{\min}(t) \leq P_{\text{NGG}i}(t) \leq P_{\text{NGG}i}^{\max}(t) \tag{12.14}$$

In the formula, $P_{\text{NGG}i}(t)$ represents the output power of the NGGi-th natural gas production unit during the t-th period. $P_{\text{NGG}i}^{\min}(t)$ and $P_{\text{NGG}i}^{\max}(t)$ represent the lower and upper bounds of the natural gas production unit's output power during the t-th period, respectively.

Considering the equipment conversion efficiency of each production unit and its own electricity demand, the actual output power sent to the energy consumption unit, energy storage unit and external power grid for each production unit is related to its maximum output power (the upper limit of output power), shown as the following formula:

$$P_{\text{NGG}i}^{\text{Actual}}(t) = \begin{cases} \eta_{\text{NGG}i}^H \times P_{\text{NGG}i}(t) & \text{if} \quad P_{\text{NGG}i}(t) > \theta_{\text{NGG}i}^H \times P_{\text{NGG}i}^{\max}(t) \\ \eta_{\text{NGG}i}^M \times P_{\text{NGG}i}(t) & \text{if} \quad \begin{pmatrix} \theta_{\text{NGG}i}^L \times P_{\text{NGG}i}^{\max}(t) < P_{\text{NGG}i}(t) \\ P_{\text{NGG}i}(t) \leq \theta_{\text{NGG}i}^H \times P_{\text{NGG}i}^{\max}(t) \end{pmatrix} \\ \eta_{\text{NGG}i}^L \times P_{\text{NGG}i}(t) & \text{if} \quad P_{\text{NGG}i}(t) \leq \theta_{\text{NGG}i}^L \times P_{\text{NGG}i}^{\max}(t) \end{cases} \tag{12.15}$$

$$P_{\text{WDG}i}^{\text{Actual}}(t) = \begin{cases} \eta_{\text{WDG}i}^H \times P_{\text{WDG}i}(t) & \text{if} \quad P_{\text{WDG}i}(t) > \theta_{\text{WDG}i}^H \times P_{\text{WDG}i}^{\max}(t) \\ \eta_{\text{WDG}i}^M \times P_{\text{WDG}i}(t) & \text{if} \quad \begin{pmatrix} \theta_{\text{WDG}i}^L \times P_{\text{WDG}i}^{\max}(t) < P_{\text{WDG}i}(t) \\ P_{\text{WDG}i}(t) \leq \theta_{\text{WDG}i}^H \times P_{\text{WDG}i}^{\max}(t) \end{pmatrix} \\ \eta_{\text{WDG}i}^L \times P_{\text{WDG}i}(t) & \text{if} \quad P_{\text{WDG}i}(t) \leq \theta_{\text{WDG}i}^L \times P_{\text{WDG}i}^{\max}(t) \end{cases} \tag{12.16}$$

$$P_{\text{PVG}i}^{\text{Actual}}(t) = \begin{cases} \eta_{\text{PVG}i}^H \times P_{\text{PVG}i}(t) & \text{if} \quad P_{\text{PVG}i}(t) > \theta_{\text{PVG}i}^H \times P_{\text{PVG}i}^{\max}(t) \\ \eta_{\text{PVG}i}^M \times P_{\text{PVG}i}(t) & \text{if} \quad \begin{pmatrix} \theta_{\text{PVG}i}^L \times P_{\text{PVG}i}^{\max}(t) < P_{\text{PVG}i}(t) \\ P_{\text{PVG}i}(t) \leq \theta_{\text{PVG}i}^H \times P_{\text{PVG}i}^{\max}(t) \end{pmatrix} \\ \eta_{\text{PVG}i}^L \times P_{\text{PVG}i}(t) & \text{if} \quad P_{\text{PVG}i}(t) \leq \theta_{\text{PVG}i}^L \times P_{\text{PVG}i}^{\max}(t) \end{cases} \tag{12.17}$$

where $P_{\text{NGG}i}(t)$, $P_{\text{WDG}i}(t)$ and $P_{\text{PVG}i}(t)$ represent the output power of the NGGi-th natural gas production unit, WDGi-th wind power generation unit and PVGi-th PV production unit during the t-th period, respectively; $P_{\text{NGG}i}^{\text{Actual}}(t)$, $P_{\text{WDG}i}^{\text{Actual}}(t)$, and $P_{\text{PVG}i}^{\text{Actual}}(t)$ represent the actual available output power for transforming each production unit, respectively; and the corresponding output power conversion coefficients between different grades are η_*^H, η_*^M, and η_*^L. θ_*^H and θ_*^L stand for the classification coefficients, which together with the maximum output power of the device $P_*^{\max}(t)$, divide the output power of each production unit $P_*^{\max}(t)$ into high, medium and low levels. The subscript "*" corresponds to the $NGGi$-th, $WDGi$-th and $PVGi$-th production units.

(2) Energy consumption unit constraints

During the whole period T, the requirement of energy consumption for the crucial loads can be fully satisfied by a relatively high electricity price, while part of the interruptible load can be interrupted during the period of demand exceeding supply; the interrupted part can not only have no effect on the normal operation but also obtains a certain percentage of economic compensation or price concessions, which depends on the amount of interrupted power and the market clearing price. In summary, the critical loads and the uninterrupted part of the interruptible loads constitute the base load of the consumption unit. The interrupted part of the interruptible loads can be adjusted within a certain range and constitutes part of the schedulable resources of the ELAN system. The interrupted rate of the interruptible loads needs to meet the requirement of Eq. 12.18:

$$R_{\text{IL}si}^{\min}(t) \leq R_{\text{IL}si}(t) \leq R_{\text{IL}si}^{\max}(t) \tag{12.18}$$

where $R_{\text{IL}si}(t)$ indicates the interruptible proportion of the ILsi-th interruptible load during the t-th period. $R_{\text{IL}si}^{\min}(t)$ and $R_{\text{IL}si}^{\max}(t)$ indicate the lower and upper bounds of the interruptible proportion, respectively.

(3) Energy storage unit constraints

The main constraint of a BESS is to suppress the frequent fluctuations from distributed renewable energy and various types of loads [20] to a certain extent, which contributes to realizing the supply and demand balance as well as full-time economic energy dispatch. Factors affecting and determining energy storage unit performance include the system supply and demand requirements, electricity power price, capacity of the energy storage unit, self-discharge rate, charging/discharging status, power, efficiency and so on. The charging or discharging power of the energy storage unit should satisfy the following requirements:

$$P_{\text{BESS}i}^{\text{Char}\,\min}(t) \leq P_{\text{BESS}i}^{\text{Char}}(t) \leq P_{\text{BESS}i}^{\text{Char}\,\max}(t) \tag{12.19}$$

$$P_{\text{BESS}i}^{\text{Dis Char}\,\min}(t) \leq P_{\text{BESS}i}^{\text{Dis Char}}(t) \leq P_{\text{BESS}i}^{\text{Dis Char}\,\max}(t) \tag{12.20}$$

The capacity of an energy storage unit is affected by its initial capacity, self-discharge rate, charging/discharging power and efficiency, which also need to meet the requirements of capacity constraints, as shown in Eqs. 12.21, 12.22. The charging/discharging times of the energy storage unit during period T satisfy the requirement of Eq. 12.23.

$$\text{Cap}_{\text{BESS}i}^{\min}(t) \leq \text{Cap}_{\text{BESS}i}(t) \leq \text{Cap}_{\text{BESS}i}^{\max}(t) \tag{12.21}$$

$$\mathrm{Cap}_{\mathrm{BESS}i}(t+1) = \left(\begin{array}{c} \mathrm{Cap}_{\mathrm{BESS}i}(t) + P_{\mathrm{BESS}i}^{\mathrm{Char}}(t) \times \Delta t \times \left(1 - LR_{\mathrm{BESS}i}^{\mathrm{Char}}(t)\right) - \\ P_{\mathrm{BESS}i}^{\mathrm{Dis\,Char}}(t) \times \Delta t / \left(1 - LR_{\mathrm{BESS}i}^{\mathrm{Dis\,Char}}(t)\right) - P_{\mathrm{BESS}i}^{\mathrm{Self\,DisChar}}(t) \times \Delta t \end{array} \right)$$

$$(12.22)$$

$$\mathrm{Num}_{\mathrm{BESS}i}^{\mathrm{Char\,2\,DisChar}}(T) + \mathrm{Num}_{\mathrm{BESS}i}^{\mathrm{DisChar\,2\,Char}}(T) \le \mathrm{Num}_{\mathrm{BESS}i}^{\mathrm{State\,Change\,max}}(T) \qquad (12.23)$$

where $P_{\mathrm{BESS}i}^{\mathrm{Char}}(t)$, $P_{\mathrm{BESS}i}^{\mathrm{DisChar}}(t)$ and $P_{\mathrm{BESS}i}^{\mathrm{Self\,DisChar}}(t)$ represent the charging power, discharging power and self-discharging power of the BESSi-th energy storage unit during the t-th period, respectively; $P_{\mathrm{BESS}i}^{\mathrm{Char\,min}}(t)$, $P_{\mathrm{BESS}i}^{\mathrm{Char\,max}}(t)$, $P_{\mathrm{BESS}i}^{\mathrm{DisChar\,min}}(t)$ and $P_{\mathrm{BESS}i}^{\mathrm{DisChar\,max}}(t)$ represent the lower and upper bounds of the corresponding charging/discharging power, respectively; $\mathrm{Cap}_{\mathrm{BESS}i}(t)$, $\mathrm{Cap}_{\mathrm{BESS}i}^{\mathrm{min}}(t)$ and $\mathrm{Cap}_{\mathrm{BESS}i}^{\mathrm{max}}(t)$ stand for the capacity and corresponding boundary values of the BESSi-th energy storage unit during the t-th period, respectively; $LR_{\mathrm{BESS}i}^{\mathrm{Char}}(t)$ and $LR_{\mathrm{BESS}i}^{\mathrm{DisChar}}(t)$ represent the charging/discharging loss rate during period t, respectively; $\mathrm{Num}_{\mathrm{BESS}i}^{\mathrm{Char\,2\,DisChar}}(T)$ and $\mathrm{Num}_{\mathrm{BESS}i}^{\mathrm{DisChar\,2\,Char}}(T)$ represent the numbers of state transitions of the BESSi-th energy storage unit, where the former means from charging to discharging and the latter means from discharging to charging during the entire period T, respectively; and $\mathrm{Num}_{\mathrm{BESS}i}^{\mathrm{State\,Change\,max}}(T)$ represents the total number of state transitions allowed during the entire period T.

(4) Constraints for interaction with the external power grid

The ELAN system can interact with multiple neighbour/non-neighbour ELANs and the utility power grid when the transmission line capacity constraints can be satisfied, as shown below:

$$P_{\mathrm{Neighbour}i}^{\mathrm{Input\,min}}(t) \le P_{\mathrm{Neighbour}i}^{\mathrm{Input}}(t) \le P_{\mathrm{Neighbour}i}^{\mathrm{Input\,max}}(t) \qquad (12.24)$$

$$P_{\mathrm{NonNeighbour}i}^{\mathrm{Input\,min}}(t) \le P_{\mathrm{NonNeighbour}i}^{\mathrm{Input}}(t) \le P_{\mathrm{NonNeighbour}i}^{\mathrm{Input\,max}}(t) \qquad (12.25)$$

$$P_{\mathrm{Neighbour}i}^{\mathrm{Output\,min}}(t) \le P_{\mathrm{Neighbour}i}^{\mathrm{Output}}(t) \le P_{\mathrm{Neighbour}i}^{\mathrm{Output\,max}}(t) \qquad (12.26)$$

$$P_{\mathrm{NonNeighbour}i}^{\mathrm{Output\,min}}(t) \le P_{\mathrm{NonNeighbour}i}^{\mathrm{Output}}(t) \le P_{\mathrm{NonNeighbour}i}^{\mathrm{Output\,max}}(t) \qquad (12.27)$$

In the formula, $P_{\mathrm{Neighbour}i}^{\mathrm{Input}}(t)$, $P_{\mathrm{Neighbour}i}^{\mathrm{Output}}(t)$, $P_{\mathrm{Neighbour}i}^{\mathrm{Output}}(t)$ and $P_{\mathrm{NonNeighbour}i}^{\mathrm{Output}}(t)$ represent the input power and the output power between the ELAN and its Neighbour$i$-th neighbour ELAN and NonNeighbour$i$-th non-neighbour ELAN during period t, respectively. $P_{\mathrm{Neighbour}i}^{\mathrm{Input\,min}}(t)$, $P_{\mathrm{Neighbour}i}^{\mathrm{Input\,max}}(t)$, $P_{\mathrm{NonNeighbour}i}^{\mathrm{Input\,min}}(t)$, $P_{\mathrm{NonNeighbour}i}^{\mathrm{Input\,max}}(t)$, $P_{\mathrm{Neighbour}i}^{\mathrm{Output\,min}}(t)$, $P_{\mathrm{Neighbour}i}^{\mathrm{Output\,max}}(t)$, $P_{\mathrm{NonNeighbour}i}^{\mathrm{Output\,min}}(t)$ and $P_{\mathrm{NonNeighbour}i}^{\mathrm{Output\,max}}(t)$, respectively, represent the lower and upper bounds of the schedulable range corresponding to the above variables during period t.

(5)   Supply and demand balance constraints

The ELAN system needs to satisfy the supply and demand balance at any time, as shown in Eq. 12.28. The left side of the equation represents the production capacity, which includes the practical output of each production unit, the discharging power of the BESS and the electricity power purchased from its neighbour/non-neighbour ELANs and the utility grid. The right side of the equation represents the consumption capacity, which includes the input power of each consumption unit, the charging power of the BESS and the electricity power sold to its neighbour/non-neighbour ELANs and the utility grid. Note that the transformation efficiency and transmission loss within the ELAN and during the exchange process between the ELAN and its externally connected grid (such as its neighbour/non-neighbour ELANs and the utility grid) are neglected.

$$
\sum_{NGGi=1}^{NGGn} P_{NGGi}^{Actual}(t) + \sum_{WDGi=1}^{WDGn} P_{WDGi}^{Actual}(t) + \sum_{PVGi=1}^{PVGn} P_{PVGi}^{Actual}(t) + \sum_{BESSi=1}^{BESSn} P_{BESSi}^{DisChar}(t)
$$

$$
+ \sum_{Neighbouri=1}^{Neighbourn} P_{Neighbouri}^{Input}(t) + \sum_{NonNeighbouri=1}^{NonNeighbourn} P_{NonNeighbouri}^{Input}(t) + P_{UGrid}^{Input}(t)
$$

$$
= \sum_{CLsi=1}^{CLsn} P_{CLsi}(t) + \sum_{ILsi=1}^{ILsn} (P_{ILsi}(t) \times (1 - R_{ILsi}(t))) + \sum_{BESSi=1}^{BESSn} P_{BESSi}^{Char}(t)
$$

$$
+ \sum_{Neighbouri=1}^{Neighbourn} P_{Neighbouri}^{Output}(t) + \sum_{NonNeighbouri=1}^{NonNeighbourn} P_{NonNeighbouri}^{Output}(t) + P_{UGrid}^{Output}(t)
$$

$$(12.28)$$

In Eq. 12.28, the first three terms on the left represent the practical output power of each production unit during period t, the fourth value represents the discharging power of each BESS during the t-th stage, and the fifth, sixth and seventh terms indicate the input power from its external grid (such as its neighbour/non-neighbour ELANs and the utility grid) during the t-th stage. On the right side of the above equation, the first two terms represent the electricity power consumed by the crucial and interruptible loads during the t-th stage, the third part represents the charging power of each battery energy storage system during the t-th stage, and the fourth, fifth and sixth terms indicate the output power to its external power grids during the t-th period.

12.2.3   Optimization Strategies

To encourage and realize the distributed renewable energy consumed locally, regional autonomy within the ELAN and collaborative optimization among different ELANs, the following optimization strategies are used to realize resource scheduling within the ELAN and power transactions with its external connected power grid.

(1) In the energy local area network, the output power generated by distributed renewable energy (such as wind and photovoltaic power) is preferentially consumed and in order is used for meeting the demands of crucial loads (CLs), interruptible loads (ILs) and the charging power of BESSs. A mechanism of demand response is introduced, and the optimization goal of minimizing the differences between the system operating costs and incomes is reached by adjusting the output capacity of the natural gas production units, the status and charging/discharging power of the BESSs, and the interrupt ratio of the interruptible loads and by satisfying the requirement of the supply and demand balance.

(2) Among different energy local area networks, an ELAN with demand exceeding its supply gives priority to buying electricity power from its neighbouring ELANs, and the buying price is slightly higher than the purchasing price for its own energy production units and battery energy storage units but lower than the price of buying electricity from its non-neighbouring ELANs and the grid. On the other hand, an ELAN with supply exceeding its demand gives priority to selling electricity power to its neighbouring ELANs, and the selling price is slightly lower than the selling price for its own energy production units and battery energy storage units but higher than the price of selling electricity power to its non-neighbouring ELANs and the utility grid.

In short, regardless of the process of purchasing or selling electricity power, the transaction order between an ELAN and its external grids always meets its internal needs first, followed by its neighbouring/non-neighbouring ELANs and finally the public power grid.

12.3 Model Solution and Results Analysis

12.3.1 Test System and Parameter Configuration

In the cooperative optimization of energy management for energy local area networks with multiple source-load-storage systems, the ELAN selected in this chapter is composed of two parts: the internal part and the external part. Among them, the internal part includes seven distributed production units (three natural gas generators, two wind power generators and two photovoltaic generators), five energy consumption units (two groups of crucial loads and three groups of interruptible loads) and one group of battery energy storage systems. The external part consists of a group of neighbouring/non-neighbouring ELANs and the utility grid. Additionally, the output power of each natural gas generator, the interrupt rate of each group of interruptible loads, the charging or discharging power of the battery energy storage units, and the interaction power with the externally connected neighbouring/non-neighbouring ELANs are regarded as decision variables, whose amount is 216. The whole time

Table 12.1 Configuration parameters of production units

Type	No.	Quadratic coefficient	Monomial coefficient	Constant coefficient	Maximum output power/kW
Natural gas generator	NGG1	$1.74e^{-5}$	0.063	0	200
	NGG2	$1.59e^{-5}$	0.06	0	200
	NGG3	$1.50e^{-5}$	0.06	0	200
Wind power generator	WDG1	–	–	–	125
	WDG2	–	–	–	125
Photovoltaic generator	PVG1	–	–	–	100
	PVG2	–	–	–	100

Table 12.2 Configuration parameters of consumption units

Type	No.	Cost compensation ratio		
Crucial load	CLs1–2	1.15		
Interruptible load	ILs1–3	0.25_H	0.15_M	0.10_L

Table 12.3 Configuration parameters of battery energy storage system

Initial capacity	Maximum capacity	Charging/discharging power	Self-discharging power
50 kW	100 kW	[0, 50] kW	0.001 kW/Δt

Table 12.4 Other configuration parameters

Type	No.	Buying/selling electricity price rate	Maximum interactive power/kW
Neighbouring ELAN	NeiELAN	1.05/0.95	50
Non-neighbouring ELAN	NonNeiELAN	1.10/0.90	50
Utility grid	UGrid	1.20/0.80	–

period is 2 h, and each time period is 5 min. All the configuration parameters and system input items are referenced in [21, 22] and shown in Tables 12.1, 12.2, 12.3, 12.4 and Fig. 12.3.

12.4 Optimization Scheme of Test System

Considering with/without the demand response and connected neighbouring/non-neighbouring ELANs, three typical optimization schemes are proposed in this chapter. The optimization results are compared and analyzed by the time-of-use

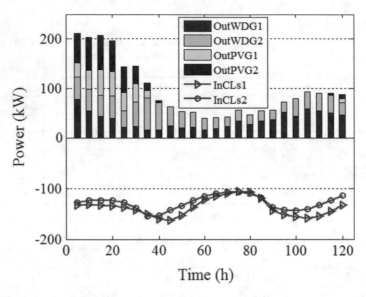

Fig. 12.3 Renewable energy output and crucial load requirements

price, as shown in Fig. 12.4, which verifies the reliability and validity of the model and method proposed in this chapter.

(1) Scenario I: Direct connection mode of source, load and utility grid. This mode considers the power exchange order inside and outside of the test system, and the ELAN is directly connected with the utility grid. This means that the supply/demand requirement in the ELAN is preferentially satisfied; then, sufficient or insufficient demands are met by exchanging power with the utility grid.

Fig. 12.4 Time-of-use price with demand response

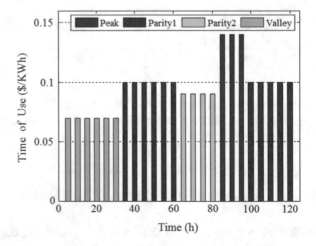

(2) Scenario II: Connection with its neighbouring/non-neighbouring ELANs. In this mode, if there are sufficient or insufficient demands after internal power exchange, this part is satisfied in order by its neighbouring ELANs first, then its non-neighbouring ELANs and finally the utility grid. Certainly, exchange with its non-neighbouring ELANs and/or the utility grid is not necessary.

(3) Scenario III: Demand response mode. This mode considers both the demand response within the system and the direct connection with its neighbouring/non-neighbouring ELANs. Among them, the demand response is mainly used to adjust the production capacity of the controllable energy production units, the charging/discharging status and power of the battery energy storage units and the interrupt ratio of the interruptible loads.

To realize the function of demand response and ordinal power transactions smoothly, the purchase price of the energy local area network is set by the following sequence: the prices among internal units are the lowest, the exchange prices with its neighbouring ELANs are slightly higher, the exchange prices with its non-neighbouring ELANs are much higher, and the exchange prices with the utility grid are the highest. In contrast, the selling price of electricity power is set as follows: the prices among internal units are the highest, the exchange prices with its neighbouring ELANs are higher than those of its non-neighbouring ELANs, and the exchange prices with the utility grid are the lowest. Additionally, the discharging price of the battery energy storage systems is much higher, which is regarded as profit. The compensation cost of the interruptible loads is changeable, which is determined by the practical interrupt ratio and regarded as an expense. In addition, all test scenarios were performed on desktops configured with Windows 7, 64-bit; Intel® Core™ i5-4460, 3.20 GHz; and 8 GB DDR2 RAM. The whole optimization time was set as two hours, and each period was five minutes.

12.4.1 Optimization Results of Different Schemes

In this chapter, both the Lagrange multiplier method and a co-evolutionary algorithm are used to optimize the above three different schemes. The constrained optimization problem is transformed into an unconstrained problem using the Lagrange multiplier method, and the co-evolutionary method is applied to solve the transformed unconstrained optimization problem. The consumption power of each energy unit in the ELAN under the above three different schemes together with the optimization results of its external grids are shown in Figs. 12.5, 12.6 and 12.7.

Figure 12.5 shows that the energy consumption of the consumption unit in Scenario III is lower than that of schemes I and II, especially when the electricity price is much higher. With the introduction of a demand response mechanism, users of interruptible loads are willing to interrupt partial loads during the corresponding periods and obtain the corresponding economic income or electricity compensation. Although the ELAN needs to pay some expense for compensating such interrupted

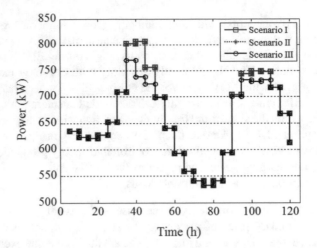

Fig. 12.5 Energy consumption curve of the consumption unit under three schemes

loads, it is much more economical to compensate the insufficient part than purchasing electricity power from external power grids, which is beneficial to decreasing the total operating cost of the system and maintaining the balance of supply and demand.

Figures 12.6 and 12.7 show the real-time power and cumulative cost curves with the utility grid and its external grids under three scenarios. Among them, positive values mean the ELAN purchases power and the corresponding expense from the utility grid and its external grids, while negative values mean the ELAN sells power and the corresponding income to the utility grid and its external grids.

The optimization results in Figs. 12.6a and 12.7a show that when the ELAN purchases power from the utility grid and its external grids under scenario III with the demand response mechanism, it is much less than that under scenarios I and II without the demand response mechanism, especially when the price is relatively high.

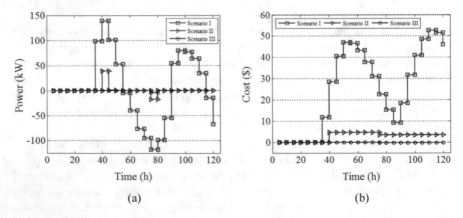

Fig. 12.6 Real-time power and cumulative cost curves with the utility grid under three scenarios

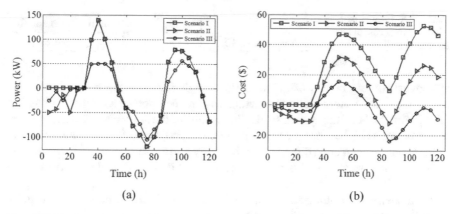

Fig. 12.7 Real-time power and cumulative cost curves with external power grids under three scenarios

Once the demand response mechanism is introduced into the optimization process of the ELAN, it will facilitate the ELAN to adjust the charging/discharging behaviour of the battery energy storage unit and the interrupted ratio of the interruptible load, which can help the ELAN to increase its internal self-regulation and self-balancing capability and alleviate or even avoid buying power from the external power grid. A further analysis of Fig. 12.6a reveals that scenario II has more ability to reduce the power purchased from the utility grid than scenario I because the ELAN can exchange power with its neighbouring/non-neighbouring ELANs prior to the utility grid.

The optimization results in Figs. 12.6b and 12.7b show the cumulative cost curves with the utility grid and external power grids under three scenarios, respectively. Compared with scenario I, which has the worst economic performance, the cumulative cost under scenario II is much smaller during period T because the latter can directly interact with the neighbouring/non-neighbouring ELANs. Compared with scenario II, scenario III with the demand response mechanism not only can meet the real-time supply and demand balance within the ELAN and the interaction demand outside of the network but also does not generate the interaction behaviour and corresponding cost with the utility grid. Figure 12.7b shows the cumulative charge curve of the ELAN interacting with its external power grids, including the utility grid, and its neighbouring/non-neighbouring ELANs. Similarly, scenario II can effectively reduce the reliance on the utility grid, reduce or avoid the fluctuations of voltage/frequency caused by frequent switching between purchasing and selling power, and contribute to realizing regional supply and demand coordination. The introduction of a demand response mechanism and the regular adjustment of the battery energy storage system can dramatically improve the supply and demand for different production and consumption units, which can also realize the economic dispatch of the whole network.

Table 12.5 shows the comparison results of the total cost under the three scenarios, which shows that during the cooperative optimization process of energy management for the energy local area networks with multiple source-load-storage systems, the

Table 12.5 Comparison
results of total cost under
three scenarios

	Scenario I	Scenario II	Scenario III
Total cost (USD)	352.57	335.90	328.66

system scheduling results of the scenario considering both the demand response mechanism and preferentially interacting with its neighbouring/non-neighbouring ELANs are the best, the system scheduling results of the scenario considering interaction priorities higher than the utility grid follow, and the system scheduling results of the scenario connecting directly with the utility grid are the worst. From the results, we can see that scenario III is the best, scenario II is second, and scenario I is the worst.

In summary, the optimization strategy in which the ELAN first interacts with its neighbouring/non-neighbouring ELANs has more economic and scheduling advantages than that of the ELAN directly interacting with the utility grid, such as contributing to the consumption of distributed renewable energy, realizing the supply and demand balance of the internal ELAN, and alleviating or even eliminating the stress and corresponding costs generated by frequent and large-scale power exchange. Additionally, the introduction of the demand response mechanism can further enhance the efficiency of power exchange and reduce the cost of internal and external scheduling for the ELAN.

12.4.2 Optimization Results Based on Optimal Strategy

To further analyze the changes in each configuration unit inside and outside the network of the ELAN, taking scenario III as an example, the exchange power and corresponding costs of production units, consumption units, battery energy storage units and external neighbouring/non-neighbouring ELANs as well as the utility grid during the whole period T are discussed and analyzed as follows:

(1) Optimization results of production units

Figure 12.8 shows the changes in the output power for each production unit and the percentage of distributed renewable energy during different periods. Among them, renewable energy resources such as wind resources and solar resources are given priority and are fully utilized according to the principle of "maximizing the utilization of renewable energy." The output power of natural gas power generation units changes with increases or decreases in the demand of energy consumption, the time-of-use price changes and the changes in its own characteristics. The ratios between the output power of renewable energy and the total capacity of all production units, shown by the red line in Fig. 12.8, fluctuate within the range of 6.09–31.49%.

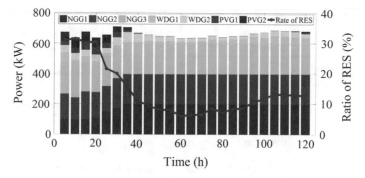

Fig. 12.8 The output power of production units and the proportion of renewable energy

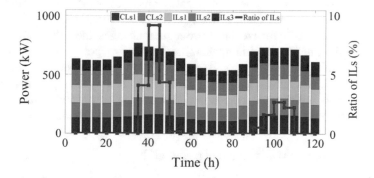

Fig. 12.9 Optimization results of consumption units and the ratio of the interrupted loads

(2)   Optimization results of consumption units

Figure 12.9 shows the optimization results of consumption units and the ratio of the interrupted loads with the change in time t. Among them, the reliability of the crucial loads (CLs1, CLs2) is highest. Even if the output power of all production units has been fully consumed, the battery energy storage system has been fully discharged, and the ratio of the interrupted load has reached the limit, the insufficient requirement will also be guaranteed by means of higher power purchases from its neighbouring/non-neighbouring ELANs and the utility grid.

Compared with the interruptible loads (IL1, IL2, and IL3), the crucial loads have obvious advantages in the unit revenue of electricity (as shown in Table 12.6). The reasons mainly include two aspects: on the one hand, the electricity price rate of the latter is higher than that of the former in any period of time. On the other hand, the former needs to pay a certain amount of revenue, which is regarded as the compensation cost for the interrupted part of interruptible loads. As shown by the red line in Fig. 12.9, the interrupt ratio (the ratio between the interrupted loads and all the interruptible loads) fluctuates from 0% to 9.22%, and the interrupted loads occur during periods of high-load ranges.

Table 12.6 Comparison results of critical loads and interruptible loads

Load type	Total electricity consumption (kW)	Total revenue (USD)	Unit electricity revenue (USD/kW)
Crucial Loads (CLs)	6364.88	− 705.12	− 0.111
Interruptible loads (ILs)	9323.25	− 897.04	− 0.096

(3) Optimization results of the battery energy storage system

The charging or discharging status and power of the BESS adjusts with changes in time-of-use price, as shown in Fig. 12.10. During the whole period T, the battery energy storage system appears to have two continuous charging processes, two continuous discharging processes and three state transitions, which meets the demand for charging at a low price (valley price, parity 2) and discharging at a high price (peak price, parity 1). To realize the above regular changes, the optimization strategies include limiting the maximum charging/discharging power, adding switching costs from charging to discharging and from discharging to charging, and setting different charging/discharging rates. The capacity of the battery energy storage system is affected by its initial capacity, charging/discharging power and efficiency, self-discharging power and other factors, which change regularly between the minimum and maximum boundary values, as shown in Fig. 12.10.

(4) Optimization results with external power grids

Figure 12.11 shows the optimization results of the ELAN with its external grids. This interaction occurs only when there is a sign of imbalance between supply and demand in the test system and when it has difficulty achieving a state of self-regulation and

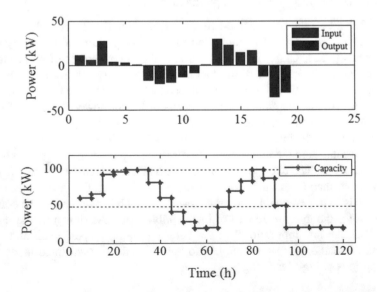

Fig. 12.10 The capacity and charging/discharging power of the battery energy storage system

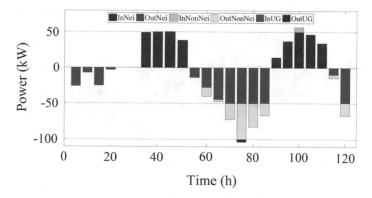

Fig. 12.11 Interaction with external power grids

self-balancing by itself. After the process of self-coordination, the ultimate state of the ELAN is demand exceeding supply or supply exceeding demand, and the power exchange sequence occurs between the ELAN and the external grid is its neighbouring ELAN first, non-neighbouring ELAN second and the utility grid last. There are differences in the transaction sequence and amount between the ELAN and its external power grids, which show that the regularity of self-regulation within the ELAN and collaboration among different ELANs can help renewable energy be used locally and the whole schedulable resources to be distributed economically.

(5)   Optimization results of the whole system

Figures 12.12 and 12.13 show the optimization results of the interaction power and corresponding cost for each unit inside/outside the network during the period T, respectively. Here, positive values represent the output power and corresponding cost, while negative values indicate the input power and corresponding benefit. Figure 12.12 shows that the total output power of all production units is equal to the total input power of all consumption units at any time, which shows that the test system can satisfy the requirements of real-time supply and demand balance. The

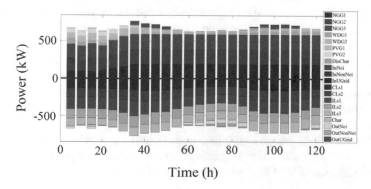

Fig. 12.12 Power optimization results of the test system

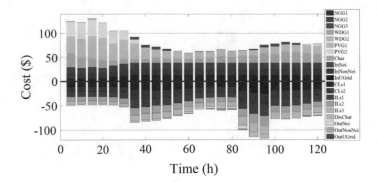

Fig. 12.13 Cost optimization results of the test system

optimization results also show that the internal units have strong self-coordination abilities, which are affected by the demand response mechanism.

Figure 12.12 also shows that both the input and output power of the external grid are much smaller than those of the internal units of the ELAN when it cannot reach the state of self-balance by means of self-coordination. Among them, the exchange frequency and quantity between the ELAN and the utility grid are much lower. The above optimization results reflect that the ELAN interacts with its external grids only when it has no self-balancing ability. Considering the optimization mechanism that the buying price from the external grid is much higher than the selling price and that the exchange sequence with its neighbouring/non-neighbouring ELANs occurs prior to that with the utility grid, the ELAN can effectively reduce or avoid disadvantages in trading price, as shown in Fig. 12.13.

12.5   Conclusions

In this chapter, a model and cooperative optimization method for energy management of energy local area networks with multiple source-load-storage systems are proposed, which can contribute to the usage of distributed renewable energy. First, considering the practical operation of an ELAN and the complex interaction relationship among different internal units, a demand response mechanism and the priority of the neighbouring/non-neighbouring ELANs being higher than that of the utility grid are introduced to construct the optimization strategy. Second, three different optimization schemes are proposed according to the idea of "autonomous within ELAN, collaboration among different ELANs and global optimization of the whole test system." Finally, the constrained problem is transformed into an unconstrained problem using the Lagrange multiplier method, and a co-evolutionary algorithm is used to solve the above different optimization schemes. Experimental results verify the reliability and validity of our proposed model and optimization method.

References

1. S.A. Janko, M.R. Arnold, N.G. Johnson, Implications of high-penetration renewables for ratepayers and utilities in the residential solar photovoltaic (PV) market. Appl. Energy **180**, 37–51 (2016)
2. I. DAdamo, P. Rosa, Current state of renewable energies performances in the European Union: a new reference framework. Energy Convers. Manag. **121**, 84–92 (2016)
3. S.R. Tito, T.T. Lie, T.N. Anderson, Optimal sizing of a wind photovoltaic battery hybrid renewable energy system considering socio-demographic factors. Sol. Energy **136**, 525–532 (2016)
4. J. Ping, Z. Xiaoxin, S. Yunting, M.A. Shiying, L. Baiqing, Review and prospect of regional renewable energy planning models. Power Syst. Technol. **37**(8), 2071–2079 (2013)
5. Z. Xueting, L. Tianqi, L. Qian, W. Fujun, G. Tieying, Short-term complementary optimal dispatch model of multi-source hybrid power system based on virtual power configuration strategy. Power Syst. Technol. **40**(5), 1379–1386 (2016)
6. L. Guojing, H. Xueshan, Y. Ming, Synergetic economic dispatch in power system operation. Proc. CSEE **34**(16), 2668–2675 (2014)
7. J. Xiaoming,C. Haoyong,X. Peizheng,Z. Xiaodong,Y. Liu, Model and strategy of source-load coordinated dispatch considering intermittent renewable energy. Southern Power Syst. Technol. **9**(5), 2–14 (2015)
8. S. Behboodi, D.P. Chassin, C. Crawford, N. Djilali, Renewable resources portfolio optimization in the presence of demandResponse. Appl. Energy **162**, 139–148 (2016)
9. Y. Qinzheng, *Research on Energy Optimization Management for Grid-Connected Micro-Grid* (North China Electric Power University, 2015)
10. P. Tianjiao, C. Naishi, W. Xiaohui et al., Application and architecture of multi-source coordinated optimal dispatch for active distribution network. Autom. Electric Power Syst. **40**(1), 17–32 (2016)
11. W.Chun, W. Ke, Z. Xiangwen, S. Haishun, Z. Cong, X. Xiaohui, Unit commitment considering coordinated dispatch of large scale electric vehicles and wind power generation. Power Syst. Protect. Control **43**(11), 41–48 (2015)
12. H. Tianxiong, X. Jiawei, Z. Xian, Z. Xiaoquan, Evolutionary programming method and its application in the locating and sizing of the distributed generator. J. Electric Power **28**(5), 370–373 (2013)
13. T. Zhang, Z. Fuxing, Z. Yan, Study on energy management system of energy internet. Power Syst. Technol. **40**(1), 146–155 (2016)
14. Z. Ming, Y. Yongqi, L. Dunnan et al., Generation-grid-load-storage coordinative optimal operation mode of energy internet and key technologies. Power Syst. Technol. **40**(1), 114–124 (2016)
15. S. Hongbin, G. Qinglai, P. Zhaoguang, Energy internet concept architecture and frontier outlook. Autom. Electric Power Syst. **39**(19), 1–8 (2015)
16. X. Fang, Q. Yang, J. Wang, W. Yan, Coordinated dispatch in multiple cooperative autonomous islanded microgrids. Appl. Energy **162**, 40–48 (2016)
17. Z. Wang, B. Chen, J. Wang, J. Kim, Decentralized energy management system fornetworked microgrids in grid-connectedand islanded modes. IEEE Trans. Smart Grid **7**(2), 1097–1105 (2016)
18. W. Jiekang, X. Yan, Establishment and solution of the complementary power generation model of wind-energy, hydro-energy and natural gas. Power Syst. Technol. **38**(3), 603–609 (2014)
19. W. Fanrong, S. Quan, L. Xiangning et al., Energy control scheduling optimization strategy for coal-wind-hydrogen energy grid under consideration of the efficiency features of hydrogen production equipment. Proc. CSEE (2017). https://doi.org/10.13334/j.0258-8013.pcsee.170044
20. S. Yu, Y. Wei, F. Jiakun, W. Jinyu, Liquid hydrogen With SMES and its application in energy internet. Power Syst. Technol. **40**(1), 172–179 (2016)

21. T. Hong, P. Pinson, S. Fan, H. Zareipour, A. Troccoli, R.J. Hyndman, Probabilistic energy forecasting: global energy forecastingcompetition 2014 and beyond. Int. J. Forecast. **32**, 896–913 (2016)
22. E.A. Farsani, H.A. Abyaneh, M. Abedi, S.H. Hosseinian, A novel policy for LMP calculation in distribution networks based on loss and emission reduction allocation using nucleolus theory. IEEE Trans. Power Syst. **31**(1), 143–152 (2016)

Chapter 13
Power Quality and Power Experience

Jie Yang and Haochen Hua

Abstract This chapter introduces the related concepts, research significance, categories, and macroscopic and microscopic problems of power quality. General rules of power quality management in the Energy Internet are proposed. Investigation monitoring and analysis of power quality in the Energy Internet are summarized. Device development for power quality control is elaborated. The effect evaluation of power quality control is described. Finally, the conclusions are given.

One of the key technological challenges that the Energy Internet needs to solve is the access of renewable energy. Renewable energy sources, for example, wind power and solar, have defects including volatility, intermittence, randomness, etc. In Energy Internet scenarios, power quality data are diversified, formats are diverse, and data are distributed widely. The Energy Internet is expected to provide green, flexible and reliable high-quality energy for customers. The electric power quality is directly linked to the stability and economic operation of the system. Therefore, the study of related power quality problems has theoretical research value and practical significance [1, 2]

13.1 Introduction to Power Quality Problems

13.1.1 The Related Concepts and Research Significance of Power Quality

The International Electrotechnical Commission (IEC) 61000 series electromagnetic compatibility standards and Institute of Electrical and Electronics Engineers (IEEE) 1159 have gradually agreed on the concept definition, index recommendation and testing methods of modern power quality. IEEE has the term 'power quality' and

J. Yang · H. Hua (✉)
Beijing National Research Center for Information Science and Technology, Tsinghua University, Beijing 100084, People's Republic of China
e-mail: hhua@tsinghua.edu.cn

© Springer Nature Switzerland AG 2020
A. F. Zobaa and J. Cao (eds.), *Energy Internet*,
https://doi.org/10.1007/978-3-030-45453-1_13

uses parameters, e.g. voltage amplitude, voltage waveform and frequency deviation, to measure the power quality. The problem of power quality is the deviation
of voltage, current or frequency that causes electrical equipment failure, including
voltage deviation, voltage fluctuation, voltage flicker, voltage sags and short-time
interruption, three-phase unbalance, temporary or transient overvoltage, waveform
distortion, frequency deviation, long-time voltage interruption, etc. [3].

The Energy Internet has put forward new demands for power quality. Distributed
power generation and energy storage can achieve a local power balance, improving
the load carrying capacity and system stability of a distribution network but can
easily lead to voltage sag, wave and flicker, and harmonic pollution. A power quality
disturbance can reduce the lifetime of a transformer, cause the failure of a relay
protection device, increase the line loss, interfere with the normal communication of
the power system, and directly affect the function of the power grid. Due to the usage
of a large number of power electronic equipment, users are very sensitive to all kinds
of electromagnetic interference. Slight voltage sags may affect the normal work of
an electronic control system, even leading to production line stop and an increase in
waste products, which directly affect the economic benefits of enterprises.

13.1.2 The Categories of Power Quality Problems

IEEE divides power quality problems into two categories: steady-state problems
and transient problems. A steady-state power quality disturbance refers to the waveform, amplitude and frequency of voltage or current, and wave distortion is a typical representative, mainly including power grid harmonics, voltage fluctuation and
flicker, voltage deviation, frequency deviation, three-phase imbalance, etc. Steady-
state power quality indicators are the differences between the real wave and the
expected wave in shape, amplitude, frequency, phase, etc., in a long period. Thus,
this disturbance is also called a continuous power quality problem. A transient power
quality disturbance mainly includes voltage sags, short-time voltage interruption,
pulse and oscillation. Thus, this disturbance is also known as an event-type power
quality problem. The transient power quality index is a measure of sudden and
transient changes in voltage and current. In addition, from a period of monitoring
statistics, the severity of transient power quality disturbances is also reflected in the
frequency of events [4].

A typical power quality single disturbance is described as follows:

1. Voltage Sag. Voltage sag is a kind of electromagnetic disturbance called voltage
   drop or voltage sag in the operation of a power system. The effective value of
   the voltage signal's power frequency voltage is reduced to 0.1–0.9 p.u., and the
   duration of time is within 0.5 T–1 min. The main reasons for the voltage drop are
   the short-circuit fault of the power system, the operation of the transformer, the
   starting of the induction motor, the access of loads with large starting currents,
   etc. [5].

2. Voltage Swell. Voltage transient is a kind of electromagnetic disturbance phenomenon in which the Root Mean Square (RMS) of the power frequency voltage signal increases to 1.1–1.8 p.u. and lasts for 0.5 T–1 min in power system operation. This electromagnetic disturbance can also be called a voltage bump or voltage rise. In general, the main reasons for the sudden rise in voltage include a single-phase grounding fault occurs in a neutral point ungrounded system, the removal of a large load, or the electricizing of a capacitor bank with large capacity.

3. Voltage Interruption. Voltage interruption refers to a kind of electromagnetic disturbance in which the power frequency voltage value of the voltage signal is reduced to less than 0.1 p.u. and its duration is within 0.5 T–1 min during the operation. The main reason for voltage interruption is transient fault caused by lightning and other matter. Power system control device misoperation and equipment failure may also lead to voltage interruption.

4. Pulse Transient and Oscillation Transient. Pulse transient is a kind of electromagnetic disturbance phenomenon in which the voltage signal or current signal appears in the stable condition during the operation of a power system. The impulse transient is mainly caused by lightning strikes. Oscillatory transient is a kind of electromagnetic disturbance phenomenon in which voltage or current signals undergo sudden and bipolar changes under stable conditions in power system operation. Pulse perturbation is unipolar, and oscillatory disturbance is bipolar. According to its spectral components, oscillatory transients are divided into three types: high, medium and low frequency.

5. Harmonic wave. A harmonic wave is a kind of electromagnetic disturbance phenomenon in which the voltage or current signal contains an integral multiple fundamental frequency in a power system. Harmonics are usually caused by power and electronic devices, such as rectifiers and electrical loads with nonlinear characteristics. The harmonic voltage total distortion rate THD_u is usually used as the index of harmonic classification. The definition of the total harmonic voltage distortion rate is shown in formula (13.1):

$$THD_u = \frac{\sqrt{\sum_{n=2}^{N} U_n^2}}{U_1},$$

(13.1)

where U_n is the value of the nth second harmonic voltage and U_1 is the effective value of the fundamental voltage.

6. Voltage Fluctuation and Voltage Flicker. Voltage fluctuation refers to an electromagnetic disturbance that has regular changes in the envelope of a power quality voltage signal or a series of random variations between the amplitude and the range of 0.9–1.1 p.u. Voltage flicker usually refers to the human eye reaction phenomenon in which the illumination of an electric light source is unstable due to voltage fluctuation.

7. Three-Phase Unbalance. Three-phase unbalance refers to the deviation of the amplitude and phase angle of three-phase voltage signals. Three-phase unbalance is mainly classified into two types: accident type and normal type. For

example, a one-phase or two-phase short circuit or ground fault will cause three-phase unbalance. Three-phase unbalance refers to the percentage of the negative sequence component relative to the positive sequence component, which is usually used as a measure index, as shown in (13.2):

$$\varepsilon_U = \frac{A_2}{A_1} \times 100\%, \tag{13.2}$$

where A_1 is the RMS value of the positive sequence components of the three-phase voltage and A_2 is the RMS value of the negative sequence components.

13.1.3 Macroscopic and Microscopic Problems of Power Quality

Within the scope of the Energy Internet, the problem of power quality has aroused great concern. From economic and market perspectives, power quality has its own value, so we must make a trade-off between cost and quality. Driven by economic interests, power users will choose a certain level of power quality to suit their operational activities. A clean electromagnetic environment is a public property that needs to be controlled and protected. This is an important resource and must be used in a sustainable way. Power retailers, power distribution operators, equipment manufacturers and users must work together to ensure power quality. Due to the high cost of investment in improvement actions, it is necessary to establish liability standards or operational guidelines for each partner on the basis of power quality contracts. Another possibility is to establish a power quality market under the Energy Internet, where the allotted power disturbance allowance can be traded if feasible. As a result, investment to reduce power disturbances has become a good business opportunity [6].

From the microscopic point of view, power quality means the quality of the alternating galvanic current energy provided for the user side via the public utility grid. The ideal public power grid should supply users with a constant frequency, standard sine wave and rated voltage. Additionally, in a three-phase AC system, the magnitudes of the voltage and galvanic current of each phase should be equal, phase symmetrical and have a phase difference of 120°. However, this ideal state does not exist because of the nonlinearity of the generators, transformers and lines in the system, the variable load nature, the imperfect control approaches, operation and external interference, and various failures.

13.2 General Rules of Power Quality Management in the Energy Internet

13.2.1 Standardization

The existing national standards and related technical specifications for monitoring and evaluation systems prepared by IEEE and IEC should be studied. Based on the actual operating conditions of the power distribution networks in cities, the established analysis and evaluation method for power quality monitoring systems should be used to perfect the non-standardized contents in the existing standards and prepare the industry specifications or local standards. The standards are divided into foundational standards, product standards, method standards and management standards by type. Several standards have been studied. Finally, the power quality work standards of the power grid in the Energy Internet are founded.

13.2.2 Integration

Power quality monitoring, analysis, governance and evaluation should not be separate from each other. An integrated operation management platform is required to share data at different steps. Monitoring data is the foundation for advanced analysis. The analysis results are the basis for preparing a governance scheme. The governance effect is evaluated based on monitoring, analysis and simulation data.

A power quality system is not a separate system and should be connected with management systems such as Energy Internet scheduling operation and customer service in the Energy Internet. The power quality system is closely associated with Energy Internet scheduling and operation. The root reasons for the problems can be known according to the scheduling operation data. If the influences of power quality are considered in scheduling, customer service will contain the power quality to improve the management service level of a power grid company.

13.2.3 Customization

Based on the above statement, the power quality problem is complicated, and no uniform governance mode is available. For economy and high efficiency, a customized solution should be used according to actual conditions.

The design, configuration and parameters of a power quality governance device can differ greatly according to different governance requirements. For example, for transient voltage decreases, the voltage grade and capacity of the corresponding governance devices are closely associated with the condition of governance loads. Related system configurations such as power storage depend on the requirements for transient decrease depth and duration, so detailed economic analysis should be performed to minimize the governance cost under the premise of meeting governance requirements.

13.2.4 Differentiation

Comprehensive governance is designed differently according to actual conditions. The difference and customization of a comprehensive governance scheme aim to meet governance requirements and minimize the governance cost.

13.3 Investigation monitoring and analysis of power quality in the Energy Internet

To solve the critical problems in the Energy Internet and improve the power quality level, monitoring and governance should be performed. For a monitoring system, the layout and selection of existing power quality monitoring sites are not reasonable. Existing terminals are installed on 110 and 10 kV buses at transformation stations without special monitoring for large pollution or sensitive loads [7, 8].

On the one hand, an imperfect layout makes the monitoring system not truly and effectively reflect the actual power quality of the users, so it affects determination of the power supply sector for authentic quality standards in the nation. On the other hand, a lack of monitoring sites for large users (including pollution source users and sensitive users) does not facilitate analyzing the reasons for power quality events and responsibility division of the power supply companies and users in power quality events and reduces the passion for both parties to carry out governance.

In addition, with increasing growth of the scale of Energy Internet monitoring systems, the existing system architecture includes weaknesses in device compatibility and network scalability. No uniform technical standards exist for monitoring device specifications, communication and networking modes. Equipment from different manufacturers is not compatible with each other. The key work is to construct a power quality monitoring system and a corresponding standard that meets the power grid application requirements of the Energy Internet.

13.3.1 Basic Functions of Energy Internet Power Quality Monitoring

The power quality problem is not only related to the reliability and safety of power equipment operation but also is closely related to standardization of the power supply and consumption markets. Only by effective monitoring of power quality can we have a clear understanding of the problem, such that a real gauge for an increase in power quality, the integration of power supply and consumption and the regulation of the power supply and consumption markets can be provided. Under such circumstances, it is of great practical significance to discuss the related theory and control technology in the field of Energy Internet power quality and to analyze its development trend.

13.3.2   Basic Form of Power Quality Observation Network

A power quality observation network is composed of the following three forms: continuous monitoring, regular or irregular monitoring and special measurement [9, 10].

(1) Continuous monitoring refers to on-line monitoring, which is generally used for monitoring public power points in important substations. The monitoring indexes include power supply frequency, voltage deviation, three-phase voltage unbalance, harmonics, etc. This monitoring can realize the continuous supervision of the power quality of both power supply and consumption. It can reflect the characteristics of all kinds of electric power quality indexes and the trend of change with time, accurately describe the actual power quality problem, and offer a foundation for comprehensive control of power quality. An on-line monitoring network is composed of several monitoring points and has a perfect network communication function. Once the power quality indicators exceed the limit, alarm signals and data recordings will be issued.

(2) The power supply quality of general public power points is usually monitored at regular intervals. There is no obvious voltage fluctuation and flicker in the Energy Internet without impulse load. These two indexes can be measured only once or twice a year. The Energy Internet with impact load is often measured once a month or quarter, and the monitoring cycle and time of measurement depend on the specific conditions. For power users with stable loads and small power consumption, power quality indicators are relatively stable, and the monitoring cycle can be once a week or month.

(3) Special monitoring, which is mainly used before and after interference source equipment, is connected to the power grid. It is used to determine the background of the energy quality index of the Energy Internet and the actual amount of interference or to verify the effect of power quality control effects. A portable power quality analyzer is usually used for testing.

13.3.3 Power Quality Monitoring for the Energy Internet

A power quality monitoring device should be able to effectively measure the parameters of power quality in real time and determine the time, position and deviation of the index deviating from the normal value in time. The functions of the power quality monitoring device should include data collection, data integration, data conversion, data storage, disturbance monitoring, fault recording and realizing intelligent management [11].

In traditional methods, different instruments and devices should be used for monitoring different power quality indicators. For example, traditional power loss measuring equipment includes active power metres and reactive power metres, voltmeters and ammeters for measuring the effective values of voltage and current, and frequency metres for measuring frequency. There are also harmonic measuring instruments, three-phase unbalance metres, voltage fluctuation and flicker metres. The disadvantages of such instruments are fewer monitoring indicators, poor universality, low accuracy and low automation [12].

As the influence of power quality transient disturbances is serious, attention for power quality is not only concentrated on steady-state indexes such as harmonic disturbance and voltage deviation but also on transient parameters such as voltage sags and transient harmonics. This requires that a power quality monitoring device can improve real-time processing and provide more intuitive analysis results, which are conducive to making decisions for power quality problems. For example, it is required that the system be able to identify faults, identify interference sources, predict faults and share information. At present, mainstream power quality monitoring devices mainly use a microprocessor based on digital signal process technology and advanced reduced instruction set computer machine technology as the core. Through analysis and processing of electrical signals, monitoring devices have made remarkable progress in measurement accuracy and reliability. At present, in addition to traditional monitoring functions, power quality monitoring emphasizes the collection, conversion, integration, classification and management of all kinds of data in functions and technologies.

13.3.4 Advanced Power Quality Analysis for the Energy Internet

An integrated platform is studied to cover the power distribution grid and user terminal monitoring, data collection, advanced application analysis, operation guides, evaluation alarms and governance effect evaluation. A transient event analysis module for the power quality based on transient decrease, transient increase and break indicators is developed. A stability analysis module for the power quality based on evaluation indicators such as frequency and voltage deviation, voltage fluctuation and flickering, three-phase unbalance, transient or instantaneous overvoltage, harmonic wave and intermittent harmonic wave should be developed.

By combining massive data mining technology, artificial intelligence methods and signal processing algorithms, the computing techniques of power quality assessment indicators and the influences on power quality should be studied based on big data analysis and computing platforms. The association rule and transfer features of the monitoring sites should be studied to provide an effective solution for evaluation, alarms and synthetic administration of power quality.

A power quality monitoring system accumulates massive monitoring data of the real-time accumulation of a power system. Compared with the primary study of power quality records, advanced analysis of power quality records is reflected in the following five aspects:

(1)   The basic analysis that is statistically oriented is changed to be prediction-oriented.
(2)   When analyzing problems, our focus changes from known problems to unknown problems.
(3)   In the analysis range, analysis of single monitoring node data changes to analysis of a regional network.
(4)   In terms of analytical methods, the single-point calculation method is changed to distributed analysis technology oriented to massive data.
(5)   For the analysis time limit, lagging bulk data analysis is shifted to real-time data analysis.

The above-advanced analysis can be realized through the application of data mining and its extended technologies.

13.4   Device Development for Power Quality Control

Currently, modern power electronic technology, with the Distribution Flexible AC Transmission System (DFACTS) as a representative, is the most effective means to settle the power quality problem of the power distribution network. The typical DFACTS equipment includes a Dynamic Voltage Recovery (DVR) device, Active Power Filter (APF), Distributed Static Synchronous Compensator (DSTATCOM) and Uniform Power Quality Control (UPQC) [13].

In the DFACTS family, the DVR is mainly used to solve the transient power quality problem in the power distribution network. The dynamic voltage recovery device is serially connected between the AC power and sensitive loads. When the power grid voltage is normal, the DVR output is under bypass status, and the power grid directly provides voltage to sensitive loads. When transient decrease or increase of the power grid voltage occurs, DVR can effectively offset the voltage increase or increase in a ms interval. DVR provides the power consumption required by the normal voltage only when the power grid voltage has a short increase or decrease, so the efficiency is high and the cost is lower than that of devices such as an uninterruptible power system/uninterruptible power supply. DVR has become one of the most economic and effective means of governance of transient power quality problems due to its better dynamic performance and higher price/performance ratio.

The APF is a special device that governs harmonic wave pollution in the DFACTS family. The APF outputs the offset current whose value is equal to the load harmonic wave current in the reverse direction and restores the current and voltage in the power grid to a 50 Hz pure sine wave. The active filter can transmit the base wave passive current and improve the load power factor.

The DSTATCOM in the power distribution grid is the main means to suppress the voltage flickering caused by the impact load. The development of DSTATCOM benefits from the development and application of dynamic passive compensation devices in the power transmission area. Large-capacity STATCOM has been developed successfully worldwide. DSTATCOM is also extensively applied.

In the DFACTS family, a serial-connected compensation device (e.g. a DVR device) is suitable for solving the voltage quality problem. A parallel-connected compensation device (e.g. APF and DSTATCOM) is suitable for solving the current problem. Two adjustment devices are needed for governance of one power quality problem. A single-function compensation device cannot meet the application requirements of comprehensive power quality governance under certain conditions.

By combining the functions of voltage and current compensation devices, the UPQC has a comprehensive adjustment function and is recognized as a serial-parallel compensation adjustment device with bright prospects.

Generally, the serial-connected part can compensate for the base wave voltage and harmonic wave voltage of the power grid as the voltage source in the UPQC compensation strategy. The parallel part can absorb the load harmonic current and adjust the DC bus voltage as the current source. With this compensation strategy, the voltage of the load can change to the standard pure since the voltage and the input current of the power grid change to the pure current whose phase is the same as that of the base wave voltage of the power grid. For the power grid, this is similar to the power supply to a pure resistor load in the power grid.

UPQC can realize serial compensation and parallel compensation of a specific load, so it can uniformly adjust and control the power supply mode and power quality in a power distribution area. When a transient decrease or short break appears in the power distribution voltage, it can maintain the load voltage at a rated value, filter the harmonic wave current generated by the nonlinear load, prevent it from entering the system and affecting other power equipment, and improve the power factor of the loads.

The development of modern power electronic technology provides powerful support for power quality governance of distributed power. The promotion and application of DFACTS technology can provide a practicable and effective solution to completely improve the power quality level.

13.5   Effect Evaluation of Power Quality Control

A power quality governance effect assessment depends not only on improvement of the power quality indicators but also on the degree of user sensitivity. Generally, the sensitive equipment of customers is affected by multiple power quality factors and is very uncertain. The influence results of these power quality problems cannot be realized by overlapping influence results of a single power quality indicator.

The main sensitive and pollution equipment of the Energy Internet are divided into the following types:

(1)   Main sensitive equipment

Based on the investigation results, the loads and industries that are sensitive to power quality disturbances are divided into the following types:

(1)   The disturbance of power quality leads to resetting the control flows of semiconductor processing equipment and digital control equipment;
(2)   The disturbance of power quality leads to errors in automatic control equipment;
(3)   A transient voltage decrease will lead to abnormal operation of equipment such as medical apparatuses and affect diagnosis, treatment and operation, even endangering patient life;
(4)   The disturbance of power quality leads to production stop, so more time is spent restarting the whole production line;
(5)   A transient voltage decrease leads to operation pauses for elevators, automatic fire control and alarm systems in commercial and civil buildings;
(6)   For industries such as the processing industry, paper-making industry and glass manufacturing industry, the disturbance of power quality leads to the sudden stop of modern production lines, which indicates that several more hours of time will be spent cleaning the waste in devices;
(7)   The disturbance of power quality leads to huge economic losses in IT and communication industries and computer centres.

(2)   Main pollution equipment

Based on the investigation results, the main pollution equipment in the Energy Internet includes the following:

(1)   AC (DC) steel-making arc furnace (power supply bus under the same voltage grade);
(2)   Roller, lifter and winch;
(3)   Motor start (very high-capacity and high-frequency start motor);
(4)   Plastic injector, welding device and smelting furnace.

Sensitive users are classified according to the customer sensitivity and severity of user complaints. Comprehensive governance users are classified by the user sensitivity level, pollution level and user complaint severity. The indicator system for user classification is described as follows:

(1) User complaint severity: This indicates the customer complaint information received by the customer service departments of power suppliers and is based on the user complaint time and event severity. User dissatisfaction with the power supply is assessed as the main basis for user governance. The user complaint severity is obtained by the customer service department according to the complaint evaluation.

(2) User load sensitivity: This indicates the user sensitivity to a specific power quality problem. When the power quality problem occurs, a higher user load sensitivity leads to easier failure occurrence. The user load sensitivity reflects the possible influences of the power quality disturbance on customers. A user may include multiple sensitive loads.

(3) User load pollution degree: This indicates the severity of different power quality disturbances in the power grid during the operation of user loads. A user may include many pollution loads.

In addition, the effect assessment method will differ with the governance scheme, and the assessment methods for whole area power quality comprehensive assessment and single-user power quality improvement are different. To evaluate power quality, it is necessary to standardize the whole evaluation process to obtain consistent and comparable evaluation results on the same platform.

13.5.1 Single Assessment of Power Quality

On the basis of different indicators of power quality, single assessment is a common method. Single assessment of power quality refers to the process of quantifying a power quality problem or a characteristic quantity to obtain a measurement value. The evaluation result is expressed in terms of the single index, and the evaluation result is the index value. When the power quality is evaluated according to the phenomena and characteristics, the monitoring instruments detect and analyze the parameters of the power quality. By decomposing diversified power quality problems, we can quickly find the main power quality problems. This contributes to seeking out the cause of a disturbance quickly and solving practical problems. However, in practical applications, all kinds of power quality disturbances usually occur simultaneously and interact with each other. A single evaluation cannot be used to judge the overall power quality [14].

13.5.2 Synthetic Estimation of Power Quality

The concentration of the research on the synthetic estimation of power quality is determining how to objectively integrate a multi-target problem into a single quantified index to assess the grade of the power quality. A comprehensive assessment is based on a single assessment. The comprehensive evaluation method quantifies some characteristic indexes of some or all power quality disturbances or a power quality disturbance by attributes. Then, a synthetic index or comprehensive grade is obtained by using an algorithm of normalization and synthesis. A number of researchers have studied the synthetic estimation of power quality based on the analytic hierarchy process, artificial neural network methods, the grey comprehensive evaluation method, etc., and achieved good results. However, due to the different mechanisms of each single comprehensive evaluation method and the different attribute levels, the evaluation conclusions are different [15].

The comprehensive evaluation method can simplify the evaluation results and enhance the operability and acceptability of the implementation. The main evaluation methods include quantitative evaluation and grade evaluation, which belong to quantitative numerical calculation and qualitative calculation, respectively.

(1)   Quantitative estimation of power quality

Quantitative assessment is the process of quantifying the power quality indicators, and the quality level obtained is a numerical value. An advantage is that it can be directly compared with the standard compatibility level, planning level or contract limit level. This is the commonly used form of evaluation results. It can directly reflect the seriousness of the problem, and the quantized value can be used as the input function of the power quality operation and the design parameters of the management facilities. This method mainly includes numerical calculation based on cannikin law and numerical calculation of weight theory.

(2)   Evaluation of power quality grade

Qualitative grade evaluation first divides the single index of power quality into a number of grade intervals, and then, on the basis of quantitative evaluation, the evaluation grade is determined according to the interval of the actual evaluation level. This approach is usually evaluated based on weight theory and cannikin law. The comprehensive grade of power quality is determined by the weighted summation result or the worst power quality index.

13.6   Conclusions

Based on the current power quality conditions of the power grid and new technologies and achievements in power quality governance of a power distribution network, the following conclusions are drawn:

(1)   Existing monitoring systems are mainly deployed at a transformation station in the power grid. The monitoring points for large pollution sources and important sensitive users are lacking, so this does not facilitate the grasping of the true power quality of the whole Energy Internet, analyzing the event causes in the power quality problem and dividing the responsibilities of the companies and users.

(2) Transient power quality events (including transient voltage decrease, transient voltage increase and short-time break), voltage flickering and line harmonic waves are the main power quality problems faced by power supply companies. The transient power quality problem is very outstanding and is one of the user complaints.
(3) Modern power electronic technologies, with DFACTS as a representative, can effectively solve the transient voltage decrease, transient voltage increase, flickering and harmonic problems in the power distribution networks in large cities.
(4) The customized and differential governance scheme is economic. The middle-voltage dynamic voltage adjuster and the power quality comprehensive governance device on the low-voltage side are practicable in the technology roadmap of comprehensive governance.

To completely improve the power quality level of the Energy Internet, on the one hand, the power supply department should identify the responsibilities between itself and users based on monitoring data, actively promote management measures for governance based on pollution makers and work hard to promote the comprehensive governance of power quality pollution sources. On the other hand, power companies should fully apply the development achievements of power electronic technologies, actively explore technologies and schemes to provide high-quality power supply services for users and work hard to improve power quality.

References

1. J. Cao, W. Zhang, Z. Xiao et al., Reactive power optimization for transient voltage stability in energy internet via deep reinforcement learning approach. Energies **12**, 1556 (2019)
2. H. Hua, J. Cao, G. Yang et al., Voltage control for uncertain stochastic nonlinear system with application to energy internet: non-fragile robust H∞ approach. J. Math. Anal. Appl. **463**(1), 93–110 (2018)
3. G.Q. Shen, D.H. Xu, L.P. Cao et al., An improved control strategy for grid-connected voltage source inverters with an LCL filter. IEEE Trans. Power Electron. **23**(4), 1899–1906 (2008)
4. S. Tsai, C. Luo, Synchronized power-quality measurement network with LAMP. IEEE Trans. Power Deliv. **24**(1), 484–485 (2009)
5. M. Wang, G.I. Rowe, A.V. Mamishev, Classification of power quality events using optimal time-frequency representations- part 2: application. IEEE Trans Power Deliv. **19**(3), 1496–1503 (2004)
6. C. Kumar, M.K. Mishra, Predictive voltage control of transformer less dynamic voltage restorer. IEEE Trans. Ind. Electron. **62**(5), 2693–2697 (2015)
7. S. Singh, B. Singh, G. Bhuvaneswari, Power factor corrected zeta converter based improved power quality switched mode power supply. IEEE Trans. Ind. Electron. **62**(9), 5422–5433 (2015)
8. A. Teke, L. Saribulut, M. Tumay, A novel reference signal generation method for power-quality improvement of unified power-quality conditioner. IEEE Trans. Power Deliv. **26**(4), 2205–2214 (2011)
9. N. Gourov, P. Tzvetkov, G. Milushev et al., Remote monitoring of the electrical power quality, in *11th International Conference on Electrical Power Quality and Utilisation* (Lisbon, Portugal, 2011)

10. A.K. Khan, Monitoring power for the future. Power Eng. J. **15**(2), 81–85 (2011)
11. L.F. Auler, D. Roberto, Power quality monitoring controlled through low-cost modules. IEEE Trans. Instrum. Meas. **58**(3), 557–562 (2009)
12. T. Lin, J.A. Domijan, F. Chu, A survey of techniques for power quality monitoring. Int. J. Power Energy Syst. **25**(3), 167–172 (2005)
13. M. Moghbel, M. Masoum, A. Fereidouni et al., Optimal sizing, siting and operation of custom power devices With STATCOM and APLC functions for real-time reactive power and network voltage quality control of smart grid. IEEE Trans. Smart Grid **9**(6), 5564–5575 (2018)
14. G.J. Lee, M.M. Albu, A power quality index based on equipment sensitivity, cost, and network vulnerability. IEEE Trans Power Deliv. **19**(3), 1504–1510 (2004)
15. M.K. Gray, W.G. Morsi, Power quality assessment in distribution systems embedded with plug-in hybrid and battery electric vehicles. IEEE Trans. Power Syst. **30**(2), 663–671 (2015)

Chapter 14
Power Restoration Approach for Resilient Active Distribution Networks in the Presence of a Large-Scale Power Blackout

Chunqiu Xia, Qiang Yang, Le Jiang, Leijiao Ge, Wei Li, and Albert Y. Zomaya

Abstract The increasing integration of different forms of distributed generations (DGs) into current medium- and low-voltage power distribution networks can result in increased system instability and protection performance degradation. Efficient power supply restoration upon a power outage is demanded to ensure resilient operation of power distribution networks consisting of DGs with stochastic generation and diverse characteristics. This work proposes an algorithmic solution of power supply restoration under the condition of a large-scale power blackout. The solution consists of DGs starting path searching and collective load restoration combining DG-based and topology reconfiguration strategies to improve network resilience under failures. In the DG starting path searching algorism, non-black-start DGs (NBDGs) can be started by black-start DGs (BDGs) nearby through the shortest path as much as possible. This solution enables simultaneous power restoration in the presence of multiple faults with maximized restored loads and minimized power loss and power flow changes. The developed solution of power supply restoration is evaluated on the basis of a 53-bus test distribution feeder penetrated with wind turbines (WTs) for a set of fault scenarios through simulations. The stochastic generation of WTs is fully considered using the heuristic moment matching (HMM) method. The proposed solution is assessed through mathematical simulations, and the results confirm that this solution can provide efficient supply restoration under a large-scale power blackout.

C. Xia · W. Li (✉) · A. Y. Zomaya
School of Information Technologies, Centre for Distributed and High Performance Computing, 2006 Sydney, Australia
e-mail: liwei@it.usyd.edu.au

C. Xia
e-mail: cxia3271@uni.sydney.edu.au

Q. Yang · L. Jiang
College of Electrical Engineering, Zhejiang University, Hangzhou 310027, China

L. Ge
School of Electrical Information and Engineering, Tianjin University, Tianjin 300072, China

© Springer Nature Switzerland AG 2020
A. F. Zobaa and J. Cao (eds.), *Energy Internet*,
https://doi.org/10.1007/978-3-030-45453-1_14

Nomenclature

F	Utility function of the power restoration model
$f_{k,n}$	kth sub-function of function F; during the nth optimization, $k \in [1, 3]$, $n \in [1, N_{change\_swi}]$
N_{change_swi}	Number of switchers to be closed in the restoration process
r_k	The base value of the kth sub-function of function F, $k \in [1, 3]$
$\mathbf{X}_n$	Switcher operation types at the nth iteration of optimization, $\mathbf{X}_n = \{x_{1,n}, x_{2,n}, \ldots, x_{N_{can\_swi},n}\}$
$x_{i,n}$	Type of operation in terms of a specific optimization iteration; when $x_{i,n} = 1$, the candidate switcher i will be turned off; otherwise, it remains open
$N_{can\_swi,i}$	The candidate switcher number for an iteration
α_m	The time cost for switcher m
m	Switcher operation types, $m \in \{1, 2, 3\}$; type 1 is bus switcher operation, type 2 is inter-switcher operation within the substation, and type 3 is across multiple substations
N_{off_node}	Number of off-line buses after faults
ρ_j	Off-line load priority of j
L_j	Off-line load of j
dis_{ij}	Restoration of the jth off-line load when set the ith candidate switcher
N_{on_swi}	The quantity of online buses after faults
$X_{i,p,n}$	Boolean value denoting if the power flow of the pth restored branch is affected by the ith off-line switcher operation at the nth iteration of optimization. If yes, then $X_{i,p,n} = 1$; else, $X_{i,p,n} = 0$
$Y_{i,q,n}$	Whether the bus voltage of the qth restored bus is affected due to the ith off-line switcher operation at the nth iteration of optimization. If yes, then $X_{i,p,n} = 1$; else, $X_{i,p,n} = 0$
$I_{p,n}$	Current of the pth restored branch at the nth iteration of optimization
$I_{limit,p}$	Current limit of the pth branch
$U_{q,n}$	Voltage of the qth restored bus at the nth iteration of optimization
$U_{lowerlimit}$	Bus voltage lower limit
$U_{upperlimit}$	Bus voltage upper limit
N_{bus}	The quantity of network buses
N_{sub}	The quantity of power substations
$path(i, j)$	The quantity of paths between buses i and j
$\gamma_{bran\_vw,n}$	If buses v and w connect with each other, $\gamma_{bran\_vw,n} = 1$; otherwise, $\gamma_{bran\_vw,n} = 0$
G_{vw}	Branch admittance from bus v to bus w
$\delta_{vw.n}$	Phase angle difference between branches of buses v and w at the nth iteration of optimization
P_v^L	Loads at bus v
P_v^D	Installed DG capacity at bus v
$I_{capacity,vw}$	Branch capacity of bus v and bus w

V_i^{BR} ith bus voltage in the failure-free section before the restoration process
V_i^{AR} ith bus voltage in the failure-free section after the restoration process
I_i^{BR} Current of the ith bus in the failure-free section before the restoration process
I_i^{AR} Current of the ith bus in the failure-free section after the restoration process
N_{normal} Number of buses in the failure-free section

Highlights

- Power supply restoration solution in the presence of a large-scale power blackout
- A DG start-up strategy and parallel power supply restoration process are used.
- Performance under uncertainties is validated using HMM-generated scenarios.

14.1 Introduction

The integration of small-scale distributed generators (DGs) in various forms, such as micro wind turbines and photovoltaic panels, effectively transforms existing electric distribution networks into active distribution networks with bidirectional power flows. Such penetrated intermittent renewable generators have brought about direct management challenges, e.g., unstable voltage profile, frequent fault occurrence, and degraded protection performance [1]. In addition, considering the starting characteristics of DGs, some DGs can be immediately utilized in the power restoration process upon large-scale power failures, known as black-start DGs (BDGs) (e.g., passive inverters, external-excited induction generators, and wind turbine/photovoltaic-based generators with storage units), while other DGs can only supply power to loads after being started, known as non-black-start DGs (NBDGs) (e.g., self-excited induction generators and wind turbine/photovoltaic-based generators without storage units). Thus, the DG start process and load restoration should be balanced during the restoration process for large-scale faults or blackouts in the distribution system. An efficient supply restoration solution upon large-scale power failures or outages (e.g., due to natural disasters and extreme weather hazards) is needed to ensure network resilience to enhance the security and reliability of electricity supply.

In general, power supply restoration under network failures aims to maximize the number of customers who receive electricity supply restoration after faults or interruptions [2] and meet the network operational constraints, e.g., within feeder/switcher ratings. In recent decades, the power restoration issue has been extensively studied, and a set of solutions are available. However, these existing solutions (e.g., [3]) addressing the power supply restoration problem have not considered DG availability and hence cannot be directly adopted in current active distribution networks with renewable distributed generation.

The significant role of DGs in guaranteeing the reliability and security of the power supply under a variety of operational scenarios in power distribution networks was highlighted in [4]. To this end, many studies of power restoration under network failures considering the presence of DGs were carried out. In [5], an analytical solution was developed based on the collective probability of power generation and loads, and the performance of a power supply service was evaluated. In [6], a load restoration optimization model and an approach were presented to investigate the benefits of operational flexibility provided by DGs through coordinating network topology reconfiguration and microgrid formation. Our previous work [7] investigated the optimal coordination among multiple autonomous microgrids and confirmed the benefit of DGs in supporting power security. The proposed approach aimed to identify appropriate DGs to supply critical loads (CLs) by determining the optimal network topologies using the minimum spanning tree (MST), which considers power damnification and stability criteria. In [8], the authors exploited multi-agent-based supply restoration through network topology reconfiguration. However, such an approach firmly relies on backup feeder capacity to restore the power demands, and hence, the performance can be significantly constrained. In [9], the authors considered the availability of distributed renewable generation and then presented a collective power supply restoration that combines both DG local restoration techniques and switcher operation-based restoration techniques. An algorithmic solution was presented in [10] to enable power dispatch through multiple microgrids to enhance power supply reliability for critical power demand. The self-healing mechanism under network failures for power distribution networks with both dispatchable and non-dispatchable DGs was exploited in [11]. In the proposed solution, the on-outage portion of the power distribution system can be optimally sectionalized to a set of connected self-supplied microgrids to maximize the network energy supply. However, the aforementioned restoration solutions did not fully consider the starting characteristics of different types of DGs and explicitly address DG uncertainties during power restoration.

In addition, topology reconfiguration has been extensively investigated for power restoration (e.g., [11–13]). A mathematical model of power restoration in radial distribution systems was exploited in [11]. This model considered multiple optimization objectives, e.g., reduction in switch operations, increase in demand satisfaction, and critical load prioritization. In [12], the authors described a graph-theoretic power restoration method aiming to augment the restored load while reducing switching operations. However, this solution was merely evaluated and validated under scenarios with a single fault. The authors in [13] developed an informed A* search-based load restoration approach based on topology reconfiguration for radial distribution networks considering single faults and multiple faults concurrently. There have also been explorations of topology reconfiguration-based restoration solutions in microgrids [14–17]. However, it is worth noting that frequent topology changes due to reconfiguration can directly lead to increased system instability due to unexpected power flow changes and the risk of cascading failure due to malfunction of switcher operations.

In summary, existing solutions (e.g., [11–20]) have not fully exploited the potential benefits of DGs and considered DG uncertainties under large-scale failures. To this end, fully considering DG availability (BDGs and NBDGs) and uncertainties, this work develops a novel restoration strategy that carries out parallel power restoration based on the strategies of local DG restoration and restoration with configurable topology in the presence of a large-scale power blackouts. In this work, the main contributions are as follows: (1) considering the DG starting characteristics (i.e., BDGs and NBDGs), the DG starting process and load restoration process are integrated during power restoration under large-scale failures; (2) DG-based and network configurable restoration strategies are collectively used to restore electricity loads in parallel when faults occur; and (3) the consideration to minimize the impacts on the non-failure sections within the network during power restoration is explicitly included.

The structure of the rest of this paper is presented as follows. The power restoration problem is proposed in Sect. 14.2, and our solution is also given in detail here. Section 14.3 presents the scenario generation approach based on the HMM method. Section 14.4 evaluates the proposed solution through simulation. Finally, the conclusions are presented in Sect. 14.5.

14.2 Formulation and Solution

Considering the DG availability in a power distribution system (including BDGs and NBDGs), the power restoration solution for the large-scale power blackout is carried out by two processes: a repetitive process of NBDG start-up by the backup of BDGs in fault sections and parallel supply restoration to power demand.

14.2.1 Strategy of DG Start-Up

In the process of NBDG start-up, the shortest power supply paths between BDGs and NBDGs can be identified based on the Dijkstra algorithm and recovered with balanced power flow constraints and safety operation constraints. As the number of started NBDGs increases, the repetitive process of NBDG start-up will not terminate until all NBDGs are restarted or there are no available NBDGs to be restarted. The detailed restoration process is illustrated in Fig. 14.1 (the shaded areas are restored regions with power supply). Upon a failure at the 11th bus, the switchers located on the downstream buses are automatically turned on to avoid further failures, as shown in Fig. 14.1a. As shown in Fig. 14.1b, NBDG1 is first started by BDG1 with the loads in bus Nos. 45, 44, 38, and 39 restored. With the limited capacity of BDG1, NBDG2 can only be started by NBDG1 in the second step with the loads in bus Nos. 33, 34, and 35 restored, as presented in Fig. 14.1c.

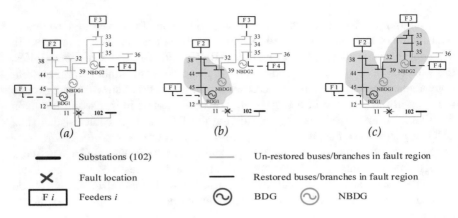

(a)   *(b)*   *(c)*

▬▬	Substations (102)	───	Un-restored buses/branches in fault region
✕	Fault location	▬▬	Restored buses/branches in fault region
☐ F *i*	Feeders *i*	⊘ BDG	⊘ NBDG

Fig. 14.1 The process of the proposed restoration solution for the electricity supply. **a** The restoration network topology; **b** Step 1 of the process; **c** Step 2 of the process

14.2.2 Parallel Power Supply Restoration

Our restoration solution for power supply is as follows: because of the generation uncertainties of BDGs and started NBDGs, the solution includes two strategies, i.e., one strategy is local restoration based on DGs, and the other strategy is topology reconfiguration-based restoration over inter-switcher operations, to collectively and simultaneously supply the loads in those areas that failed previously. In different time slots, the proposed solution can determine the optimal supply restoration approach using one or both restoration strategies as follows:

Strategy 1 **(DG-based local restoration)**: the DGs can support the network management to supply the loads locally during network failures. Figure 14.2d shows that NBDG2 is identified with sufficient power generation to supply bus Nos. 35 and 36 simultaneously upon the fault at bus No. 11. As a result, the network loads can be supplied without network topology changes through inter-switcher operations;

Strategy 2 **(Topology reconfiguration-based restoration)**: Network reconfiguration operates the switchers (opens or closes the switchers) to redistribute loads across the network to restore power supply while meeting the operational objectives, e.g., minimization of power loss and voltage violations. Figure 14.2d shows that bus Nos. 12, 44, and 45 can be restored by closing the inter-switcher between Feeders 1 and 12.

Figure 14.2 schematically illustrates the overall process of our power supply restoration strategy, where we assume that a fault exists at the 11th bus of the network with DG generation. After the DG start-up process, the available DGs can perform load restoration locally and collectively through multiple switcher operations (Fig. 14.2b); under the condition that the DG generation is insufficient, additional actions of switcher operation need to be taken for load restoration, including 2 inter-switchers and 7 intra-switchers, as illustrated in Fig. 14.2b–d.

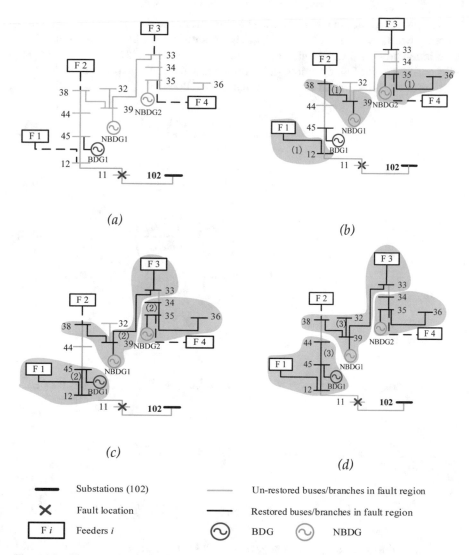

Fig. 14.2 The step-by-step load restoration process: **a** Network topology before restoration; **b** restoration step 1; **c** restoration step 2; and **d** restoration step 3

14.2.3 Formulation and Implementation

Here, the electricity supply restoration under the condition of a large-scale power blackout considering the availability of NBDGs and BDGs can be considered as a multi-objective integer programming problem, including DG start-up sequences and load restoration optimization.

The proposed solution is outlined in (14.1)–(14.12) and is carried out sequentially; in other words, there is exactly one switcher operation within each iteration of the restoration process. In addition, such a restoration process could be performed in parallel and at various failed sections of the power distribution network. The restoration process in the nth iteration can be formulated as follows:

$$\text{Min } F(\mathbf{X_n}) = \sum_{k=1}^{3} \phi\left(f_{k,n}\right) \tag{14.1}$$

$$f_{1,n} = \sum_{i=1}^{N_{\text{can\_swi}}} x_{i,n} \times \alpha_m \tag{14.2}$$

$$f_{2,n} = \sum_{i=1}^{N_{\text{can\_swi}}} x_{i,n} \times \left(\sum_{j=1}^{N_{\text{off\_node}}} \frac{\rho_j \times L_j}{\text{dis}_{ij,n}} \right) \tag{14.3}$$

$$f_{3,n} = \sum_{i=1}^{N_{\text{can\_swi}}} \left\{ x_{i,n} \times \left(\min_{p=1}^{N_{\text{swi}}} \left(X_{i,p,n} \times \frac{I_{p,n} - I_{\text{limit},p}}{I_{\text{limit},p}} \right) + \min_{q=1}^{N_{\text{swi}}} \left(Y_{i,q,n} \times \frac{U_{q,n} - U_{\text{lowerlimit}}}{U_{\text{lowerlimit}}} \right) \right) \right\} \tag{14.4}$$

Subject to:

$$U_{v,n} \sum_{w=1}^{N_{\text{bus}}} \gamma_{\text{bran\_}vw,n} U_{w,n} (G_{vw} \cos \delta_{vw.n} + B_{vw} \sin \delta_{vw.n}) = P_v^L - P_v^D, \quad v \in [1, N_{\text{bus}}] \tag{14.5}$$

$$U_{v,n} \sum_{w=1}^{N_{\text{bus}}} \gamma_{\text{bran\_}vw,n} U_{w,n} (G_{vw} \sin \delta_{vw.n} - B_{vw} \cos \delta_{vw.n}) = Q_v^L - Q_v^D \quad v \in [1, N_{\text{bus}}] \tag{14.6}$$

$$U_{\text{lowerlimit}} \leq U_{v,n} \leq U_{\text{upperlimit}} \quad v \in [1, N_{\text{bus}}] \tag{14.7}$$

$$I_{vw,n}^2 \times \gamma_{\text{bran\_}vw,n} \leq I_{\text{capacity},vw}^2 \times \gamma_{\text{bran\_}vw,n} \quad v \in [1, N_{\text{bus}}], w \in [1, N_{\text{bus}}] \tag{14.8}$$

$$\sum_{i=1}^{N_{\text{cam\_swi}}} x_{i,n} = 1 \tag{14.9}$$

$$\sum_{v=1}^{N} \sum_{w=1}^{N} \gamma_{\text{bran\_}vw,n+1} - \sum_{v=1}^{N} \sum_{w=1}^{N} \gamma_{\text{bran\_}vw,n} = 1 \tag{14.10}$$

$$\text{path}(v, w) \leq 1 \quad v \in [1, N_{\text{bus}}], w \in [1, N_{\text{bus}}] \tag{14.11}$$

$$\sum_{p=1}^{N_{\text{sub}}} \text{path}(p, q) = 1, q \in [1, N_{\text{bus}}] \tag{14.12}$$

Here, the power restoration simultaneously considers three different operational objectives, i.e., the time consumption of switcher operations, the load restoration capacity of switchers, and the adverse influences on non-failure sections, in a utility function, as given in (14.1). In (14.13)–(14.14), the penalty function $\phi(x)$, a nonlinear and piecewise function, is used for different objectives. For each objective with a base value r, $\phi(x)$ increases by a metric value x (e.g., $\frac{x}{r} < 2/3$), and the slope of $\phi(x)$ increases quickly with a heavy penalty (e.g., $\frac{x}{r} \geq 1$). Thus, three objectives can be assigned with different weights with appropriate r_k.

$$\phi(x) = \sum \phi_a(x), x \geq 0 \tag{14.13}$$

$$\phi_a'(x) = \begin{cases} 1, & 0 \leq \frac{x}{r} < \frac{1}{3} \\ 3, & \frac{1}{3} \leq \frac{x}{r} < \frac{2}{3} \\ 10, & \frac{2}{3} \leq \frac{x}{r} < \frac{9}{10} \\ 70, & \frac{9}{10} \leq \frac{x}{r} < 1 \\ 500, & 1 \leq \frac{x}{r} < \frac{11}{10} \\ 5000, & \frac{11}{10} \leq \frac{x}{r} \end{cases} \tag{14.14}$$

In addition, the costs of different types of switcher operations are explicitly described in (14.2) as follows: bus switcher operation consumption, $\alpha_1 = 1$; inter-switcher operation within the scope of the same substation, $\alpha_2 = 2$; and inter-switcher operation over various substations, $\alpha_3 = 4$. In (14.3), the restoration capacity of switcher i can be defined as $\sum_{j=1}^{N_{\text{off\_node}}} \frac{\rho_j \times L_j}{\text{dis}_{ij,n}}$, where $\text{dis}_{ij,n} = D_{ij,n} \times N_{ij,n}^{\text{open}}$. Here, $D_{ij,n}$ is the optimal supply path between switcher i and off-state load j. The unsupplied load is determined based on the Dijkstra algorithm (the branch weight is associated with the line impedance), and $N_{ij,n}^{\text{open}}$ indicates the number of active switchers on the restoration path. Finally, the topology reconfiguration adverse influence on the network is described in (14.4).

Additionally, a collection of operational constraints must be strictly met during the power supply restoration: the power balance equations are illustrated in (14.5) (active power) and (14.6) (reactive power); constraints (14.7) and (14.8) indicate the voltage and current limits, respectively; and constraints (14.9) and (14.10) ensure only one switcher operation during individual iterations. Constraint (14.11) guarantees that the network topology is loop-free during the supplemental restoration process. Additionally, constraint (14.12) guarantees that each bus can only be powered by one power substation. The $(n + 1)$th iteration of the optimization process can only be carried out subject to constraints (14.5)–(14.12) after the close of the switcher in iteration n. Figure 14.3 illustrates the overall procedure of the restoration solution in the presence of large-scale failures.

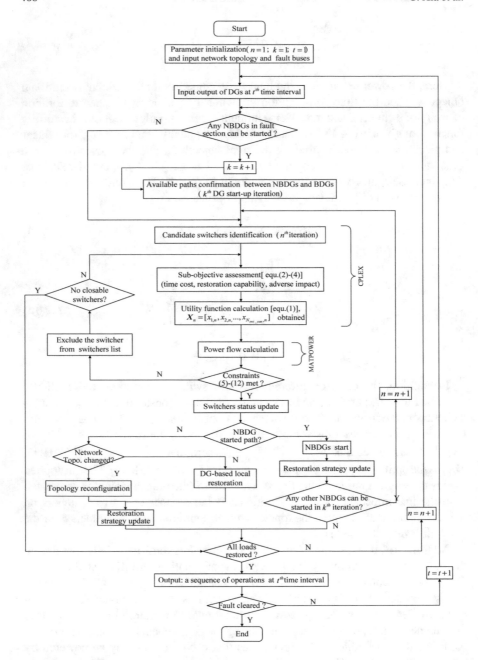

Fig. 14.3 Process of the proposed power distribution network supply restoration solution

14.3 DG Uncertainty Characterization Using the HMM Method

In this study, a set of scenarios is generated using the HMM approach [21] to capture the generation uncertainties of wind turbines in the power distribution system [22]. The scenario generation includes 2 steps: matrix and cubic transformations, as presented in detail as follows.

(1) *Matrix transformation*: with a lower triangular matrix L, an n-dimensional matrix Y can be obtained. Y is correlated with matrix $R = LL^T$. It starts with an n-dimensional random matrix X that follows the normal distribution, consisting of independent column vector X_i, $i = 1, \ldots, n$. Y can be formulated as

$$Y = L \times X = \sum_{j=1}^{i} L_{ij} \times X_i \tag{14.15}$$

(2) *Cubic transformation*: the univariate column vector Z_i is transformed from the column vector Y_i following the normal distribution through cubic transformation. The transformation process considers four different moments, including expectation, standard deviation, skewness, and kurtosis, as formulated in (14.16):

$$Z_i = a_i + b_i Y_i + c_i Y_i^2 + d_i Y_i^3 \tag{14.16}$$

where a_i, b_i, c_i, and d_i are the coefficients during the transformation, which can be derived through (14.17):

$$M_{i,k}(Z_i) = M_{i,k}^T \tag{14.17}$$

where $M_{i,k}^T$ represents the given kth moment of the ith column vector. $M_{i,k}(Z_i)$ is the kth moment of the column vector Z_i.

Step 1 *Initialization*: the target moments and correlation matrix R of the collected statistics of wind power generation are calculated, and the target moments are normalized based on (14.18):

$$M_{i,1}^{NT} = 0, \quad M_{i,2}^{NT} = 1,$$
$$M_{i,3}^{NT} = \frac{M_{i,3}^T}{\left(\sqrt{M_{i,2}^T}\right)^3}, \quad M_{i,4}^{NT} = \frac{M_{i,4}^T}{\left(M_{i,4}^T\right)^2} \tag{14.18}$$

where $M_{i,k}^{NT}$ is the normalized moment of column vector i and $M_{i,k}^T$ is the target moment of column vector i.

Step 2 *Random scenario generation:* once the number of wind turbines N_w and the expected number of scenarios N_h are given, the matrix $X_{N_w \times N_h}$ can be generated subject to $N(0, 1)$.

Step 3 *Matrix transformation:* matrix $X_{N_w \times N_h}$ is transformed to $Y_{N_w \times N_h}$ based on matrix transformation, as given in (14.1), to meet the correlation of historical generation statistics.

Step 4 *Cubic transformation:* the coefficients a_i, b_i, c_i, and d_i in (14.16) are calculated by solving (14.17), and then matrix $Y_{N_w \times N_h}$ is transformed into $Z_{N_w \times N_h}$ based on cubic transformation using (14.2) to meet the moments of historical statistics.

Step 5 *Verification:* The moment (ε_m) and correlation (ε_c) errors are calculated using (14.19) and (14.20), respectively, to verify the eligibility of the generated scenarios, where $\varepsilon_m = 0.1$ and $\varepsilon_c = 0.1$ are the predefined thresholds in this study.

$$\varepsilon_m = \sum_{i=1}^{N_w} \left(|M_{i1}^G - M_{i1}^{N-T}| + \sum_{k=2}^{4} |M_{ik}^G - M_{ik}^{N-T}| / M_{ik}^{N-T} \right) \qquad (14.19)$$

$$\varepsilon_c = \sum_{i=1}^{N_h} \sqrt{\frac{2}{N_w(N_w - 1)} \sum_{i=1}^{N_w} \sum_{i=1}^{N_w} \left(R_{il}^G - R_{il}^T \right)^2} \qquad (14.20)$$

where M_{ik}^G is the moment of column vector i that is generated using HMM. R_{il}^G and R_{il}^T are the correlation matrix of the HMM-generated matrices and the target interaction matrix, respectively.

Step 6 *Inversion:* the normalized scenarios $Z_{N_w \times N_h}$ are inverted to meet the target moments $M_{i,k}^T$ by (14.21).

$$Z_i^T = \sqrt{M_{i,2}^N} \times Z_i + M_{i,1}^N. \qquad (14.21)$$

14.4 Performance Assessment and Simulation Results

In this work, the efficiency of our algorithmic solution is evaluated through simulation experiments based on a 53-bus distribution test feeder [11] that consists of three power substations, 101, 102, and 104, with a substation voltage of 1.0 p.u., as illustrated in Fig. 14.4. The nominal bus voltage and the network demand are 13.8 kV and $45.67 + j22.12$ MVA, respectively. The buses connected with the critical demand and DGs (a total of 9 BDGs and 6 NBDGs) are highlighted. Here, the demand profile is considered to be predictable and remains unchanged during the power supply restoration process. It is supposed that the capacities and daily output of DGs in the distribution system are almost the same, which is 5 MW and consists of a

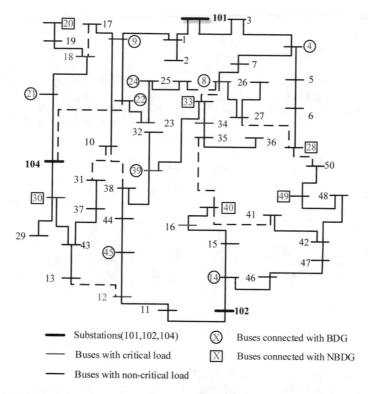

Fig. 14.4   53-bus test network topology and configurations

group of WTs. In Fig. 14.5, the scenarios of half-hourly changed DG outputs are generated using the HMM method based on the wind generation scaled down from the realistic statistics. The proposed arithmetic solution is simulated using MATLAB (version 8.3) and CPLEX (version 12.5). In this work, the solution is validated under two fault conditions: large-scale fault restoration with high and low wind power generation in comparison with existing solutions.

14.4.1   Experiment 1: Large-Scale Fault Restoration

The proposed supply restoration solution is first assessed for a multi-fault scenario (faults occur simultaneously in bus Nos. 1, 3, 11, and 14) at 21:00. Here, the time duration of power supply restoration is assumed to be 2 h consisting of 4 consecutive time slots (30 min per slot). The switcher operations and load restoration in individual time slots are shown in Fig. 14.6, and the detailed switcher operations can be found in Table 14.1.

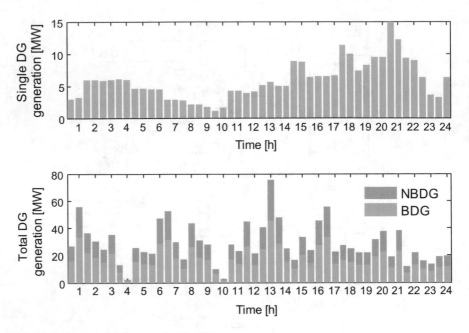

Fig. 14.5 Generated daily scenarios for WTs

With sufficient DG outputs, all NBDGs in fault sections can be started in time slot #1 step-by-step through the DG starting process, as presented in Table 14.1 (step 1–step 4). After restoring some loads in the DG starting process, most loads can be restored in the load restoration process. Even with a sharp decrease in DG outputs, the load restoration in time slot #2 is slightly lower than that in time slot #1 with the backup of feeders in failure-free sections. In time slot #3, where the outputs of DGs are close to those in time slot #2, only two switcher operations (one switcher closing and one switcher opening) are needed for topology reconfiguration to utilize the DG outputs. For the scenario with the lowest DG generation (i.e., time slot #4, 2.10 MW), a portion of the noncritical loads adjusts and goes back to unrestored, while most critical loads in fault sections are guaranteed to be supplied by topology reconfiguration with limited load restoration capacity of failure sections and DGs in fault sections.

Here, Table 14.2 presents the numerical results of power loss; power flow changes, e.g., the bus voltage drop and current increase that are defined in (14.22) and (14.23), respectively; and the unrestored power load in the four different time slots.

$$\Delta V_{\max} = \max\left\{ \frac{\left|V_i^{AR} - V_i^{BR}\right|}{V_i^{BR}} \right\}, i = 1, \ldots, N_{\text{normal}} \qquad (14.22)$$

$$\Delta I_{\max} = \max\left\{ \frac{\left|I_i^{AR} - I_i^{BR}\right|}{I_i^{AR}} \right\}, i = 1, \ldots, N_{\text{normal}} \qquad (14.23)$$

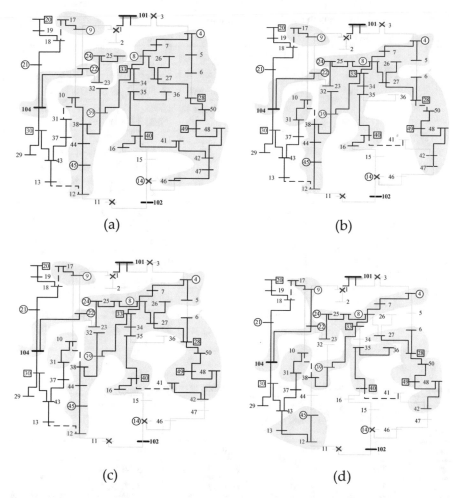

Fig. 14.6 The restored power supply in different time slots (considering DG generation dynamics). **a** 21:00–21:30; **b** 21:30–22:00; **c** 22:00–22:30; **d** 22:30–23:00

Table 14.2 shows that the power loss, influences on non-fault sections and unrestored loads vary with the DG scenarios. For the scenario with the highest DG generation (i.e., time slot #1, 4.06 MW), most loads can be supplied by DGs in islanding mode, which can minimize the influence on failure-free sections. Thus, the maximum values of bus voltage drop and current increase in time slot #1 are the smallest among all time slots. Though the number of unrestored buses is similar for time slots #1–#3, the influence on the failure-free sections in time slots #2–#3 is much higher than that in time slot #1 due to the limited DG outputs. In time slot #4, more loads return to unrestored status, causing relatively lower power losses and impact on the fault-free sections.

In detail, Fig. 14.7 presents the DG utilization during load restoration for each

Table 14.1 The restoration path and switcher operations

Time slot	DG scenario	Operations	Steps	Parallel switcher operations
#1 (21:00–21:30)	4.06 MW	DG restoration	Steps 1	39–33 (NBDG start)
			Steps 2	33–8; 8–27; 27–28 (NBDG start)
			Steps 3	28–50; 50–49 (NBDG start)
			Steps 4	49–48; 48–42; 42–41; 41–40 (NBDG start)
		Load restoration		12–45; 9–17; 40–16; 24–23; 33–34; 4–7; 38–39; 4–5; 38–10; 42–47; 40–35; 24–25; 27–26; 104–22; 38–44; 39–32; 45–44; 47–46; 5–6; 35–36; 7–8
#2 (21:30–22:00)	3.09 MW	Closing		6–28; 34–35; 10–31; 25–8
		Opening		6–5; 35–36; 46–47; 41–42; 10–38; 40–41
#3 (22:00:22:30)	3.00 MW	Closing		42–41
		Opening		42–47
#4 (22:30–23:00)	2.10 MW	Closing		35–36; 12–13
		Opening		8–27; 27–26; 4–5; 44–45; 44–38; 39–32; 33–34; 34–35; 40–16; 42–41

Table 14.2 Restoration performance for multiple faults at different buses

Time slot	Power loss (MW)	Maximum bus voltage drop	Maximum current increase	Unrestored buses
#1	0.0181	1.68×10^{-16}	1.08×10^{-15}	2,15
#2	0.0191	0.0179	0.0157	2,15,36,41,46
#3	0.0188	0.0179	0.0157	2,15,36,46,47
#4	0.0057	0.0095	0.0155	2,5,15,16,26,32,34,41,44,46,47

time slot. All NBDGs can be started in time slot #1. The variation trend of restored loads is approximately consistent with the variation trend of available DGs. Restored loads always remain high with sufficient DG outputs, as shown in time slots #1–#3 in Fig. 14.7. Owing to the limited capacity of feeders in the failure-free sections, some loads in the fault sections cannot be supplied, as presented in time slot #4 in Fig. 14.7.

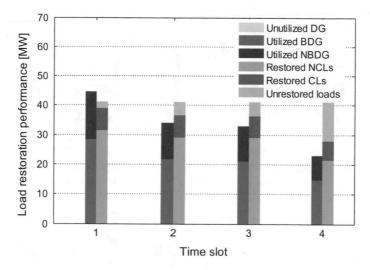

Fig. 14.7   Condition of DG utilization and load restoration

14.4.2   Experiment 2: Large-Scale Faults with Different DG Penetration Rates

The performance of the proposed solution can be further demonstrated for different DG penetration rates upon large-scale faults. Given the same DG placement and fault conditions (fault locations, fault duration, and fault time) as in experiment 1, the DG capacity is gradually increased from 4.5 MW to 6.0 MW. Thus, the comparison of load restoration performance is presented in Fig. 14.8.

As shown in Fig. 14.8a, with the lowest DG capacity in the distribution network, only one NBDG (NBDG at bus No. 36) is started, resulting in the lowest load restoration among the four DG penetration cases. However, even with low DG outputs, some loads can be resupplied by feeders in the failure-free sections, as shown in time slots #1–#3 in Fig. 14.8a, where the supplied loads are larger than the available DG generation. As shown in Fig. 14.8b–d, all NBDGs can be started in time slot #1 with sufficient BDG outputs. In addition, except for time slot #4 shown in Fig. 14.8b, the available DGs are always larger than the loads that need to be restored. However, the DGs may not be fully utilized in the restoration process due to the bus and branch constraints, and hence, some loads cannot be supplied in each slot. This implies that the appropriate capacity of DGs needs to be carefully considered to enhance network resilience in power distribution network planning to meet the need for NBDG start-up as well as load restoration.

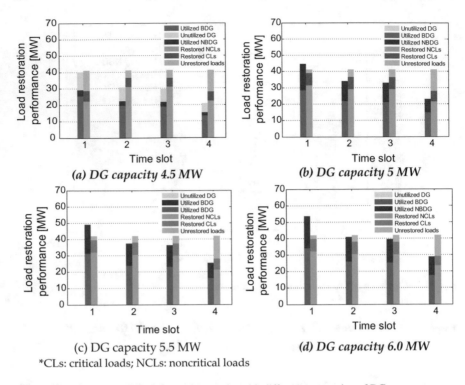

Fig. 14.8 Comparison of load restoration results with different penetration of DGs

14.4.3 Experiment 3: Large-Scale Faults with Different BDG/NBDG Proportions

The impact of the proportion of BDGs/NBDGs on load restoration under large-scale faults is studied in experiment 3. The fault locations, fault moment, and fault time are the same as in experiment 1. In comparative experiments, the total number of DGs in the fault section remains unchanged (11 DGs in the fault section), and the proportion of NBDGs/BDGs in the fault section is varied from 2:9 to 8:3. The comparative results of load restoration are presented in Table 14.3.

With the high DG generation output (4.06 MW) in time slot #1, all NBDGs can be started in the first three proportion cases, where the number of NBDGs is smaller or close to the number of BDGs. Owing to the limited capacity of BDGs, one NBDG (bus No. 4) stays unstarted and unutilized during the restoration process with 8 NBDGs and 3 BDGs in the fault section.

For the load restoration, all critical loads can be restored during time slots #1–#3, and only one critical load (bus No. 34) cannot be supplied in time slot #4 with low DG outputs and limited capacity of feeders in the failure-free sections. The recovery of noncritical loads decreases with the DG outputs in the four cases. In addition, the loads restored in the same time slot for the four different NBDG/BDG

Table 14.3 Comparison results of load restoration and NBDG starting

Time slot	NBDGs/BDGs = 2:9			NBDGs/BDGs = 4:7			NBDGs/BDGs = 6:5			NBDGs/BDGs = 8:3		
	Restored CL	Restored NCL	Unstarted NBDGs	Restored CL	Restored NCL	Unstarted NBDGs	Restored CL	Restored NCL	Unstarted NBDGs	Restored CL	Restored NCL	Unstarted NBDGs
1	6.44	25.78	0	6.44	27.23	0	6.44	27.23	0	6.44	24.39	1
2	6.44	25.16	0	6.44	25.16	0	6.44	25.16	0	6.44	23.56	1
3	6.44	25.09	0	6.44	25.09	0	6.44	25.09	0	6.44	22.45	1
4	5.61	18.57	0	5.61	18.57	0	5.61	19.40	0	5.61	20.44	1

proportions are close but not the same in the first three cases, which results from the fact that the total DG outputs are the same (all NBDGs are started) and the loads can be fully supplied by different DGs or feeders in the failure-free sections considering the safety constraints. However, the restored noncritical loads in the last case (NBDGs/BDGs = 8/3) are lower than those in time slots #1–#3, with one NBDG unutilized. Thus, this observation indicates that it is vital to determine the appropriate locations (near critical loads) and capacity of DGs as well as the NBDG/BDG proportion to strengthen the self-repair ability of distribution networks.

In addition, the influences on failure-free sections may change with the variation in the BDG/NBDG proportion, as shown in Table 14.4. By comparing the results of the first three cases, the power losses in the same time slot are close to each other, but the influence on the failure-free sections (including voltage drop and current increase) increases with the proportion of NBDGs/BDGs during time slot #1 but remains in the other three-time slots, which may result from the final power supply path in time slot #1 being influenced by the path of starting NBDGs. With lower restored loads, the restoration process is carried out with lower power loss and influence on failure-free sections.

14.5   Conclusions and Remarks

In this study, an algorithmic solution of power supply restoration was developed to enhance the resilience of power distribution systems under large-scale faults. The proposed solution effectively integrates DG-based local restoration and restoration based on switcher operations considering the intermittency and randomness intro-duced by distributed renewable generation. The restoration strategy was designed to carry out power restoration to maximize the restored loads and minimize the power loss and adverse impacts on network power flows. The algorithmic solution was evaluated using a 53-bus power distribution feeder penetrated with wind turbines, and the numerical results clearly demonstrated its effectiveness.

In future work, additional research can be carried out in the following aspects. The performance of the suggested solution can be further exploited and assessed in the presence of different types of distributed green energy generation and installed diesel generators. In addition, because the appropriate capacity and locations of DG instal-lations can affect network resilience during power blackouts, additional research efforts need to be made from the aspect of distribution network planning, e.g., opti-mal siting and capacity configuration of DGs, to improve the network resilience and self-repair capability.

Table 14.4 Comparison of the influence of restoration under faults at multiple buses

Time slot	NBDGs/BDGs = 2:9			NBDGs/BDGs = 4:7			NBDGs/BDGs = 6:5			NBDGs/BDGs = 8:3		
	Power loss (MW)	Maximum bus voltage drop	Maximum current increase	Power loss (MW)	Maximum bus voltage drop	Maximum current increase	Power loss (MW)	Maximum bus voltage drop	Maximum current increase	Power loss (MW)	Maximum bus voltage drop	Maximum current increase
1	0.016	1.2×10^{-16}	1.1×10^{-15}	0.018	1.7×10^{-16}	1.1×10^{-15}	0.013	0.0080	0.0073	0.013	0	0
2	0.019	0.0179	0.0157	0.019	0.0179	0.0157	0.019	0.0179	0.0157	0.006	0.0088	0.0080
3	0.019	0.0179	0.0157	0.019	0.0179	0.0157	0.019	0.0179	0.0157	0.020	0.0179	0.0157
4	0.006	0.0095	0.0155	0.006	0.0095	0.0155	0.006	0.0095	0.0155	0.003	0.0079	0.0069

References

1. A. Ehsan, Q. Yang, Optimal integration and planning of renewable distributed generation in the power distribution networks: a review of analytical techniques. Appl. Energy **210**, 44–59 ()2018
2. L. Yutian, F. Rui, V.J. Terzija, Power system restoration: a literature review from 2006 to 2016. J. Modern Power Syst. Clean Energy **4**(3), 332–341 (2016)
3. F. Yang, V. Donde, Z. Wang, J. Stoupis, D. Lubkeman, *Load restoration for feeder automation in electric power distribution systems*. Google Patents (2015)
4. V.N. Coelho, M.W. Cohen, I.M. Coelho, N. Liu, F.G. Guimarães, Multi-agent systems applied for energy systems integration: State-of-the-art applications and trends in microgrids. Appl. Energy **187**, 820–832 (2017)
5. M. Abdullah, A. Agalgaonkar, K. Muttaqi, Assessment of energy supply and continuity of service in distribution network with renewable distributed generation. Appl. Energy **113**, 1015–1026 (2014)
6. T. Ding, Y. Lin, Z. Bie, C. Chen, A resilient microgrid formation strategy for load restoration considering master-slave distributed generators and topology reconfiguration. Appl. Energy **199**, 205–216 (2017)
7. X. Fang, Q. Yang, J. Wang, W. Yan, Coordinated dispatch in multiple cooperative autonomous islanded microgrids. Appl. Energy **162**, 40–48 (2016)
8. A. Elmitwally, M. Elsaid, M. Elgamal, Z. Chen, A fuzzy-multiagent service restoration scheme for distribution system with distributed generation. IEEE Trans. Sustain. Energy **6**(3), 810–821 (2015)
9. Q. Yang, L. Jiang, A. Ehsan, Y. Gao, S. Guo, Robust power supply restoration for self-healing active distribution networks considering the availability of distributed generation. Energies **11**(1), 210 (2018)
10. X. Fang, Q. Yang, Cooperative energy dispatch for multiple autonomous microgrids with distributed renewable sources and storages, in *Smart Power Distribution Systems* (Elsevier, 2019), pp. 127–160
11. Z. Wang, J. Wang, Self-healing resilient distribution systems based on sectionalization into microgrids. IEEE Trans. Power Syst. **30**(6), 3139–3149 (2015)
12. R. Romero, J.F. Franco, F.B. Leão, M.J. Rider, E.S. de Souza, A new mathematical model for the restoration problem in balanced radial distribution systems. IEEE Trans. Power Syst. **31**(2), 1259–1268 (2016)
13. J. Li, X.-Y. Ma, C.-C. Liu, K.P. Schneider, Distribution system restoration with microgrids using spanning tree search. IEEE Trans. Power Syst. **29**(6), 3021–3029 (2014)
14. A. Botea, J. Rintanen, D. Banerjee, Optimal reconfiguration for supply restoration with informed A * Search. IEEE Trans. Smart Grid **3**(2), 583–593 (2012)
15. B. Zhao, X. Dong, J. Bornemann, Service restoration for a renewable-powered microgrid in unscheduled island mode. IEEE Trans. Smart Grid **6**(3), 1128–1136 (2015)
16. C. Moreira, F. Resende, J.P. Lopes, Using low voltage microgrids for service restoration. IEEE Trans. Power Syst. **22**(1), 395–403 (2007)
17. C. Yuan, M.S. Illindala, A.S. Khalsa, Modified Viterbi algorithm based distribution system restoration strategy for grid resiliency. IEEE Trans. Power Deliv. **32**(1), 310–319 (2017)
18. A. Zidan, E.F. El-Saadany, A cooperative multiagent framework for self-healing mechanisms in distribution systems. IEEE Trans. Smart Grid **3**(3), 1525–1539 (2012)
19. Y. Kumar, B. Das, J. Sharma, Multiobjective, multiconstraint service restoration of electric power distribution system with priority customers. IEEE Trans. Power Deliv. **23**(1), 261–270 (2008)
20. M.R. Kleinberg, K. Miu, H.-D. Chiang, Improving service restoration of power distribution systems through load curtailment of in-service customers. IEEE Trans. Power Syst. **26**(3), 1110–1117

21. K. Høyland, M. Kaut, S.W. Wallace, A heuristic for moment-matching scenario generation. Comput. Optim. Appl. **24**(2–3), 169–185 (2003)
22. A. Ehsan, Q. Yang, M.J.I.G. Cheng, A scenario-based robust investment planning model for multi-type distributed generation under uncertainties. Transm. Distrib. **12**(20), 4426–4434 (2018)

Chapter 15
Internet Thinking for Layered Energy Infrastructure

Haochen Hua, Chuantong Hao, and Yuchao Qin

Abstract Huge shifts in the structure and functionality are brewing in the sector of power and energy with the wide deployment of renewable energy and rapid development of electricity market. Growing demand for intelligent appliances and autonomous devices poses a great challenge to the existing power-energy systems. With inspirations from the Internet, in this chapter, a layered infrastructure for the future Energy Internet system is introduced. In the meantime, the functionalities and typical application scenarios of each layer in this structure are analyzed and discussed in depth.

15.1 Internet Thinking and Technology in Energy Infrastructure

15.1.1 Introduction of Energy System Infrastructure

Electricity is the premise of the energy system that covers all kinds of primary energy. It is electricity that closely combines various scattered and systematic energy systems. Therefore, the new generation of energy systems includes a type of network system that centres on electricity and uses the power grid as its backbone. In this system, the production, transmission, usage, storage and conversion devices, as well as the information, communication, and control protection devices of primary and secondary energy sources, are connected directly or indirectly [1]. Based on the characteristics and applications of existing power grid construction, especially the unparalleled advantages in energy transmission and conversion efficiency, the power grid remains a central infrastructure in the energy and utility industry. In the future, the

H. Hua · Y. Qin (✉)
Beijing National Research Center for Information Science and Technology, Tsinghua University, Beijing 100084, People's Republic of China
e-mail: qinyc17@mails.tsinghua.edu.cn

C. Hao
School of Engineering, University of Edinburgh, Edinburgh EH9 3DW, UK

© Springer Nature Switzerland AG 2020
A. F. Zobaa and J. Cao (eds.), *Energy Internet*,
https://doi.org/10.1007/978-3-030-45453-1_15

main energy transmission infrastructure will inevitably be the power grid, in addition to a variety of energy storage devices, power electronic control devices, charging piles and other renewable energy infrastructures. Broadly speaking, the infrastructures in energy systems cover numerous equipment, machines, systems and energy transport pipelines, covering the chain of energy generation, energy transmission, and energy consumption.

15.1.2 The Internet-Styled Next Generation of Energy Systems

Constructed with the ideas of the Internet, the Energy Internet is a new type of comprehensive energy system that realizes the fusion of information and energy in a wide area. It takes large-scale power utilities as the backbone and renewable energy sources, distributed microgrids, and intelligent communities as a Local Area Network (LAN). An information-energy fusion architecture with open and peer-to-peer features truly realizes two-way and on-demand transmission and dynamic balance of energy, maximizing the access of renewable energy [2]. The Energy Internet adopts a decentralized bottom-up management mode to realize the effective combination of traditional centralized energy systems and distributed energy systems. The Energy Internet is the specific realization form of new generation energy systems. Compared with traditional energy systems, the main characteristics of these new generation systems include a multi-energy structure that allows for renewable energy with priority; centralized, distributed, coordinated and reliable energy production and supply modes; the synthetic utilization of all kinds of energy sources; the efficient use of energy with the interaction of supply and demand; and platform-oriented, commercial and user-oriented services for the whole society.

In view of the electric utility system as the heart of the next-generation energy system, the "Internet thinking" towards the traditional power system is mainly reflected in the following aspects:

There are two main points in the understanding of "Internet thinking" in the way of social production: decentralization of the production factor configuration and flattening of the production management mode. Based on the characteristics of the Internet, which include openness, equality, collaboration and sharing, the main form of the production factor configuration of various systems is decentralized and distributed, and the management of enterprises will also move from a traditional multi-level style to a flat and networked mode.

The power supply structure of all kinds of decentralized primary energy generation in a traditional power system (backbone power supply) is connected to thousands of families through a large-scale interconnected transmission and distribution network. Power users do not need to know which power plants are used for power generation, and they can simply take electricity from the grid network according to their requirements, which has typical Internet features: open and shared. This is the idea that has been realized through years of investigation of computer information systems (network computing, Internet and cloud computing, etc.). The traditional power system does not support the mutual conversion and complementarity of multiple primary and secondary energy sources and cannot support the access of a high proportion of distributed renewable energy sources. The efficiency of comprehensive energy utilization is limited. The centralized and unified management, scheduling and control system of the traditional power system is not adapted to the development trend of building bottom-up energy infrastructures in the future.

There are three levels using Internet thinking to construct and transform the power system. The first level is multi-energy interconnection and interoperability, such that the integrated utilization efficiency of primary and secondary energy can be improved. The second level refers to learning from the Internet and its extended network technology, realizing the support of plug-and-play, energy routers, big data, and cloud computing for energy systems, especially for distributed energy, microgrids, and customers in energy production, consumption, market management and services. The third level includes new formats and business modes of energy production, consumption and services. The Energy Internet cannot completely abandon the existing power grid. In the structure of the Energy Internet, we should consider integrating micro-energy networks into the large-scale power grid and constructing a new Internet-styled power architecture.

With the renewable energy revolution, the power grid is also facing a new mission. For example, the power grid will become a large-scale renewable energy transmission and distribution network. Furthermore, the power grid should be integrated with distributed generation, energy storage devices and comprehensive energy-efficient utilization systems to become a flexible and efficient energy network. In addition, the power grid must have high power supply reliability and basically eliminate the risk of large-scale power outages. It should be widely integrated with information communication systems, such that a comprehensive service system of electric power, energy and information can be built [3–5].

15.1.3   Typical Scenario Analysis

Drawing on the thinking and architecture of the Internet, we consider a typical Energy Internet scenario based on energy routers. As the core element of the Energy Internet, energy routers are designed to integrate power electronic control devices, energy storage and data centres. The "object" in the energy infrastructure is connected through a sensor, and the user is connected through a mobile terminal. Through an energy router, the energy layer and information layer can be effectively fused. On the one hand, an energy router provides a power electronic interface for information-based energy control, which realizes the routing of energy in the networked model and the fine control of energy flow, e.g. power quality control. In addition, an energy router is also the main channel for information collection based on energy infrastructure. Taking the data centre as the core and by using the collection devices and sensors from customers and electrical devices, the energy router obtains relevant information of the energy infrastructure, users, running status, etc., and sends it back to the data centre to realize information interconnection within the Energy Internet.

Data centres and energy routers need to be integrated; transmission lines and optical fibre communications need to be integrated; and sensors and local/wide area networks need to be integrated. A data centre can be powered by energy routers or even the entire Energy Internet. Due to the integration of communication infrastructure and power infrastructure, the cable-laying cost of a data centre can be saved. Theoretically, the operating state of power equipment can be better monitored by means of communication network and data centre facilities. Through the combination of power network topology and communication network topology, a fault can be better located, repair work can be performed in time, and system security and efficiency can be improved. Each energy router should be equipped with a certain capacity of energy storage equipment that can be used for backup power in the data centre and smooth transmission of energy. The communication link can utilize power line communication or optical fibre communication, both of which can be integrated with transmission lines. Based on existing facilities, the process of integration is an information process, a process of embedding the Internet into the energy network.

15.1.4　A Summary of the Infrastructure Level

The transformation of traditional power systems by Internet thinking is a key step in the construction of a new generation of energy systems. It is necessary to learn from Internet thinking and technology to further achieve the broad integration of energy. This approach is embodied in the following aspects [6–8]:

(1) Openness

For the Internet, the realization of access to information at anytime and anywhere mainly depends on the open architecture. The new generation of Internet-based energy systems requires plug-and-play features for renewable energy, energy storage equipment and loads to achieve the access and acquisition of energy anywhere and anytime.

(2) Interconnection

Based on a concise and standard protocol, the Energy Internet achieves the interconnection of information, establishes energy communication protocols to achieve interconnection of multiple energy sources, maximally develops and utilizes renewable energy, and improves the overall utilization efficiency of primary and secondary energy resources.

(3) Equal

The Internet architecture ensures its security and stability. The overall reliability is ensured by means of large amounts of redundancy. In addition, the dynamic backup of equipment and lines can be realized by decentralized routing to maintain a certain utilization rate. Different from the top-down structure of the traditional power grid, the formation of the Energy Internet is peer-to-peer interconnection among energy-autonomous units. The connections between any units must be based on decentralized routing. The transmission of energy should be the result of multiple routings that are decoupled, thus avoiding a series of safety and stability problems. Additionally, the transmission routing path can be dynamically alternated.

(4) Sharing

Distribution, decentralization and sharing are the main characteristics of the Energy Internet. Using the information-sharing mechanisms of social networks in Internet applications, the energy exchange and routing between LANs in the Energy Internet are real-time dynamic, using decentralized optimal schedules to achieve global energy management and optimization.

15.2 Internet Thinking and Technology in Cyber-Physical Systems

15.2.1 Introduction of Information-Physics Fusion

The Energy Internet, as a typical Cyber-Physical System (CPS), is the inevitable consequence of integrating the concepts and structures in the information field into the energy system. Due to the characteristics of the Energy Internet, its CPS has put forward new capability requirements in different aspects, including sensing, communication, computation and control; see, e.g. [9, 10].

(1) Comprehensive situational awareness. The Energy Internet CPS involves various facilities and terminals. For realization of information management and precise control of this system, comprehensive and accurate situation awareness and prediction ability are required. Therefore, it is necessary to design an integrated intelligent monitoring device and to build an advanced metering infrastructure for the Energy Internet.

(2) Highly reliable and secure communication capability. The Energy Internet CPS has a complex communication structure and multi-scale dynamics. Such systems are required to have highly reliable and secure communication capabilities. Proper communication architectures and protocols should be analyzed and designed for the Energy Internet to realize the integration and security protection of heterogeneous communication systems.

(3) Analysis and computing ability based on big data processing. With the maturity of the Energy Internet application mode, the scale of data will also increase along with the data types, measurement terminals, frequency of adoption, etc. We need to build an Energy Internet big data processing and analysis platform to expand the application of big data.

(4) Distributed cooperative control capability. With the extensive access of distributed energy sources, the Energy Internet CPS will contain numerous heterogeneous and autonomous energy LAN systems. This requires the Energy Internet system to have strong distributed coordination control capability. Therefore, it is necessary to build a centralized and distributed system control framework.

15.2.2 Interaction Between Energy and Information in the Energy Internet

In the Energy Internet, energy interconnection and sharing should be realized like information sharing in the Internet. The energy system architecture based on a CPS provides a new way to solve these problems. That is, the energy is discretized (or fragmented) in physics so that the discrete energy can be scheduled. Next, based on the discretization of energy, the energy system and the load system are integrated into the information system, and then network management and control shall be realized through network technology.

The future Energy Internet has a similar interconnection structure with the information infrastructure. Some scholars have proposed that the information infrastructure should be integrated into the Energy Internet as a whole to achieve the fusion of energy and information infrastructures. Existing concepts and models of the Energy Internet and energy routers are proposed from the perspectives of power electronic technology, energy storage, information acquisition and decision support, multiple forms of energy conversion and direct current transmission. In this note, we focus on the idea of the Internet to integrate information infrastructure (e.g. data centres and cloud computing platforms) and energy infrastructure (e.g. energy storage, power electronic control, and the power grid).

Decentralized cooperative scheduling and control of the Energy Internet needs the support of real-time dynamic information collection, transmission, analysis and decision making. It mainly includes power information acquisition and control systems, power quality monitoring and analysis systems, power grid energy management systems, user-side energy management systems, etc. Load information inaccuracy and parameter uncertainty are always important problems in power system simulation analysis and energy management. The close combination of the energy system and information infrastructure can provide the most powerful technical support for real-time dynamic collection and processing of massive load information. Intelligent information processing and decision support capabilities can achieve power quality control and other advanced energy management functions. This approach is able to flexibly control the energy exchange between renewable energy generation and the power grid according to the conditions of energy demand, market information and operation constraints. In the era of the Energy Internet, energy flow and information flow should be reconstructed, the bi-directional flows of energy and information would be realized, and a high degree of integration with business flow could be achieved.

15.2.3   Typical Energy Internet Information Service Mode

The Energy Internet information system forms an end-to-end data connection, realizing data acquisition, transmission, storage, analysis and processing. The typical Energy Internet information service mode is mainly big data analysis and optimization service. Based on the analysis and mining of data, the operation, planning, construction and service modes of the power grid can be changed, and the decision management of energy enterprises can be effectively supported. A typical case is a smart city project in Austin, Texas. This project constructed a comprehensive service platform for energy data, including intelligent electrical appliances; electric vehicles; solar photovoltaics; and other types of detailed electrical, gas and water data.

In the era of the Energy Internet, data mining and analysis capabilities are essential. Based on big data analysis, energy and electricity companies can achieve in-depth insights into consumers, providing accurate services, obtaining scientific decision-making capacity, maximizing asset efficiency, and minimizing pollution and greenhouse gas emissions. Big data records the power curves of all electrical appliances and the habits of consumers, helping enterprises to manage each electrical appliance meticulously. Through optimization, the maximum output of each power station, even that of each wind turbine generator, can be realized, and the maximum return on investment can be realized. In addition, the use of big data also makes accurate weather forecasting and electricity demand forecasting possible. Relying on big data forecasts, power generation enterprises and power grid management enterprises can better manage and dispatch power to realize a more rational utilization of resources. In fact, the massive energy and electricity consumption data generated by the Energy Internet also have unimaginable commercial value, but it needs to be realized through the applications of other industries. To illustrate, for families, turning on the lights often means going home, whereas turning off the lights means going out. If it is possible to use a power data collector to monitor the light-switching time of each family, it is potentially feasible to provide a travel plan and a home entertainment scheme for the user. Similar applications of energy and power big data in other fields also have numerous possibilities.

15.2.4 Summary of the Information Level

Using the idea of full interaction of network information, we can realize the intelligent interaction of energy flow in the Energy Internet. This is not only the interaction among the supply, output and use of various kinds of energy, such as electricity, thermoelectricity, gas and refrigeration, but also the cross-boundary interaction of all kinds of energy sources. The relationship among production, transmission and consumption of energy would be rebuilt, as the Internet has caused the subversion of information release, transmission and reception. Based on the mass information processing mechanism of the Internet, as well as the types, characteristics and analysis requirements of big data for the Energy Internet, key technologies such as big data modelling, storage and analysis need to be utilized comprehensively to construct a big data processing and analysis platform for the Energy Internet.

15.3 Internet Thinking and Applications in the Energy Internet

15.3.1 Future Typical Energy Internet Applications

Typical Energy Internet applications include the following major sections: the family Energy Internet level (smart facility, electric vehicle, etc.), community Energy Internet level (smart buildings, microgrids), and regional Energy Internet level (distribution network, distributed generation, demand-side management, etc.). Energy management at all levels needs to reflect the dynamic deployment of electricity supply and demand. Virtual power plants (VPPs), as a new form of demand-side management, would emerge in large numbers. From a micro perspective, a VPP can be considered as an integrated and coordinated optimization method of controllable loads, electric vehicles and other distributed energy sources with advanced information and communication technology. In this sense, VPP is regarded as a power supply coordination management system for a special power plant to participate in the power market and power grid operation. From a macro perspective, a VPP acts as a traditional power plant in power systems and markets.

Trading is an indispensable element of the Internet, and it is also the core of the Energy Internet. Energy trading electronic commerce platforms will provide open and shared trading environments for Energy Internet participants, supporting various types of transactions, such as energy transactions and carbon trading. In the Internet era, the traditional modes of purchase and sale will be changed, and electronic commerce platforms based on electricity fees will drive the integration of energy, finance and services to realize the transformation of power grid service modes, marketing modes and even management strategies.

15.3.2   Characteristics of Internet-Based Energy Applications

The specific embodiments of applying Internet thinking and technology to the field of electric power include

(1) Information transparency: Breaking the asymmetric information structure and doing everything we can to ensure the transparency of the information.
(2) Open and shared: Energy Internet participants can conduct various types of transactions in an open and shared environment.
(3) Equality and win-win: All energy units can participate in an energy market, ancillary service market, and carbon trading market regardless of size.
(4) Self-adaption: The Internet has a self-regulation mechanism.
(5) Democratization: Internet-based energy is more democratic, and energy is no longer monopolized by power companies. People can generate electricity through renewable energy equipment and earn profit by selling electricity to the grid.

15.3.3   Typical Scenarios of the Energy Internet

A typical scenario of the Energy Internet is composed of energy LANs and the main power grid. Within the Energy Internet, an energy microgrid is the basic unit in the Energy Internet power supply model. The microgrid includes home users, commercial users, factories and all kinds of electricity suppliers. The power supply of the microgrid is supported by distributed energy and the utility grid. For example, power generation is mainly provided by distributed energy sources and is supplemented by the utility grid. The microgrid consists of basic distributed energy units, such as wind turbine generators, photovoltaic panels, energy storage devices and electric vehicles (the basic elements of the microgrid), and information infrastructure construction, such as optical fibre communications, mobile communications, sensors and data centres. The information layer collects power consumption information, generation information and environmental information, referring to historical data. According to the current user demand, power supply and dispatching decisions are made.

The main difference between the Energy Internet and the microgrid is that the open and peer-to-peer information and energy exchange among all the elements in the Energy Internet is not available in the microgrid. At the information level, through the added information layer, the microgrid itself is no longer a simple energy supply link and becomes an intelligent energy supply entity; that is, the coordination and balance of various sources of energy is realized. Moreover, a more important point is that the existence of the information layer makes the microgrid no longer an independent entity. The information layer has established information channels for interworking between microgrids, making networking possible for microgrids and enabling energy to flow between microgrids.

Compared with the traditional top-down transmission and distribution modes, the Energy Internet in a certain area has the functions of flexible access of distributed energy and demand-side response with user interaction. Additionally, the Energy Internet largely shields the dynamics of power generation and consumption. It enhances the security and reliability of the large-scale power grid by using large redundancies and improves the overall utilization efficiency of the large-scale power grid.

Democratization of energy has forced power companies to reconsider their business practices. Ten years ago, the four largest power generation groups in Germany, EON, RWE, EnBW, and Vattenfall, provided almost all power for Germany. Currently, these companies no longer have monopolies on power generation. In recent years, farmers, urban residents and small- and medium-sized enterprises have established various power generation cooperatives in Germany. In fact, all of these cooperatives are very successful in obtaining low-interest loans from banks and installing solar, wind, or other renewable energy equipment. Banks are also willing to provide these loans. They are sure that the principal and interest can be recovered because, through an Internet price subsidy mechanism, the power cooperatives are earning a profit from the power grid for selling green electricity back to the power grid at a higher price than the market. Today, the vast majority of German green electricity is provided by small power generation cooperatives. By contrast, the four big power companies provide less than seven percent of green electricity, while German industry is leading the third industrial revolution. Although these traditional power companies have proven to be quite successful in the use of traditional fossil and nuclear energy to produce cheap electricity, they cannot effectively compete with local power cooperatives. The horizontal operation of these power cooperatives enables them to better manage the use of electricity from a cooperative platform composed of thousands of small participants. The widespread use of the Internet is accompanied by a huge shift in energy from centralized to distributed generation. The larger electricity and utility companies must adjust to one reality; that is, in the long run, the scale of power companies' profits from traditional energy will be much lower than what we have seen previously. Therefore, we must make use of the Internet to reconstruct and enlarge the tiny gains that have been achieved.

15.3.4 Summary of the Application Level

The end users of the information Internet or Energy Internet are human beings. The reason why the information Internet is widely popularized is that it provides a new platform for information exchange and cooperation for people, which is in line with the needs of human nature itself. As a more widely used network, the Energy Internet should pay more attention to human factors. It needs to take full consideration of the power generation of distributed energy sources, the feeling of security generated by energy routers, the happiness of sharing energy, and the sense of social belonging of energy cooperation, finally building a network that takes user experience as a priority.

15.4   Internet Thinking in the Business Model of the Energy Internet

15.4.1   Introduction of a New Business Model Under Internet Thinking

The business model under Internet thinking uses the spirit of the Internet (equality, openness, collaboration and sharing) to yield user value and provide innovative services, thus subverting and reconstructing the whole business value chain. At present, there are six main types of business models:

(1)   Tools + Community + Crowdsourcing

With the development of the Internet, information exchange is becoming more convenient. Like-minded people are more likely to get together and form communities. The Internet has brought scattered needs on a platform, forming new common needs.

The Energy Internet is an energy network defined by crowdsourcing from the very beginning. Unlike traditional centralized power generation networks, in the Energy Internet, each distributed generation project and microgrid are a crowdsourcing unit. These units form an intelligent, robust and green power network through collaboration and complementarity. Behind this network are large numbers of enterprises and family participants, involving power plant development, power station operation, microgrid operation, management platform, demand-side response, etc. In contrast to the traditional power network, every participant in the Energy Internet has an equal role, and cooperation will replace the dispatching commands of the traditional power grid. In addition, in the era of the Energy Internet, individuals will be able to participate in energy system management and investment for the first time. Individuals can not only build distributed power station projects through Internet finance but also participate in demand-side management as an important body of the Energy Internet and become the most important terminal cells of the power network.

(2)   Long-tail Business Model

The long-tail concept describes the transformation of the media industry from selling a few products to a large number of users to selling a large number of niche products. Although each niche product produces relatively small sales in relative terms, the total sales volume of the niche products can be comparable to that of traditional sales models for a large number of users to sell a few products. The core of mass customization through customer to business is "more styles with fewer volumes". Thus, the long-tail model requires low inventory costs and strong platforms and makes it easier for interested buyers to acquire niche products.

(3)   Transboundary Business Model

The subversion of the Internet to traditional industry is essentially the use of high efficiency to integrate low efficiency, redistribute the core elements of traditional industry and restructure the production relationship to improve the overall system efficiency. By reducing the unnecessary losses of all channels and the links that products need to experience from production to users, an Internet enterprise can increase efficiency and reduce costs. Therefore, for Internet companies, as long as the links of low efficiency or high profit in the traditional industry value chain are seized, by using Internet tools and Internet thinking and reconstructing a business value chain, there is a chance to succeed.

Transboundary operation within the energy industry is not uncommon. In China, a large number of traditional manufacturing enterprises are evolving into comprehensive energy supply enterprises and have even transformed into energy service and management enterprises. The convergence of the energy industry with other industries will emerge. For example, Internet companies will become power selling companies; communication equipment enterprises are engaged in operation and maintenance for photovoltaic power generators; the everyday operation of microgrid business by electric vehicle companies has become a reality; and even telecom operators may become important players in the energy industry. There will be more similar cross-border integrations in the future. If energy and power enterprises cannot take the initiative to participate in the process of integration, it is likely that they will be passively integrated by the dominant enterprises in other industries. In addition, energy and power enterprises cannot be confined to their own circles. Entrepreneurs with spare efforts and ideals should also jump out of the energy sector to see the world based on customers or products.

(4)   Free Business Model

The Internet industry never raises price wars. The transformation of traditional enterprises to Internet enterprises requires a deep understanding of the essence of the business logic of "free". The "Internet +" era is an era of "information surplus" and an era of "attention scarcity". Determining how to obtain "limited attention" in the "infinite information" becomes the core proposition of the "Internet +" era. Lack of attention has led many Internet entrepreneurs to try their best to compete for attention resources. Nevertheless, the most important aspect of Internet products is network flow. With network flow, we can build our own business mode on this basis. Thus, the Internet economy is based on attracting public attention to create value and then turn it into profit.

Many Internet companies attract many users with free and good products and then build business models on the basis of new products or services to different users. The common way that the Internet subverts traditional enterprises is by changing the areas that traditional enterprises can make money into free to completely take away the customer base of traditional enterprises and turn them into flow and then make use of an extended value chain or value-added services to achieve profits.

If there is a business model that can dominate the future market, it can also crush the current market. That is the free mode. In the book "Free: The Past and the Future of a Radical Price" by Chris Anderson, the spiritual leader of the information age, a completely free business model based on core services is concluded as follows. The first is direct cross-subsidy; the second is the third-party market; the third is roughly free; and the fourth is purely free. Strictly speaking, free can be used as a marketing tool, but it is by no means a business mode. In Energy Internet applications, traffic thinking can be implemented in three ways. The first is that basic services are free, and value-added services are charged. Second, short-term services are free, and long-term services are charged. The third is free for users, whereas the third parties pay. To illustrate, a PV operation and maintenance enterprise can provide free data monitoring services to power stations while selling data reports to banks with power plant financing and transaction demands or financial institutions such as insurance companies.

(5)   O2O Business Model

O2O is short for online to offline. The location information of the mobile Internet brings a brand-new opportunity. This opportunity is O2O. O2O, in a narrow sense, is the business mode of online trading and offline experience consumption and mainly contains two scenarios. First, in online to offline, users purchase or book services online and then enjoy the services offline. At present, there are more examples of this type. Second, in offline to online, users experience and select products through offline stores and then buy goods through online purchase orders. The broad sense of O2O is the integration of Internet thinking and traditional industries. The development of O2O in the future will break through the boundaries between online and offline and achieve deep integration between online and offline. The core of this model is based on equality, openness, interaction, iteration and sharing, using high-efficiency and low-cost Internet information technologies to transform inefficient links in traditional industrial chains.

(6)   Platform Business Model

The world of the Internet is boundless, and the market is the whole country and even the whole world. The core of the platform-based business model is to create enough large platforms where products are more diversified and more attention is paid to user experience and closed-loop design of products.

The biggest platform in the energy industry is the power grid. However, due to the public utility nature of the grid, the platform thinking of the power grid super platform is weaker. However, it also provides opportunities to other platforms, for example, photovoltaic power station operation and maintenance platforms and intelligent building energy management platforms. In the future, the competition in the Energy Internet industry must be the competition between platforms and even the ecosphere because a single platform does not have systematic competitiveness.

However, when traditional energy companies want to transform into Internet companies or new Energy Internet companies, if they do not have an eco-platform, it is necessary to think about how to use the existing platform. Not all companies have the capability to become an Energy Internet platform, and the number of platforms that market space can accommodate is not large. In a symbiotic and win-win open platform, excellent enterprises in the subdivision area can still make considerable progress. A real Energy Internet company must not be fragmented and monopolized.

15.4.2   The Embodiment of Internet Thinking in the Business Mode of the Energy Internet

The Energy Internet under the Internet thinking mode will eventually move towards the consumer side, such as smart homes, smart communities, electric vehicles and family energy management. The Energy Internet entering the consumer side will have greater imagination and innovative business models. In the process of transforming the traditional power grid system into the Energy Internet system, a number of interesting business models will emerge. In the future, power companies will gradually evolve from the investment and operation management enterprises of transmission and distribution assets to information service enterprises, and many new service patterns will emerge.

Power grid enterprises can cooperate with power generation companies to provide energy management services to downstream electricity consumers and provide energy packaging services according to their energy use characteristics. Power grid enterprises are also able to identify high-quality electricity consumers based on big data and recommend high-quality electric consumers to power generation companies to achieve satisfactory sales agreements between the two sides. Future exchange can provide more and better diversified electricity futures to financial investors and power consumers according to the large historical data provided by power grid enterprises. Power grid enterprises will be disaster warning centres. When equipment is short circuited, the power grid will be informed the first time and notify users or the fire brigade in time.

With copious amounts of user demands and smart terminal access requirements, from the perspective of big data in the Internet, the operation of information and data will become a core competitiveness of electric power enterprises. The future business model will be transformed into a diversified energy service provider close to power users. Data mining and analysis capabilities are essential. Through the data collection and data analysis of power users, the habits of energy use of different power consumption groups can be analyzed. Then, users can be provided with rich product solutions regarding accurate energy supply, power demand-side management and low-carbon energy savings. Energy and power companies can achieve in-depth insights into consumers, provide accurate services and marketing, obtain scientific management decision-making capacity, maximize asset efficiency, and minimize pollution and greenhouse gas emissions.

15.4.3   A Summary of the Business Model

The Energy Internet ecosystem under the Internet thinking mode supports energy exchange, energy information sharing and energy value-added services; provides a platform for the free choice of both suppliers and users; and is the basis of the innovative business model of the power industry. The energy business mode will be transformed from the original B2B to more user-centric B2C and C2C modes.

15.5   The Development of the Energy Internet System

15.5.1   The Development of the Internet

Since the late twentieth century, the fifth scientific and technological revolution, with information technology as the core, has greatly changed the form of social development and has greatly promoted the development of the economy and society. The occurrence and development of the IT era have gone through three important milestone events. The first milestone was that in 1942, Pennsylvania State University successfully developed the first computer "ENIAC", signifying the emergence of a digitalized and programmed information processing mode. The second milestone was the invention of semiconductor transistors by Alcatel-Lucent Bell Labs in 1947, which laid the physical foundation for the development of the whole IT era. The miniaturization of IT equipment has brought computers into every household. The third milestone was that, in the late 1970s, the Internet appeared and has become popular around the world. It can be clearly seen that these three milestones in the development of information technology have brought four fundamental technical features in the IT era, namely, digitalization, programming, miniaturization and computer-machine interconnectivity.

In the post-IT era, information technology has been increasingly integrated with nanomaterials, brain and cognition, materials, energy, manufacturing, biology, oceans and the environment. The four fundamental technological characteristics in the IT era gradually developed into big data, intellectualization, nanotechnology and human-machine interfaces. Big data will open a major era of transformation, changing our understanding of life and the world, and become the source of new inventions and new services. The high intellectualization of the development of information science, brain science and cognitive science is trying to make robots have the ability and creativity to solve problems and even make robots develop their own characters, which is beyond the original idea of the designer.

15.5.2   Future Evolution of the Energy Internet

In view of the similarity shared by the Energy Internet and the information Internet, it can be envisaged that the development of the Energy Internet will experience a similar route to the development of the information Internet.

In the early stage, the construction of network infrastructure should be carried out. In this period, breakthroughs will be made in energy exchange technology and energy storage technology, and a scalable and flexible distributed energy sharing network will be established.

Then, with the expansion of commercial applications, a rich network application development will be started, and energy network applications such as information publishing; information search; business transactions; and point-to-point real-time energy exchange, B2B and B2C energy source trading platforms will appear.

In the end, the Energy Internet will be widely popularized, and an Energy Internet economic model that is completely different from the traditional economy will be created; like the information Internet, the Energy Internet will change human life and work.

15.5.3   A Summary of the System and Mechanism Level

In the development of the Energy Internet, we should not only learn from the successful experience of the information Internet but also fully consider the uniqueness of the Internet. The integration of renewable energy technology and advanced communication and information technology in the Energy Internet is a complex engineering issue that should be solved from the following aspects.

(1) Issues such as Energy Internet architecture, related standard protocols and distributed collaborative control should be studied.
(2) Focus on renewable energy, information, intelligent control, system management, network security and other technical fields; address the key technologies of the Energy Internet and enhance the innovation capability of supporting Energy Internet technology.
(3) Active development of high-tech products as well as independent intellectual property rights. Carry out typical demonstrations of experimental validation.

15.6   Conclusion

The Energy Internet breaks the information asymmetry in the industry and tremendously improves the efficiency of the traditional energy power system. Nevertheless, the value of the Energy Internet goes far beyond that. Its core is to reconstruct the thinking mode of energy enterprises by a "full link", in the sense that power consumers and power generation enterprises, power generation and the grid, electricity consumers and the grid, and service enterprises and consumers are all connected. The conventional operation modes of energy enterprises must be restructured with the characteristics of the Internet as the starting point. Only those enterprises that can think and use the Internet actively can stand out in the era of the Energy Internet.

References

1. J. Rifkin, *The Third Industrial Revolution: How Lateral Power is Transforming Energy, the Economy, and the World* (Palgrave Macmillan, New York, 2013), pp. 31–46
2. J. Cao, H. Hua, G. Ren, *"Energy Use and the Internet," The SAGE Encyclopedia of the Internet* (Sage, Newbury Park, CA, USA, 2018), pp. 344–350
3. M. Simonov, Event-Driven communication in smart grid. IEEE Commun. Lett. **17**(6), 1061–1064 (2013)
4. J.M. Selga, A. Zaballos, J. Navarro, Solutions to the computer networking challenges of the distribution smart grid. IEEE Commun. Lett. **17**(3), 588–591 (2013)
5. C.J. Mozina, Impact of smart grid and green power generation on distribution systems. IEEE Trans. Ind. Appl. **49**(3), 1079–1090 (2013)
6. K. Wang, Y. Wang, Y. Sun, Green industrial internet of things architecture: an energy-efficient perspective. IEEE Commun. Mag. **54**(12), 48–54 (2016)
7. W. Ejaz, M. Naeem, A. Shahid, Efficient energy management for the internet of things in smart cities. IEEE Commun. Mag. **55**(1), 84–91 (2017)
8. P. Kamalinejad, C. Mahapatra, Z. Sheng, Wireless energy harvesting for the internet of things. IEEE Commun. Mag. **53**(6), 102–108 (2015)
9. H. Hua, Y. Qin, C. Hao, J. Cao, Optimal energy management strategies for energy internet via deep reinforcement learning approach. Appl. Energy **239**, 598–609 (2019)
10. H. Hua et al., Stochastic optimal control for energy internet: a bottom-up energy management approach. IEEE Trans. Ind. Inform. **15**(3), 1788–1797 (2019)

Index

© Springer Nature Switzerland AG 2020
A. F. Zobaa and J. Cao (eds.), *Energy Internet*,
https://doi.org/10.1007/978-3-030-45453-1

Printed in the United States
by Baker & Taylor Publisher Services